Information Science and Statistics

Information Science and Statistics

Rubinstein/Kroese: The Cross-Entropy Method: A Unified Approach to Combinatorial Optimization, Monte-Carlo Simulation, and Machine Learning.

Reuven Y. Rubinstein   Dirk P. Kroese

The Cross-Entropy Method

A Unified Approach to Combinatorial Optimization, Monte-Carlo Simulation, and Machine Learning

With 38 Illustrations

Springer

Reuven Y. Rubinstein
Department of Industrial Engineering
and Management
Technion
Technion City
Haifa 32 000
Israel
ierrr01@ie.technion.ac.il

Dirk P. Kroese
Department of Mathematics
University of Queensland
Brisbane
Queensland 4072
Australia
kroese@maths.uq.edu.au

Library of Congress Cataloging-in-Publication Data
Rubinstein, Reuven Y.
The cross-entropy method : a unified approach to combinatorial optimization, Monte-Carlo simulation, and machine learning / Reuven Y. Rubinstein, Dirk P. Kroese.
p. cm. — (Information in sciences and statistics)
Includes bibliographical references and index.
ISBN 0-387-21240-X (hardback : alk. paper)
1. Cross-entropy method. 2. Combinatorial optimization. 3. Monte-Carlo simulation. 4. Machine learning. I. Kroese, Dirk P. II. Title. III. Series.
QA402.5.R815 2004
519.6′4—dc22 2004048202

ISBN 0-387-21240-X e-ISBN 0-387-21240-X Printed on acid-free paper.

Printed in the United States of America. (MVY)

9 8 7 6 5 4 3 2 1 SPIN 10966174

Springer is a part of Springer Science+Business Media
springeronline.com

To the memory of my teacher and friend Leonard Rastrigin

Reuven Rubinstein

For Lesley, Elise, and Jessica

Dirk Kroese

Preface

This book is a comprehensive and accessible introduction to the *cross-entropy* (CE) method. The CE method started life around 1997 when the first author proposed an adaptive algorithm for rare-event simulation using a cross-entropy minimization technique. It was soon realized that the underlying ideas had a much wider range of application than just in rare-event simulation; they could be readily adapted to tackle quite general combinatorial and multi-extremal optimization problems, including many problems associated with the field of learning algorithms and neural computation.

The book is based on an advanced undergraduate course on the CE method, given at the Israel Institute of Technology (Technion) for the last three years. It is aimed at a broad audience of engineers, computer scientists, mathematicians, statisticians and in general anyone, theorist or practitioner, who is interested in smart simulation, fast optimization, learning algorithms, image processing, etc. Our aim was to write a book on the CE method which was accessible to advanced undergraduate students and engineers who simply want to *apply* the CE method in their work, while at the same time accentuating the unifying and novel mathematical ideas behind the CE method, so as to stimulate further research at a postgraduate level.

The emphasis in this book is placed on concepts rather than on mathematical completeness. We assume that the reader has some basic mathematical background, such as a basic undergraduate course in probability and statistics. We have deliberately tried to avoid the formal "definition – lemma – theorem – proof" style of many mathematics books. Instead we embed most definitions in the text and introduce and explain various concepts via examples and experiments. In short, our goal is to promote a new unified way of thinking on the connection between rare events in simulation and optimization of complex systems in general, rather than burden the reader with too much technical detail.

Most of the combinatorial and continuous multi-extremal optimization case studies in this book are benchmark problems taken from the World Wide Web, and CE was compared with the best known solutions. In *all* examples tested so far the relative error of CE was within the limits of 1-2% of the best known solution. For some instances CE produced even more accurate solutions. It is crucial to emphasize that for the "noisy" counterparts of these test problems, which are obtained by adding random noise to the objective function, CE still performs quite accurately, provided the sample size is increased accordingly. Since our extensive numerical experience with different case studies suggests that CE is quite reliable (by comparing it with the best known solution), we purposely avoided comparing it with other heuristics such as simulated annealing and genetic algorithms. This of course does not imply that one will not find a problem where CE performs poorly and therefore will be less accurate than some other methods. However, when such problems do occur, our FACE (fully adaptive CE) algorithm (see Chapter 5) should identify it reliably.

Chapter 1 starts with some background on the cross-entropy method. It provides a summary of mathematical definitions and concepts relevant for this book, including a short review of various terms and ideas in probability, statistics, information theory, and modern simulation. A good self-contained *entry point* to this book is the tutorial Chapter 2, which provides a gradual introduction to the CE method, and shows immediately the elegance and versatility of the CE algorithm. In Chapter 3 we discuss the state of the art in efficient simulation and adaptive importance sampling using the CE concept. Chapter 4 deals with CE optimization techniques, with particular emphasis on combinatorial optimization problems. In Chapter 5 we apply CE to continuous optimization problems, and give various modifications and enhancements of the basic CE algorithm. The contemporary subject of noisy (stochastic) optimization is discussed in Chapter 6. Due to its versatility, tractability, and simplicity, the CE method has great potential for a diverse range of new applications, for example in the fields of computational biology, graph theory, and scheduling. Various applications, including DNA sequence alignment, are given in Chapter 7. A connection between the CE method and machine learning — specifically with regard to optimization — is presented in Chapter 8. A wide range of exercises is provided at the end of each chapter. Difficult exercises are marked with a $*$ sign. Finally, example CE programs, written in Matlab, are given in the appendix.

Acknowledgement. We thank all of those who contributed to this work. Sergey Porotsky provided a thorough review of an earlier version of this book and supplied an effective modification to the basic CE algorithm for continuous optimization. Arkadii Nemirovskii and Leonid Mytnik made a valuable contribution to the convergence proofs of the CE method. Søren Asmussen, Pieter-Tjerk the Boer, Peter Glynn, Tito Homem-de-Melo, Jonathan Keith, Shie Mannor, Phil Pollett, Ad Ridder, and Perwez Shahabuddin read and commented upon sections of this book.

We are especially grateful to the many undergraduate and graduate students at the Technion and the University of Queensland, who helped make this book possible, and whose valuable ideas and experiments where extremely encouraging and motivating. Yohai Gat, Uri Dubin, Leonid Margolin, Levon Kikinian, and Alexander Podgaetsky ran CE on a huge variety of combinatorial and neural-based problems. Rostislav Man ran extensive experiments on heavy-tailed distributions and supplied useful convergence results. He also provided the Matlab programs in Appendix A.6. Ron Zohar provided the second generation algorithm for the buffer allocation problem. Joshua Ross helped establish the material in Appendix 3.13. Sho Nariai pointed out various errors in an earlier version. Zdravko Botev conscientiously went through and checked the exercises. Thomas Taimre provided various Matlab programs in the appendix and meticulously carried out the clustering experiments. This book was supported by the Israel Science Foundation (grant no. 191-565).

Haifa and Brisbane,
January 2004

Reuven Y. Rubinstein
Technion

Dirk P. Kroese
The University of Queensland

Acronyms

ACO	Ant Colony Optimization
ARM	Acceptance Rejection Method
ASP	Associated Stochastic Problem
BAP	Buffer Allocation Problem
cdf	cumulative distribution function
CE	Cross-Entropy
CLT	Central Limit Theorem
CMC	Crude Monte Carlo
CoM	Change of Measure
COP	Combinatorial Optimization Problem
CVRP	Capacitated Vehicle Routing Problem
DES	Discrete Event System
DP	Dynamic Programming
ECM	Exponential Change of Measure
FACE	Fully Adaptive Cross-Entropy
FKM	Fuzzy K-Means
i.i.d.	independent identically distributed
ITLR	Inverse Transform Likelihood Ratio
IS	Importance Sampling
LPP	Longest Path Problem
LR	Likelihood Ratio
LVQ	Linear Vector Quantization
max-cut	Maximal Cut

MC	Maximum Clique
MCMC	Markov Chain Monte Carlo
MCWR	Markov Chain With Replacement
MDP	Markov Decision Process
MLE	Maximum Likelihood Estimate (or Estimator)
MME	Method of Moments Estimate (or Estimator)
NEF	Natural Exponential Family
NOP	Noisy Optimization Problem
ODTM	Optimal Degenerate Transition Matrix
pdf	probability density function
pmf	probability mass function
PFSP	Permutation Flow Shop Problem
QAP	Quadratic Assignment Problem
RE	Relative Error
RL	Reinforcement Learning
SA	Stochastic Approximation
SF	Score Function
SCV	Squared Coefficient of Variation
SEN	Stochastic Edge Network
SMCDDP	Single Machine Common Due Date Problem
SMTWTP	Single Machine Total Weighted Tardiness Problem
SNN	Stochastic Node Network
SPP	Shortest Path Problem
TL	Table Look
TSP	Traveling Salesman Problem
VM	Variance Minimization
VRP	Vehicle Routing Problem
WIP	Work In Process

List of Symbols

$\gg$	much greater than
$\sim$	is distributed according to
$\approx$	approximately
∇	∇f is the gradient of f
∇^2	$\nabla^2 f$ is the Hessian of f
$\mathbb{E}$	expectation
$\mathbb{N}$	set of natural numbers $\{0, 1, \ldots\}$
$\mathbb{P}$	probability measure
$\mathbb{R}$	the real line = 1-dimensional Euclidean space
$\mathbb{R}^n$	n-dimensional Euclidean space
$\mathcal{D}$	Kullback-Leibler cross-entropy
$\mathcal{H}$	Shannon entropy
$\mathcal{I}$	Fisher information
$\mathcal{L}$	likelihood function
$\mathcal{M}$	mutual information
$\mathcal{S}$	score function
Ber	Bernoulli distribution
Beta	beta distribution
Bin	binomial distribution
DExp	double exponential distribution

DU	discrete uniform distribution
Exp	Exponential distribution
Gam	gamma distribution
G	geometric distribution
N	normal or Gaussian distribution
Pareto	Pareto distribution
Po	Poisson distribution
SExp	shifted exponential distribution
U	uniform distribution
Weib	Weibull distribution

α	smoothing parameter
γ	level parameter
ε	relative experimental error
ψ	parameterization; $\theta = \psi(u)$
ζ	cumulant function (log of moment generating function)
ϱ	rarity parameter

$D(\mathbf{v})$	Objective function for CE minimization
f	probability density function
I_A	indicator function of event A
ln	(natural) logarithm
N	sample size
o	Little-o order symbol
$\mathcal{O}$	Big-O order symbol
S	performance function
$S_{(i)}$	i-th order statistic
$\mathbf{u}$	nominal reference parameter (vector)
$\mathbf{v}$	reference parameter (vector)
$\widehat{\mathbf{v}}$	estimated reference parameter
$\mathbf{v}^*$	CE optimal reference parameter
${}_*\mathbf{v}$	VM optimal reference parameter
$V(\mathbf{v})$	Objective function for VM minimization
W	likelihood ratio

$\mathbf{x}, \mathbf{y}$	vectors
$\mathbf{X}, \mathbf{Y}$	random vectors/matrices
$\mathcal{X}, \mathcal{Y}$	sets

Contents

1

Preliminaries

The purpose of this preparatory chapter is to provide some initial background for this book, including a summary of the relevant definitions and concepts in probability, statistics, simulation and information theory. Readers familiar with these ideas may wish to skip through this chapter and proceed to the tutorial on the cross-entropy method in Chapter 2.

1.1 Motivation for the Cross-Entropy Method

Many everyday tasks in operations research involve solving complicated optimization problems. The travelling salesman problem (TSP), the quadratic assignment problem (QAP) and the max-cut problem are a representative sample of combinatorial optimization problems (COP) where the problem being studied is completely known and static. In contrast, the buffer allocation problem (BAP) is a *noisy* estimation problem where the objective function needs to be estimated since it is unknown. Discrete event simulation is one method for estimating an unknown objective function. An increasingly popular approach is to tackle these problems via *stochastic (or randomized) algorithms*, specifically via the simulated cross-entropy or simply the *cross-entropy* (CE) method pioneered in [145]. The CE method derives its name from the cross-entropy (or Kullback-Leibler) distance, which is a fundamental concept of modern information theory [39, 95]. The method was motivated by an adaptive algorithm for estimating probabilities of *rare events* in complex stochastic networks [144], which involves variance minimization. It was soon realized [145, 146] that a simple modification of [144], involving CE minimization, could be used not only for estimating probabilities of rare events but for solving difficult combinatorial optimization and continuous multi-extremal problems as well. This is done by translating the original "deterministic" optimization problem into a related "stochastic" estimation problem and then applying the rare-event simulation machinery similar to [144]. In a nutshell, the CE method involves an iterative procedure where each iteration can be broken down into two phases:

1. Generate a random data sample (trajectories, vectors, etc.) according to a specified mechanism.
2. Update the parameters of the random mechanism based on the data to produce a "better" sample in the next iteration.

The power and generality of this new method lie in the fact that the updating rules are often simple and explicit (and hence fast), and are "optimal" in some well-defined mathematical sense. Moreover, the CE method provides a unifying approach to simulation and optimization and has great potential for opening up new frontiers in these areas.

An increasing number of applications is being found for the CE method. Recent publications on applications of the CE method include: buffer allocation [9]; static simulation models [85]; queueing models of telecommunication systems [43, 46]; neural computation [50]; control and navigation [83]; DNA sequence alignment [98]; signal processing [110]; scheduling [113]; vehicle routing [36]; reinforcement learning [117]; project management [37]; heavy-tail distributions [15, 103]; CE convergence [114]; network reliability [87]; repairable systems [137]; and max-cut and bipartition problems [147].

It is important to note that the CE method can deal successfully with both deterministic and *noisy* problems. In the later case it is assumed that the objective function is "corrupted" with an additive "noise." Such situation typically occurs in simulation-based problems, for example while solving the buffer allocation problem [9].

The CE home page, featuring many links, articles, references, tutorials and computer programs on CE, can be found at

`http://www.cemethod.org`

1.2 Random Experiments and Probability Distributions

This section will review some important concepts in probability theory. For more details we refer to standard texts such as [55] and [73].

Probability theory is about modeling and analyzing *random experiments*: experiments whose outcome cannot be determined in advance, but are nevertheless still subject to analysis. Mathematically, we can model a random experiment by specifying an appropriate *probability space* $(\Omega, \mathcal{F}, \mathbb{P})$ where Ω is the set of all possible outcomes of the experiment (the *sample space*), $\mathcal{F}$ the collection of all *events*, and $\mathbb{P}$ the *probability (measure)*. However, in practice the probability space only plays a "back-stage" role; often the most convenient way to describe random experiments is by introducing *random variables*, which are usually represented by capital letters such as X, Y, and Z. One could think of these as the *measurements* that will be obtained if we carry out our experiment tomorrow. A vector $\mathbf{X} = (X_1, \ldots, X_n)$ of random variables is often called a *random vector*. A collection $\{X_t, t \in \mathcal{T}\}$ of random variables indexed with some index set $\mathcal{T}$ is called a *stochastic process*.

In this book we have kept the measure theoretic foundations of probability to a minimum. The only notable "concession" is in the definition of probability densities. In particular, it is customary in elementary probability to make the distinction between *discrete* and *continuous* random variables, whose distributions are, in turn, specified by the *probability mass function* (pmf) and the *probability density function* (pdf) of X. However, this creates a great deal of redundancy: every idea and definition is repeated for both the discrete and continuous case. A more elegant and unifying approach is to define the *probability distribution* of a random variable X as the measure ν defined by

$$\nu(B) = \mathbb{P}(X \in B)\,, \quad B \in \mathcal{B}\,,$$

where $\mathcal{B}$ is the collection of Borel sets in $\mathbb{R}$ (containing, for example, countable unions of intervals). If X is a discrete random variable with pmf f, then

$$\nu(B) = \sum_B f(x)\,,$$

and if X is a continuous random variable with pdf f, then

$$\nu(B) = \int_B f(x)\,dx\,.$$

In both cases f is the *density* of ν with respect to some *base measure* μ, that is

$$\nu(B) = \int_B f(x)\,\mu(dx)\,, \quad B \in \mathcal{B}\,.$$

In the discrete case f is the density of ν with respect to the *counting measure*, in the continuous case f is the density of ν with respect to the *Lebesgue measure* on $\mathbb{R}$. From now on we will call f, in *both* the discrete and continuous case, the pdf or density of X. The function F defined by

$$F(x) = \int_{-\infty}^{x} f(x)\,\mu(dx) = \mathbb{P}(X \leqslant x)$$

is the well-known *(cumulative) distribution function* (cdf) of X. The *expectation* of $g(X)$ for some (measurable) function g can now be written as

$$\mathbb{E}g(X) = \int g(x)\,\nu(dx) = \int g(x)f(x)\,\mu(dx) = \int g(x)\,dF(x)\,.$$

Tables 1.1 and 1.2 list a number of important continuous and discrete distributions, to which we will refer throughout this book.

We will use the notation $X \sim f$, $X \sim F$ or $X \sim \mathsf{Dist}$ to signify that X has a pdf f, cdf F or a distribution Dist. Note that in Table 1.1, Γ is the gamma function:

$$\Gamma(a) = \int_0^\infty \mathrm{e}^{-x} x^{a-1}\, dx\,, \quad a > 0\,.$$

Table 1.1. Commonly used continuous distributions.

Name	Notation	$f(x)$	$x \in$	Params.
Uniform	$\mathsf{U}[a,b]$	$\dfrac{1}{b-a}$	$[a,b]$	$a<b$
Normal	$\mathsf{N}(a,b^2)$	$\dfrac{1}{b\sqrt{2\pi}}\,\mathrm{e}^{-\frac{1}{2}\left(\frac{x-a}{b}\right)^2}$	$\mathbb{R}$	$b>0$
Gamma	$\mathsf{Gam}(a,b)$	$\dfrac{b^a x^{a-1}\mathrm{e}^{-bx}}{\Gamma(a)}$	$\mathbb{R}_+$	$a,b>0$
Exponential	$\mathsf{Exp}(a)$	$a\,\mathrm{e}^{-ax}$	$\mathbb{R}_+$	$a>0$
Double exponential	$\mathsf{DExp}(a,b)$	$\dfrac{b}{2}\,\mathrm{e}^{-b\lvert x-a\rvert}$	$\mathbb{R}$	$b>0$
Shifted exponential	$\mathsf{SExp}(a,b)$	$b\mathrm{e}^{-b(x-a)}$	$[a,\infty)$	$b>0$
Truncated exponential	$\mathsf{TExp}(a,b)$	$\dfrac{a\,\mathrm{e}^{-ax}}{1-\mathrm{e}^{-ab}}$	$[0,b]$	$a,b>0$
Beta	$\mathsf{Beta}(a,b)$	$\dfrac{\Gamma(a+b)}{\Gamma(a)\Gamma(b)}\,x^{a-1}(1-x)^{b-1}$	$[0,1]$	$a,b>0$
Weibull	$\mathsf{Weib}(a,b)$	$ab\,(bx)^{a-1}\,\mathrm{e}^{-(bx)^a}$	$\mathbb{R}_+$	$a,b>0$
Pareto	$\mathsf{Pareto}(a,b)$	$ab\,(1+bx)^{-(a+1)}$	$\mathbb{R}_+$	$a,b>0$

Light- and Heavy-Tail Distributions

In Chapter 3 we will deal extensively with heavy- and light-tail distributions in the context of rare-event simulation. A random variable X with distribution function F is said to have a *light-tail* distribution if the *moment generation function* $s \mapsto \mathbb{E}\,\mathrm{e}^{sX}$ is finite for some $s > 0$. That is,

$$\mathbb{E}\,\mathrm{e}^{sX} \leqslant c < \infty\,. \tag{1.1}$$

Since for every x

$$\mathbb{E}\,\mathrm{e}^{sX} \geqslant \mathbb{E}\,\mathrm{e}^{sX} I_{\{X>x\}} \geqslant \mathrm{e}^{sx}\,\mathbb{P}(X>x)\,,$$

Table 1.2. Commonly used discrete distributions.

Name	Notation	$f(x)$	$x \in$	Params.
Bernoulli	$\mathsf{Ber}(p)$	$p^x(1-p)^{1-x}$	$\{0,1\}$	$0 \leqslant p \leqslant 1$
Binomial	$\mathsf{Bin}(n,p)$	$\binom{n}{x} p^x(1-p)^{n-x}$	$\{0,1,\ldots,n\}$	$0 \leqslant p \leqslant 1$, $n \in \mathbb{N}$
Discrete uniform	$\mathsf{DU}\{1,\ldots,n\}$	$\dfrac{1}{n}$	$\{1,\ldots,n\}$	$n \in \{1,2,\ldots\}$
Geometric	$\mathsf{G}(p)$	$p(1-p)^{x-1}$	$\{1,2,\ldots\}$	$0 \leqslant p \leqslant 1$
Poisson	$\mathsf{Po}(a)$	$\mathrm{e}^{-a}\dfrac{a^x}{x!}$	$\{0,1,\ldots\}$	$a > 0$

it follows that for any $s > 0$ and c satisfying (1.1) the following inequality holds:

$$\mathbb{P}(X > x) \leqslant c\,\mathrm{e}^{-sx}\ .$$

In other words, if X has a light tail, then $\bar{F}(x) = 1 - F(x)$ decays at an exponential rate or faster. Examples of light-tailed distributions are the exponential, normal and Weibull distribution — the latter with increasing failure rate, that is $\bar{F}(x) = \mathrm{e}^{-x^a}$ with $a \geqslant 1$ — and the geometric, Poisson and any distribution with bounded support.

When $\mathbb{E}\,\mathrm{e}^{sX} = \infty$ for all $s > 0$, X is said to have a *heavy-tail* distribution. Examples of heavy-tail distributions are the Weibull distribution with decreasing failure rate ($\bar{F}(x) = \mathrm{e}^{-x^a}$, $a < 1$), the log-normal ($X = \mathrm{e}^Y$, with $Y \sim \mathsf{N}(a, b^2)$) and any regularly varying distribution: $\bar{F}(x) = L(x)/x^\alpha$, where $L(tx)/L(x) \to 1$ as $x \to \infty$ for all $t > 0$, such as the Pareto distribution.

A particularly important class of heavy-tailed distributions is the class of subexponential distributions. A distribution with cdf F on $(0, \infty)$ is said to be *subexponential* if, with $X_1, X_2, \ldots, X_n$ a random sample from F, we have

$$\lim_{x\to\infty} \frac{\mathbb{P}(X_1 + \cdots + X_n > x)}{\mathbb{P}(X_1 > x)} = n\ , \tag{1.2}$$

for all n. Examples are the Pareto and Log-normal distribution and the Weibull with decreasing failure rate. See [54] for additional properties of this class of distributions.

Unbounded and Finite Support Distributions

In this book we classify distributions into two categories: the ones with *unbounded support*, like the exponential, Poisson and normal distributions, and the ones with *bounded support*, such as uniform $\mathsf{U}(a, b)$, truncated exponential and discrete n-point distributions. More formally, we say that a random variable X has *bounded support* if $\mathbb{P}(|X| > a) = 0$ for some a large enough. Note that bounded support distributions can be viewed as *zero tail* distributions as compared to their counterparts with infinite tail, which belong to the category of either *light*- or *heavy*-tail distributions; see above and [148]. Notice also that, in particular, *finite* support distribution, (only a finite number of values can be attained) have zero tail.

1.3 Exponential Families

Exponential families play an important role in statistics. Let $\mathbf{X}$ be a random variable or vector (in this section vectors will always be interpreted as *column* vectors) with pdf $f(\mathbf{x};\boldsymbol{\theta})$ (with respect to some base measure μ), where $\boldsymbol{\theta} = (\theta_1, \ldots, \theta_m)^T$ is an m-dimensional parameter vector. $\mathbf{X}$ is said to belong to an m-parameter *exponential family* if there exist real-valued functions $t_i(\mathbf{x})$ and $h(\mathbf{x}) > 0$ and a (normalizing) function $c(\boldsymbol{\theta}) > 0$, such that

$$f(\mathbf{x};\boldsymbol{\theta}) = c(\boldsymbol{\theta})\, \mathrm{e}^{\boldsymbol{\theta}\cdot\mathbf{t}(\mathbf{x})}\, h(\mathbf{x}) , \tag{1.3}$$

where $\mathbf{t}(\mathbf{x}) = (t_1(\mathbf{x}), \ldots, t_m(\mathbf{x}))^T$ and $\boldsymbol{\theta}\cdot\mathbf{t}(\mathbf{x})$ is the inner product $\sum_{i=1}^m \theta_i t_i(\mathbf{x})$. The representation of an exponential family is in general not unique.

Table 1.3 displays the functions $c(\boldsymbol{\theta})$, $t_k(x)$ and $h(x)$ for several commonly used distributions (a dash means that the corresponding value is not used).

Table 1.3. The functions $c(\boldsymbol{\theta})$, $t_k(x)$ and $h(x)$ for commonly used distributions.

Distr.	$\mathbb{E}X$	$t_1(x),\ t_2(x)$	$c(\boldsymbol{\theta})$	$\theta_1,\ \theta_2$	$h(x)$
$\mathsf{Gam}(a,b)$	$\frac{a}{b}$	$x,\ \ln x$	$\frac{b^a}{\Gamma(a)}$	$-b,\ a-1$	1
$\mathsf{N}(a,b^2)$	a	$x,\ x^2$	$\frac{\mathrm{e}^{-\frac{1}{2}\left(\frac{a}{b}\right)^2}}{b\sqrt{2\pi}}$	$\frac{a}{b^2},\ -\frac{1}{2b^2}$	1
$\mathsf{Weib}(a,b)$	$b^{-1}\Gamma\left(1+\frac{1}{a}\right)$	$\ln x,\ -x^a$	ab^a	$a-1,\ b^a$	1
$\mathsf{Bin}(n,p)$	np	$x,\ -$	$(1-p)^n$	$\ln\left(\frac{p}{1-p}\right),\ -$	$\binom{n}{x}$
$\mathsf{Po}(a)$	a	$x,\ -$	e^{-a}	$\ln a,\ -$	$\frac{1}{x!}$
$\mathsf{G}(p)$	$\frac{1}{p}$	$x-1,\ -$	p	$\ln(1-p),\ -$	1

Consider in particular the 1-dimensional case where (dropping the boldface font) X is a random variable, $m = 1$ and $t(x) = x$. The one-parameter family of distributions with densities $\{f(x;\theta), \theta \in \Theta \subset \mathbb{R}\}$ given by

$$f(x;\theta) = c(\theta)\,\mathrm{e}^{\theta x}\,h(x) \tag{1.4}$$

is called a *natural exponential family* (NEF) [121].

If $h(x)$ is a pdf itself, then $c^{-1}(\theta)$ is the corresponding *moment generating function*:

$$c^{-1}(\theta) = \int \mathrm{e}^{\theta x}\,h(x)\,\mu(dx)\;.$$

It is sometimes convenient to introduce instead the logarithm of the moment generating function:

$$\zeta(\theta) = \ln \int \mathrm{e}^{\theta x} h(x)\,\mu(dx)\;,$$

which is called the *cumulant function.* We can now write (1.4) in the following convenient form

$$f(x;\theta) = \mathrm{e}^{\theta x - \zeta(\theta)}\,h(x)\;. \tag{1.5}$$

Example 1.1. If we take h as the density of the $\mathsf{N}(0,\sigma^2)$-distribution, $\theta = \lambda/\sigma^2$ and $\zeta(\theta) = \sigma^2\theta^2/2$, then the class $\{f(\cdot;\theta), \theta \in \mathbb{R}\}$ is the class of $\mathsf{N}(\lambda,\sigma^2)$ densities, where σ^2 is fixed and $\lambda \in \mathbb{R}$.

Similarly, if we take h as the density of the $\mathsf{Gam}(a,1)$-distribution, and let $\theta = 1-\lambda$ and $\zeta(\theta) = -a\ln(1-\theta) = -a\ln\lambda$ we obtain the class of $\mathsf{Gam}(a,\lambda)$ distributions, with a fixed and $\lambda > 0$. Note that in this case $\Theta = (-\infty, 1)$.

There are many NEFs. In fact, starting from any pdf f_0 we can easily generate a NEF $\{f_\theta, \theta \in \Theta\}$ in the following way: Let Θ be the largest interval for which the cumulant function ζ of f_0 exists. This includes $\theta = 0$, since f_0 is a pdf. Now define

$$f_\theta(x) = \mathrm{e}^{\theta x - \zeta(\theta)}\,f_0(x)\;. \tag{1.6}$$

Then $\{f_\theta, \theta \in \Theta\}$ thus defined, is a NEF. We say that f_θ is obtained from f_0 by an *exponential twist* with *twisting parameter* θ. Exponential twisting plays an important role in rare-event simulation as we shall see in Chapter 3.

In some cases it is useful to reparameterize a NEF. In particular, suppose that a random variable X has a distribution in some NEF $\{f_\theta\}$. It is not difficult to see that

$$\mathbb{E}_\theta X = \zeta'(\theta) \quad \text{and} \quad \mathrm{Var}_\theta(X) = \zeta''(\theta)\;. \tag{1.7}$$

Since $\zeta'(\theta)$ is increasing in θ, its derivative $\zeta''(\theta) = \mathrm{Var}_\theta(X)$ is always greater than 0, and thus we can reparameterize the family using the mean $v = \mathbb{E}_\theta X$. In particular, to the above NEF there is a corresponding family $\{g_v\}$ such that for each pair (θ, v) satisfying $\zeta'(\theta) = v$ we have $g_v = f_\theta$.

Example 1.2. Consider the second case in Example 1.1. Note that we constructed in fact a NEF $\{f_\theta, \theta \in (-\infty,1)\}$ by exponentially twisting the $\mathsf{Gam}(a,1)$ distribution, with density $f_0(x) = x^{a-1}\mathrm{e}^{-x}/\Gamma(a)$. We have $\zeta'(\theta) = a/(1-\theta) = v$. This leads to the reparameterized density

$$g_v(x) = \exp\left(\theta x + a \ln(1-\theta)\right) f_0(x) = \frac{\exp\left(-\frac{a}{v}\,x\right)\left(\frac{a}{v}\right)^a x^{a-1}}{\Gamma(a)} ,$$

corresponding to the $\mathsf{Gam}(a, av^{-1})$ distribution, $v > 0$.

1.4 Efficiency of Estimators

In this book we will frequently use

$$\widehat{\ell} = \frac{1}{N} \sum_{i=1}^{N} Z_i , \tag{1.8}$$

which presents an *unbiased* estimator of the unknown quantity $\ell = \mathbb{E}\widehat{\ell} = \mathbb{E}Z$, where $Z_1, \ldots, Z_N$ are independent replications of some random variable Z.

By the central limit theorem, $\widehat{\ell}$ has approximately a $\mathsf{N}(\ell, N^{-1}\mathrm{Var}(Z))$ distribution, for large N. We shall estimate $\mathrm{Var}(Z)$ via the *sample variance*

$$S^2 = \frac{1}{N-1} \sum_{i=1}^{N} (Z_i - \widehat{\ell})^2 .$$

By the law of large numbers, S^2 converges with probability 1 to $\mathrm{Var}(Z)$. Consequently, for $\mathrm{Var}(Z) < \infty$ and large N, the approximate $(1-\alpha)$ confidence interval for ℓ is given by

$$\left(\widehat{\ell} - z_{1-\alpha/2}\,\frac{S}{\sqrt{N}},\ \widehat{\ell} + z_{1-\alpha/2}\,\frac{S}{\sqrt{N}}\right) ,$$

where $z_{1-\alpha/2}$ is the $(1-\alpha/2)$-quantile of the standard normal distribution. For example, for $\alpha = 0.05$ we have $z_{1-\alpha/2} = z_{0.975} = 1.96$. The quantity

$$\frac{S/\sqrt{N}}{\widehat{\ell}}$$

is often used in the simulation literature as an accuracy measure for the estimator $\widehat{\ell}$. For large N it converges to the *relative error* (RE) of $\widehat{\ell}$, defined as:

$$\kappa = \frac{\sqrt{\mathrm{Var}(\widehat{\ell})}}{\mathbb{E}\,\widehat{\ell}} = \frac{\sqrt{\mathrm{Var}(Z)/N}}{\ell} . \tag{1.9}$$

The square of the relative error

$$\kappa^2 = \frac{\mathrm{Var}(\widehat{\ell})}{\ell^2} \tag{1.10}$$

is called the *squared coefficient of variation* (SCV).

Example 1.3 (Estimation of Rare-Event Probabilities). Consider estimation of the tail probability $\ell = \mathbb{P}(X \geqslant \gamma)$ of some random variable X, for a *large* number γ. If ℓ is very small, then the event $\{X \geqslant \gamma\}$ is called the *rare event* and the probability $\mathbb{P}(X \geqslant \gamma)$ is called the *rare-event probability.*

We may attempt to estimate ℓ via (1.8) as

$$\widehat{\ell} = \frac{1}{N} \sum_{i=1}^{N} I_{\{X_i \geqslant \gamma\}} , \tag{1.11}$$

which involves drawing a random sample $X_1, \ldots, X_N$ from the pdf of X and defining the indicators $Z_i = I_{\{X_i \geqslant \gamma\}}$, $i = 1, \ldots, N$. The estimator ℓ thus defined is called the *crude Monte Carlo* (CMC) estimator. For small ℓ the RE of the CMC estimator is given by

$$\kappa = \frac{\sqrt{\mathrm{Var}(\widehat{\ell})}}{\mathbb{E}\,\widehat{\ell}} = \sqrt{\frac{1-\ell}{N\,\ell}} \approx \sqrt{\frac{1}{N\,\ell}} . \tag{1.12}$$

As a numerical example, suppose $\ell = 10^{-6}$. In order to estimate ℓ accurately with relative error (say) $\kappa = 0.01$ we need to choose a sample size

$$N \approx \frac{1}{\kappa^2 \ell} = 10^{10} .$$

This shows that estimating small probabilities via CMC is computationally meaningless.

Complexity

The theoretical framework in which one typically examines rare-event probability estimation is based on *complexity theory*, as introduced in [18, 102]. In particular, the estimators are classified either as *polynomial-time* or as *exponential-time*. It is shown in [18, 148] that for an arbitrary estimator, $\widehat{\ell}$ of ℓ, to be polynomial-time as a function of some γ, it suffices that its *squared coefficient of variation* κ^2 or its *relative error*, κ, be bounded in γ by some polynomial function, $p(\gamma)$. For such polynomial-time estimators the required sample size to achieve a fixed relative error does not grow too fast as the event becomes rarer.

Consider the estimator (1.11) and assume that ℓ becomes very small as $\gamma \to \infty$. Note that

$$\mathbb{E}Z^2 \geqslant (\mathbb{E}Z)^2 = \ell^2 .$$

Hence, the best one can hope from such estimator is that its second moment of Z^2 decreases proportionally to ℓ^2 as $\gamma \to \infty$. We say that the rare-event estimator (1.11) has *bounded relative error* if for all γ

$$\mathbb{E}Z^2 \leqslant c\,\ell^2 , \tag{1.13}$$

for some fixed $c \geqslant 1$. Because bounded relative error is not always easy to achieve, the following weaker criterion is often used. We say that the estimator (1.11) is *logarithmically efficient* (sometimes called "asymptotically optimal") if

$$\lim_{\gamma \to \infty} \frac{\ln \mathbb{E} Z^2}{\ln \ell^2} = 1 \,. \tag{1.14}$$

Example 1.4 (The CMC Estimator is not Logarithmically Efficient). Consider the CMC estimator (1.11). We have

$$\mathbb{E} Z^2 = \mathbb{E} Z = \ell \,,$$

so that

$$\lim_{\gamma \to \infty} \frac{\ln \mathbb{E} Z^2}{\ln \ell^2(\gamma)} = \frac{\ln \ell}{\ln \ell^2} = \frac{1}{2} \,.$$

Hence, the CMC estimator is not logarithmically efficient and therefore alternative estimators must be found to estimate small ℓ.

1.5 Information

In this section we discuss briefly various measures of information in a random experiment. Suppose we describe the measurements on a random experiment via a random vector $\mathbf{X} = (X_1, \ldots, X_n)$ with pdf f. Then, all the information about the experiment (all our probabilistic knowledge) is obviously contained in the pdf f. However, in most cases we wish to characterize our information about the experiments with just a few key numbers. Well-known examples are the *expectation* and the *covariance matrix* of $\mathbf{X}$, which provide information about the mean measurements and the variability of the measurements, respectively. Another informational measure comes from coding and communications theory, where the *Shannon entropy* characterizes the average amount of bits needed to transmit a message $\mathbf{X}$ over a (binary) communication channel. Yet another approach to information can be found in statistics. Specifically, in the theory of point estimation the pdf f depends on a parameter vector $\boldsymbol{\theta}$. The question is how well $\boldsymbol{\theta}$ can be estimated via an outcome of $\mathbf{X}$, in other words, how much information about $\boldsymbol{\theta}$ is contained in the "data" $\mathbf{X}$. Various measures for this type of information are associated with the *maximum likelihood*, the *score* and the *(Fisher) information matrix*. Finally, the amount of information in a random experiment can often be quantified via a *"distance"* concept; the most notable for this book is the *Kullback-Leibler* "distance" (divergence), also called the *cross-entropy*.

Shannon Entropy

One of the most celebrated measures of uncertainty in information theory is the *Shannon entropy*, or simply *entropy*. A good reference is [39], where the entropy of a discrete random variable X with density f is defined as

$$\mathbb{E} \log_2 \frac{1}{f(X)} = -\mathbb{E} \log_2 f(X) = -\sum_{\mathcal{X}} f(x) \log_2 f(x) \, .$$

Here X is interpreted as a random character from an alphabet $\mathcal{X}$, such that $X = x$ with probability $f(x)$. We will use the convention $0 \ln 0 = 0$.

It can be shown that the most efficient way to transmit characters sampled from f over a binary channel is to encode them such that the number of bits required to transmit x is equal to $\log_2(1/f(x))$. It follows that $-\sum_{\mathcal{X}} f(x) \log_2 f(x)$ is the expected bit length required to send a random character $X \sim f$; see [39].

A more general approach, which includes continuous random variables, is to define the entropy of a random variable X with density f (with respect to some measure μ) by

$$\mathcal{H}(X) = -\mathbb{E} \ln f(X) = -\int f(x) \ln f(x) \, \mu(dx) \, . \tag{1.15}$$

Definition (1.15) can easily be extended to random vectors $\mathbf{X}$ as

$$\mathcal{H}(\mathbf{X}) = -\mathbb{E} \ln f(\mathbf{X}) = -\int f(\mathbf{x}) \ln f(\mathbf{x}) \, \mu(d\mathbf{x}) \, . \tag{1.16}$$

Often $\mathcal{H}(\mathbf{X})$ is called the *joint* entropy of the random variables $X_1, \ldots, X_n$, and is also written as $\mathcal{H}(X_1, \ldots, X_n)$. When μ is the Lebesgue measure we have

$$\mathcal{H}(\mathbf{X}) = -\int f(\mathbf{x}) \ln f(\mathbf{x}) \, d\mathbf{x} \, ,$$

which is frequently referred to as the *differential entropy*, to distinguish it from the discrete case.

Example 1.5. Let X have a $\mathsf{Ber}(p)$ distribution, for some $0 \leqslant p \leqslant 1$. The density f of X is given by $f(1) = \mathbb{P}(X = 1) = p$ and $f(0) = \mathbb{P}(X = 0) = 1 - p$, so that the entropy of X is

$$\mathcal{H}(X) = -p \ln p - (1-p) \ln(1-p) \, .$$

Note that the entropy is maximal for $p = 1/2$, which gives the "uniform" density on $\{0, 1\}$. Next, consider a sequence $X_1, \ldots, X_n$ of i.i.d. $\mathsf{Ber}(p)$ random variables. Let $\mathbf{X} = (X_1, \ldots, X_n)$. The density of $\mathbf{X}$, g say, is simply the product of the densities of the X_i, so that

$$\mathcal{H}(\mathbf{X}) = -\mathbb{E} \ln g(\mathbf{X}) = -\mathbb{E} \ln \prod_{i=1}^{n} f(X_i) = \sum_{i=1}^{n} -\mathbb{E} \ln f(X_i) = n \, \mathcal{H}(X) \, .$$

The properties of $\mathcal{H}(\mathbf{X})$ in the continuous case are somewhat different from the discrete one. In particular,

1. The differential entropy can be negative whereas the discrete entropy is always positive.
2. The discrete entropy is insensitive to invertible transformations, whereas the differential entropy is not. Specifically, if $\mathbf{X}$ is discrete, $\mathbf{Y} = g(\mathbf{X})$ and g is an invertible mapping, then $\mathcal{H}(\mathbf{X}) = \mathcal{H}(\mathbf{Y})$, because $f_{\mathbf{Y}}(\mathbf{y}) = f_{\mathbf{X}}(g^{-1}(\mathbf{y}))$. However, in the continuous case we have an additional factor due to the Jacobian of the transformation.

It is not difficult to see that of any density f, the one that gives the maximum entropy is the uniform density on $\mathcal{X}$. That is,

$$\mathcal{H}(\mathbf{X}) \text{ is maximal } \Leftrightarrow f(\mathbf{x}) = \frac{1}{\mu(\mathcal{X})} \text{ (constant)} . \tag{1.17}$$

This of course up to a set of μ-measure 0.

For two random vectors $\mathbf{X}$ and $\mathbf{Y}$ with joint pdf f we define the *conditional entropy* of $\mathbf{Y}$ given $\mathbf{X}$ as

$$\mathcal{H}(\mathbf{Y} \mid \mathbf{X}) = -\mathbb{E} \ln \frac{f(\mathbf{X}, \mathbf{Y})}{f_{\mathbf{X}}(\mathbf{X})} = \mathcal{H}(\mathbf{X}, \mathbf{Y}) - \mathcal{H}(\mathbf{X}) , \tag{1.18}$$

where $f_{\mathbf{X}}$ is the pdf of $\mathbf{X}$ and $\frac{f(\mathbf{x},\mathbf{y})}{f_{\mathbf{X}}(\mathbf{x})}$ is the conditional density of $\mathbf{Y}$ (at $\mathbf{y}$), given $\mathbf{X} = \mathbf{x}$. It follows that

$$\mathcal{H}(\mathbf{X}, \mathbf{Y}) = \mathcal{H}(\mathbf{X}) + \mathcal{H}(\mathbf{Y} \mid \mathbf{X}) = \mathcal{H}(\mathbf{Y}) + \mathcal{H}(\mathbf{X} \mid \mathbf{Y}) . \tag{1.19}$$

It is reasonable to impose that any sensible additive measure describing the average amount of uncertainty should satisfy at least (1.19) and (1.17). It follows that the uniform density carries the least amount of information, and the entropy (average amount of uncertainty) of $(\mathbf{X}, \mathbf{Y})$ is equal to the sum of the entropy of $\mathbf{X}$ and the amount of entropy in $\mathbf{Y}$ after the information in $\mathbf{X}$ has been accounted for.

It is argued in [99] that any concept of entropy that includes the general properties (1.17) and (1.19) must lead to the definition (1.16).

The *mutual information* of $\mathbf{X}$ and $\mathbf{Y}$ is defined as

$$\mathcal{M}(\mathbf{X}, \mathbf{Y}) = \mathcal{H}(\mathbf{X}) + \mathcal{H}(\mathbf{Y}) - \mathcal{H}(\mathbf{X}, \mathbf{Y}) , \tag{1.20}$$

which as the name suggests can be interpreted as the amount of information shared by $\mathbf{X}$ and $\mathbf{Y}$. An alternative expression, which follows from (1.19) and (1.20), is

$$\mathcal{M}(\mathbf{X}, \mathbf{Y}) = \mathcal{H}(\mathbf{X}) - \mathcal{H}(\mathbf{X} \mid \mathbf{Y}) = \mathcal{H}(\mathbf{Y}) - \mathcal{H}(\mathbf{Y} \mid \mathbf{X}) , \tag{1.21}$$

which can be interpreted as the reduction of the uncertainty of one random variable due to the knowledge of the other. It is not difficult to show that the mutual information is always positive. It is also related to the cross-entropy concept, which follows.

Kullback-Leibler Cross-Entropy

Let g and h be two densities with respect to the measure μ on $\mathcal{X}$. The Kullback-Leibler cross-entropy between g and h (compare with (1.16)) is defined as

$$\begin{aligned}\mathcal{D}(g,h) &= \mathbb{E}_g \ln \frac{g(\mathbf{X})}{h(\mathbf{X})} \\ &= \int g(\mathbf{x}) \ln g(\mathbf{x})\, \mu(d\mathbf{x}) - \int g(\mathbf{x}) \ln h(\mathbf{x})\, \mu(d\mathbf{x})\,.\end{aligned} \tag{1.22}$$

$\mathcal{D}(g,h)$ is also called the Kullback-Leibler divergence, the *cross-entropy* and the *relative entropy.* If not stated otherwise, we shall call $\mathcal{D}(g,h)$ the *cross-entropy* (CE) between g and h. Notice that $\mathcal{D}(g,h)$ is not a "distance" between g and h in the formal sense, since in general $\mathcal{D}(g,h) \neq \mathcal{D}(h,g)$. Nonetheless, it is often useful to think of $\mathcal{D}(g,h)$ as a distance because

$$\mathcal{D}(g,h) \geqslant 0$$

and $\mathcal{D}(g,h) = 0$ if and only if $g(x) = h(x)$. This follows from Jensen's inequality (if ϕ is a convex function, such as $-\ln$, then $\mathbb{E}\phi(X) \geqslant \phi(\mathbb{E}X)$). Namely,

$$\mathcal{D}(g,h) = \mathbb{E}_g \left[-\ln \frac{h(\mathbf{X})}{g(\mathbf{X})} \right] \geqslant -\ln \left[\mathbb{E}_g \frac{h(\mathbf{X})}{g(\mathbf{X})} \right] = -\ln 1 = 0\,.$$

It can be readily seen that the mutual information $\mathcal{M}(\mathbf{X}, \mathbf{Y})$ of vectors $\mathbf{X}$ and $\mathbf{Y}$ defined in (1.20) is related to the CE in the following way:

$$\mathcal{M}(\mathbf{X}, \mathbf{Y}) = \mathcal{D}(f, f_{\mathbf{X}} f_{\mathbf{Y}}) = \mathbb{E}_f \ln \frac{f(\mathbf{X}, \mathbf{Y})}{f_{\mathbf{X}}(\mathbf{X})\, f_{\mathbf{Y}}(\mathbf{Y})}\,.$$

where f is the (joint) pdf of $(\mathbf{X}, \mathbf{Y})$, and $f_{\mathbf{X}}$ and $f_{\mathbf{Y}}$ are the (marginal) pdfs of $\mathbf{X}$ and $\mathbf{Y}$, respectively. In other words the mutual information can be viewed as the CE that measures the "distance" between the joint pdf f of $\mathbf{X}$ and $\mathbf{Y}$ and the product of their marginal pdfs $f_{\mathbf{X}}$ and $f_{\mathbf{Y}}$, that is under assumption that the vectors $\mathbf{X}$ and $\mathbf{Y}$ are *independent.*

The CE distance is a particular case of the Ali–Silver "distance" [8] between two probability densities g and h, which is defined as

$$d(g,h) = \xi \left[\mathbb{E}_g \phi \left(\frac{g(\mathbf{X})}{h(\mathbf{X})} \right) \right], \tag{1.23}$$

where $\xi(\cdot)$ is a continuous convex function on $(0, +\infty)$ and $\phi(\cdot)$ is an increasing, real-valued function of a real variable. For the CE distance we have $\phi = \ln$ and $\xi(y) = y$. In Chapter 3 we will consider also the variance minimization (VM) "distance," where $\phi(y) = y^2$ and $\xi(y) = y$.

Finally, we mention that the CE is often used in *Bayesian statistics* to measure the "distance" between *prior* and the *posterior* distributions. In particular, CE enables selection of the most "uninformative" prior distribution, which is crucial in Bayesian statistics and is associated with the *maximum entropy* principle [23]. It should be noted that in Bayesian inference the cross-entropy is usually written with the *opposite sign*, so that maximization of the Bayesian cross-entropy corresponds to minimization of the CE distance.

The MLE and the Score Function

We introduce here the notion of the *score function* via the classical *maximum likelihood estimator* (MLE). Consider a random vector $\mathbf{X} = (X_1, \ldots, X_n)$, which is distributed according to a fixed pdf $f(\cdot; \boldsymbol{\theta})$ with unknown parameter (vector) $\boldsymbol{\theta} \in \Theta$. Assume that we wish to estimate $\boldsymbol{\theta}$ on the basis of a given outcome $\mathbf{x}$ (the data) of $\mathbf{X}$. For a given $\mathbf{x}$, the function $\mathcal{L}(\boldsymbol{\theta}; \mathbf{x}) = f(\mathbf{x}; \boldsymbol{\theta})$ is called the *likelihood function*. Note that $\mathcal{L}$ is a function of $\boldsymbol{\theta}$ for a fixed parameter $\mathbf{x}$, whereas for the pdf f it is the other way around. The maximum likelihood *estimate* MLE $\widehat{\boldsymbol{\theta}} = \widehat{\boldsymbol{\theta}}(\mathbf{x})$ of $\boldsymbol{\theta}$ is defined as

$$\widehat{\boldsymbol{\theta}} = \operatorname*{argmax}_{\boldsymbol{\theta} \in \Theta} \mathcal{L}(\boldsymbol{\theta}; \mathbf{x}) \,. \tag{1.24}$$

Because the function ln is monotone increasing we also have

$$\widehat{\boldsymbol{\theta}} = \operatorname*{argmax}_{\boldsymbol{\theta} \in \Theta} \ln \mathcal{L}(\boldsymbol{\theta}; \mathbf{x}) \,. \tag{1.25}$$

The random variable $\widehat{\boldsymbol{\theta}}(\mathbf{X})$ with $\mathbf{X} \sim f(\cdot; \boldsymbol{\theta})$ is the corresponding maximum likelihood *estimator*, which is also abbreviated as MLE and again written as $\widehat{\boldsymbol{\theta}}$. Note that often the data $X_1, \ldots, X_n$ form a random sample from some pdf $f_1(\cdot; \boldsymbol{\theta})$, in which case $f(\mathbf{x}; \boldsymbol{\theta}) = \prod_{i=1}^N f_1(x_i; \boldsymbol{\theta})$ and

$$\widehat{\boldsymbol{\theta}} = \operatorname*{argmax}_{\boldsymbol{\theta} \in \Theta} \sum_{i=1}^N \ln f_1(X_i; \boldsymbol{\theta}) \,. \tag{1.26}$$

If $\mathcal{L}(\boldsymbol{\theta}; \mathbf{x})$ is a continuously differentiable concave function with respect to $\boldsymbol{\theta}$ and the maximum is attained in the interior of Θ, then we can find the MLE of $\boldsymbol{\theta}$ by solving

$$\nabla_{\boldsymbol{\theta}} \ln \mathcal{L}(\boldsymbol{\theta}; \mathbf{x}) = \mathbf{0} \,.$$

The function $\mathcal{S}(\cdot; \mathbf{x})$ defined by

$$\mathcal{S}(\boldsymbol{\theta}; \mathbf{x}) = \nabla_{\boldsymbol{\theta}} \ln \mathcal{L}(\boldsymbol{\theta}; \mathbf{x}) = \frac{\nabla_{\boldsymbol{\theta}} f(\mathbf{x}; \boldsymbol{\theta})}{f(\mathbf{x}; \boldsymbol{\theta})} \tag{1.27}$$

is called the *score function*. For the exponential family (1.3) it is easy to see that

$$\mathcal{S}(\boldsymbol{\theta}; \mathbf{x}) = \frac{\nabla c(\boldsymbol{\theta})}{c(\boldsymbol{\theta})} + \mathbf{t}(\mathbf{x}) \,. \tag{1.28}$$

The *random vector* $\mathcal{S}(\boldsymbol{\theta}) = \mathcal{S}(\boldsymbol{\theta}; \mathbf{X})$ with $\mathbf{X} \sim f(\cdot; \boldsymbol{\theta})$ is called the *(efficient) score*. The expected score is always equal to the zero vector, that is

$$\mathbb{E}_{\boldsymbol{\theta}}\mathcal{S}(\boldsymbol{\theta}) = \int \nabla_{\boldsymbol{\theta}} f(\mathbf{x};\boldsymbol{\theta})\,\mu(d\mathbf{x}) = \nabla_{\boldsymbol{\theta}} \int f(\mathbf{x};\theta)\,\mu(d\mathbf{x}) = \nabla_{\boldsymbol{\theta}} 1 = \mathbf{0}\,,$$

where the interchange of differentiation and integration is justified via the bounded convergence theorem.

Fisher Information

The covariance matrix $\mathcal{I}(\boldsymbol{\theta})$ of the score $\mathcal{S}(\boldsymbol{\theta})$ is called the *Fisher information matrix.* Since the expected score is always $\mathbf{0}$, we have

$$\mathcal{I}(\boldsymbol{\theta}) = \mathbb{E}_{\boldsymbol{\theta}}\mathcal{S}(\boldsymbol{\theta})\mathcal{S}(\boldsymbol{\theta})^T\,. \tag{1.29}$$

In the 1-dimensional case we thus have

$$\mathcal{I}(\theta) = \mathbb{E}_\theta \left(\frac{\partial \ln f(X;\theta)}{\partial \theta}\right)^2 .$$

Because

$$\frac{\partial^2}{\partial\theta^2}\ln f(x;\theta) = \frac{\frac{\partial^2}{\partial\theta^2} f(x;\theta)}{f(x;\theta)} - \left(\frac{\frac{\partial}{\partial\theta} f(x;\theta)}{f(x;\theta)}\right)^2 ,$$

we see that (under straightforward regularity conditions) the Fisher information is also given by

$$\mathcal{I}(\theta) = -\mathbb{E}_\theta \frac{\partial^2 \ln f(X;\theta)}{\partial\theta^2} .$$

In the multidimensional case we have similarly

$$\mathcal{I}(\boldsymbol{\theta}) = -\mathbb{E}_{\boldsymbol{\theta}}\,\nabla\mathcal{S}(\boldsymbol{\theta}) = -\mathbb{E}_{\boldsymbol{\theta}}\,\nabla^2 \ln f(\mathbf{X};\boldsymbol{\theta})\,, \tag{1.30}$$

where $\nabla^2 \ln f(\mathbf{X};\boldsymbol{\theta})$ denotes the *Hessian* of $\ln f(\mathbf{X};\boldsymbol{\theta})$, that is the (random) matrix

$$\left(\frac{\partial^2 \ln f(\mathbf{X};\boldsymbol{\theta})}{\partial\theta_i\partial\theta_j}\right) .$$

The importance of the Fisher information in statistics is corroborated by the famous *Cramér-Rao inequality* which (in a simplified form) states that the variance of any unbiased estimator Z of $g(\boldsymbol{\theta})$ is bounded from below via

$$\mathrm{Var}(Z) \geqslant (\nabla g(\boldsymbol{\theta}))^T\,\mathcal{I}^{-1}(\boldsymbol{\theta})\,\nabla g(\boldsymbol{\theta})\,. \tag{1.31}$$

For more details see [107].

1.6 The Score Function Method *

Many real-world complex systems in science and engineering can be modeled as *discrete-event systems* (DES). The behavior of such systems is identified via a sequence of discrete "events," which causes the system to change from one "state" to another. Because of their complexity, the *performance evaluation* of DES is usually studied by simulation and it is often associated with the estimation of the following *response* function

$$\ell = \ell(\mathbf{u}) = \mathbb{E}_{\mathbf{u}} H(\mathbf{X}) = \int H(\mathbf{x})\, f(\mathbf{x};\mathbf{u})\, \mu(d\mathbf{x}) \,. \tag{1.32}$$

Here $\mathbf{X}$ is assumed to be distributed $f(\cdot;\mathbf{u})$, $\mathbf{u} \in \mathcal{V}$ with respect to some base measure μ, and $H(\mathbf{X})$ is called the sample performance, for example, the steady-state waiting time process in a queueing network. *Sensitivity analysis* is concerned with evaluating sensitivities (gradients, Hessians, etc.) of the response function $\ell(\mathbf{u})$ with respect to parameter vector $\mathbf{u}$ and it is based on the score function and the Fisher information. It provides guidance for design and operational decisions and plays an important role in selecting system parameters that optimize certain performance measures.

In this section we give a brief review of *the score function (SF) method*, a powerful approach to sensitivity analysis and optimization of DES via the score function. This approach was independently discovered in the late 1960s by Aleksandrov, Sysoyev and Shemeneva [7]; Mikhailov [119]; Miller [120] and Rubinstein [140]. For relevant references see [16, 17, 67, 106, 136, 141, 143, 149, 148, 156].

The SF approach permits estimation of *all* sensitivities (gradients, Hessians, etc.) from a *single simulation run* (experiment), in the course of evaluating performance (output) measures. Moreover, it can handle sensitivity analysis and optimization with hundreds of decision parameters [148].

To proceed let us write (1.32) as

$$\ell(\mathbf{u}) = \mathbb{E}_{\mathbf{v}} H(\mathbf{X})\, W(\mathbf{X};\mathbf{u},\mathbf{v}) = \int H(\mathbf{x})\, \frac{f(\mathbf{x};\mathbf{u})}{f(\mathbf{x};\mathbf{v})} f(\mathbf{x};\mathbf{v})\, \mu(d\mathbf{x}) \,, \tag{1.33}$$

where

$$W(\mathbf{X};\mathbf{u},\mathbf{v}) = \frac{f(\mathbf{X};\mathbf{u})}{f(\mathbf{X};\mathbf{v})}$$

is the *likelihood ratio* (LR) of $f(\cdot;\mathbf{u})$ and $f(\cdot;\mathbf{v})$, $\mathbf{X} \sim f(\cdot;\mathbf{v})$ and it is assumed that all densities $f(\cdot;\mathbf{v})$, $\mathbf{v} \in \mathcal{V}$ have the same support. For a fixed $\mathbf{v} \neq \mathbf{u}$ the density $f(\cdot;\mathbf{v})$ is called the *importance sampling* density, which will play an important role in this book.

An unbiased estimator of $\ell(\mathbf{u})$ is $\widehat{\ell}(\mathbf{u};\mathbf{v}) = \frac{1}{N}\sum_{i=1}^{N} H(\mathbf{X}_i)\, W(\mathbf{X}_i;\mathbf{u},\mathbf{v})$, which is called the *likelihood ratio estimator*. Here $\mathbf{X}_1, \ldots, \mathbf{X}_N$ is a random

*This section can be omitted at first reading.

sample from $f(\cdot;\mathbf{v})$. Note that $\widehat{\ell}(\mathbf{u};\mathbf{v})$ is an unbiased estimator of $\ell(\mathbf{u})$ for all $\mathbf{v}$. Therefore, by varying $\mathbf{u}$ and keeping $\mathbf{v}$ fixed we can estimate unbiasedly the whole *response surface* $\{\ell(\mathbf{u}), \mathbf{u} \in \mathcal{V}\}$ via a *single simulation.* Moreover, we can estimate from that single simulation (single sample $\mathbf{X}_1,\ldots,\mathbf{X}_N$) also the *sensitivities* of ℓ, meaning the gradient $\nabla\ell(\mathbf{u})$, the Hessian $\nabla^2\ell(\mathbf{u})$ and higher order derivatives.

Let us first consider the gradient of ℓ. Assume first that the parameter u is a scalar and the parameter set $\mathcal{V}$ is an open interval on the real line. Suppose that for all $\mathbf{x}$, the function $f(\mathbf{x};u)$ is continuously differentiable in u and that there exists an integrable function h, such that

$$\left| H(\mathbf{x}) \frac{\mathrm{d} f(\mathbf{x};u)}{\mathrm{d}u} \right| \leqslant h(\mathbf{x}) \tag{1.34}$$

for all $u \in \mathcal{V}$. Then by the Lebesgue Dominated Convergence theorem, the differentiation and expectation (integration) operators are interchangeable, so that differentiation of (1.32) yields

$$\begin{aligned}\frac{\mathrm{d}\ell(u)}{\mathrm{d}u} &= \frac{\mathrm{d}}{\mathrm{d}u} \int H(\mathbf{x}) f(\mathbf{x};u)\mu(\mathrm{d}\mathbf{x}) = \int H(\mathbf{x}) \frac{\mathrm{d}f(\mathbf{x},u)}{\mathrm{d}u}\,\mu(\mathrm{d}\mathbf{x}) \\ &= \int H(\mathbf{x}) \frac{\frac{\mathrm{d}f(\mathbf{x};u)}{\mathrm{d}u}}{f(\mathbf{x};u)} f(\mathbf{x};u)\,\mu(\mathrm{d}\mathbf{x}) = \mathbb{E}_u H(\mathbf{X}) \frac{\mathrm{d}\ln f(\mathbf{X};u)}{\mathrm{d}u} \\ &= \mathbb{E}_u H(\mathbf{X})\, \mathcal{S}(u;\mathbf{X})\,.\end{aligned}$$

Consider next the multidimensional case. Similar arguments allow us to represent the gradient and the higher order derivatives of $\ell(\mathbf{u})$ in the form

$$\nabla^k \ell(\mathbf{u}) = \mathbb{E}_{\mathbf{u}} H(\mathbf{X})\, \mathcal{S}^{(k)}(\mathbf{u};\mathbf{X})\,, \tag{1.35}$$

where

$$\mathcal{S}^{(k)}(\mathbf{u};\mathbf{x}) = \frac{\nabla^k f(\mathbf{x};\mathbf{u})}{f(\mathbf{x};\mathbf{u})} \tag{1.36}$$

is the *k-th order score function*, $k = 0,1,2,\ldots$. In particular, $\mathcal{S}^{(0)}(\mathbf{u};\mathbf{x}) = 1$ (by definition), $\mathcal{S}^{(1)}(\mathbf{u};\mathbf{x}) = \mathcal{S}(\mathbf{u};\mathbf{x})$ and $\mathcal{S}^{(2)}(\mathbf{u};\mathbf{x})$ can be represented as

$$\begin{aligned}\mathcal{S}^{(2)}(\mathbf{u};\mathbf{x}) &= \nabla\mathcal{S}(\mathbf{u};\mathbf{x}) + \mathcal{S}(\mathbf{u};\mathbf{x})\,\mathcal{S}(\mathbf{u};\mathbf{x})^T \\ &= \nabla^2 \ln f(\mathbf{x};\mathbf{u}) + \nabla \ln f(\mathbf{x};\mathbf{u})\, \nabla \ln f(\mathbf{x};\mathbf{u})^T\,.\end{aligned} \tag{1.37}$$

All partial derivatives are taken with respect to the components of the parameter vector $\mathbf{u}$.

Applying likelihood ratios to (1.35) yields

$$\nabla^k \ell(\mathbf{u}) = \mathbb{E}_{\mathbf{v}} H(\mathbf{X})\, \mathcal{S}^{(k)}(\mathbf{u};\mathbf{X})\, W(\mathbf{X};\mathbf{u},\mathbf{v})\,. \tag{1.38}$$

Table 1.4 displays the score functions $\mathcal{S}(\mathbf{u};x)$ for the commonly used distributions in Table 1.3.

Table 1.4. Score functions for commonly used distributions.

Distr.	$f(x;\mathbf{u})$	$\mathbf{u}$	$\mathcal{S}(\mathbf{u};x)$
$\mathsf{Gam}(a,b)$	$\dfrac{b^a x^{a-1} \mathrm{e}^{-bx}}{\Gamma(a)}$	(a,b)	$\left(\ln(bx) - \frac{\Gamma'(a)}{\Gamma(a)},\ ab^{-1} - x\right)$
$\mathsf{N}(a,b^2)$	$\dfrac{1}{b\sqrt{2\pi}}\,\mathrm{e}^{-\frac{1}{2}\left(\frac{x-a}{b}\right)^2}$	(a,b)	$\left(b^{-2}(x-a),\ -b^{-1} + b^{-3}(x-a)^2\right)$
$\mathsf{Weib}(a,b)$	$ab\,(bx)^{a-1}\,\mathrm{e}^{-(bx)^a}$	(a,b)	$\left(a^{-1} + \ln(bx)[1-(bx)^a],\ \frac{a}{b}[1-(bx)^a]\right)$
$\mathsf{Bin}(n,p)$	$\dbinom{n}{x} p^x (1-p)^{n-x}$	p	$\dfrac{x-np}{p(1-p)}$
$\mathsf{Po}(a)$	$\dfrac{a^x \mathrm{e}^{-a}}{x!}$	a	$\dfrac{x}{a} - 1$
$\mathsf{G}(p)$	$p(1-p)^{x-1}$	p	$\dfrac{1-px}{p(1-p)}$

In general, the quantities $\nabla^k \ell(\mathbf{u}),\ k = 0,1,\ldots,$ are not available analytically, since the response $\ell(\mathbf{u})$ is not. They can be evaluated, however, either by conventional deterministic numerical methods [164] or via simulation. Simulation is particularly convenient, as the response, $\ell(\mathbf{u})$, and *all* the sensitivities, $\nabla^k \ell(\mathbf{u}),\ k = 1,2,\ldots,$ are expressed as expectations with respect to the same pdf, $f(\mathbf{x};\mathbf{v})$. The LR estimator for $\nabla^k \ell(\mathbf{u})$ is

$$\widehat{\nabla}^k \ell(\mathbf{u}) = \frac{1}{N}\sum_{i=1}^{N} H(\mathbf{X}_i)\, \mathcal{S}^{(k)}(\mathbf{u};\mathbf{X}_i)\, W(\mathbf{X}_i;\mathbf{u},\mathbf{v})\,. \tag{1.39}$$

It is important to note that the LR estimators $\widehat{\nabla}^k \ell(\mathbf{u}),\ k = 0,1,\ldots,$ allow us to estimate the corresponding $\nabla^k \ell(\mathbf{u})$ at virtually any point $\mathbf{u} \in \mathcal{V}$, provided the interchangeability of integration and differentiation is valid. This fact renders the above estimators particularly suitable for solving optimization problems.

The variance of $\widehat{\ell}(\mathbf{u},\mathbf{v}) = \widehat{\nabla}^0 \ell(\mathbf{u},\mathbf{v})$ is determined by the second moment of $H(\mathbf{X})W(\mathbf{X};\mathbf{u},\mathbf{v})$, with $\mathbf{X} \sim f(\cdot;\mathbf{v})$. For exponential families of the form (1.3) explicit formulas can be derived. Specifically, with $\boldsymbol{\theta}$ taking the role of $\mathbf{u}$ and $\boldsymbol{\eta}$ the role of $\mathbf{v}$ we have

$$\begin{aligned}
\mathbb{E}_{\boldsymbol{\eta}}\left\{H(\mathbf{X})W(\mathbf{X};\boldsymbol{\theta},\boldsymbol{\eta})\right\}^2 &= \mathbb{E}_{\boldsymbol{\theta}} H^2(\mathbf{X}) W(\mathbf{X};\boldsymbol{\theta},\boldsymbol{\eta}) \\
&= \int H^2(\mathbf{x}) \frac{c(\boldsymbol{\theta})}{c(\boldsymbol{\eta})}\, \mathrm{e}^{(\boldsymbol{\theta}-\boldsymbol{\eta})\cdot \mathbf{t}(\mathbf{x})}\, c(\boldsymbol{\theta})\, \mathrm{e}^{\boldsymbol{\theta}\cdot\mathbf{t}(\mathbf{x})}\, h(\mathbf{x})\, \mu(d\mathbf{x}) \\
&= \frac{c^2(\boldsymbol{\theta})}{c(\boldsymbol{\eta})} \int H^2(\mathbf{x})\, \mathrm{e}^{(2\boldsymbol{\theta}-\boldsymbol{\eta})\cdot\mathbf{t}(\mathbf{x})}\, h(\mathbf{x})\, \mu(d\mathbf{x}) \\
&= \frac{c^2(\boldsymbol{\theta})}{c(\boldsymbol{\eta})\, c(2\boldsymbol{\theta}-\boldsymbol{\eta})}\, \mathbb{E}_{2\boldsymbol{\theta}-\boldsymbol{\eta}} H^2(\mathbf{X}) \\
&= \mathbb{E}_{\boldsymbol{\eta}} W^2(\mathbf{X};\boldsymbol{\theta},\boldsymbol{\eta})\ \mathbb{E}_{2\boldsymbol{\theta}-\boldsymbol{\eta}} H^2(\mathbf{X})\,.
\end{aligned} \tag{1.40}$$

Confidence regions for $\nabla^k \ell(\mathbf{u})$ can also be obtained, by standard techniques; see [149].

Table 1.5 displays the second moments $\mathbb{E}_v W^2(X; u, v)$, for common exponential families in Tables 1.3 and 1.4. Note that in Table 1.5 we change *one* parameter only, which is denoted by u and is changed to v. For the Gamma and Weibull we have used a reparameterization that will be convenient in the following chapters. The values of $\mathbb{E}_v W^2(X; u, v)$ are calculated via (1.3) and (1.40). In particular, we first reparameterize the distribution in terms of (1.3), with $\theta = \psi(u)$ and $\eta = \psi(v)$, and then calculate

$$\mathbb{E}_\eta W^2(X; \theta, \eta) = \frac{c^2(\theta)}{c(\eta)\, c(2\theta - \eta)} \,. \tag{1.41}$$

At the end we substitute u and v back in order to obtain the desired $\mathbb{E}_v W^2(X; u, v)$.

Table 1.5. $\mathbb{E}_v W^2$ for commonly used distributions.

Distr.	$f(x;u)$	$\theta = \psi(u)$	$c(\theta)$	$\mathbb{E}_v W^2(X;u,v)$
$\mathsf{Gam}(a, u^{-1})$	$\dfrac{u^{-a} x^{a-1} \mathrm{e}^{-x/u}}{\Gamma(a)}$	$-u^{-1}$	$\dfrac{(-\theta)^{2a}}{\Gamma(a)}$	$\left(\dfrac{v^2}{u(2v-u)}\right)^a$
$\mathsf{N}(u, b^2)$	$\dfrac{1}{b\sqrt{2\pi}}\, \mathrm{e}^{-\frac{1}{2}\left(\frac{x-u}{b}\right)^2}$	$\dfrac{u}{b^2}$	$\dfrac{\mathrm{e}^{-\frac{1}{2}\theta^2 b^2}}{b\sqrt{2\pi}}$	$\mathrm{e}^{\left(\frac{u-v}{b}\right)^2}$
$\mathsf{Weib}(a, u^{-1})$	$a u^{-1} (x/u)^{a-1}\, \mathrm{e}^{-(x/u)^a}$	u^{-a}	$a\,\theta$	$\dfrac{(v/u)^{2a}}{2\,(v/u)^a - 1}$
$\mathsf{Bin}(n, u)$	$\dbinom{n}{x} u^x (1-u)^{n-x}$	$\ln \dfrac{u}{1-u}$	$(1+\mathrm{e}^\theta)^{-n}$	$\left(\dfrac{u^2 - 2uv + v}{(1-u)v}\right)^n$
$\mathsf{Po}(u)$	$\dfrac{u^x \mathrm{e}^{-u}}{x!}$	$\ln u$	$\mathrm{e}^{-\mathrm{e}^\theta}$	$\mathrm{e}^{\left(\frac{(u-v)^2}{v}\right)}$
$\mathsf{G}(u)$	$u(1-u)^{x-1}$	$\ln(1-u)$	$1 - e^\theta$	$\dfrac{u^2(v-1)}{v(u^2 - 2u + v)}$

1.7 Generating Random Variables

This section we briefly describe several methods for generating random variables from a prescribed distribution.

1.7.1 The Inverse-Transform Method

Let X be a random variable with cdf F. Since F is a nondecreasing function, the inverse function F^{-1} may be defined as

$$F^{-1}(y) = \inf\{x : F(x) \geqslant y\}\,, \quad 0 \leqslant y \leqslant 1\,. \tag{1.42}$$

It is easy to prove that if $U \sim \mathsf{U}(0,1)$, then

$$X = F^{-1}(U) \tag{1.43}$$

has cdf $F(x)$. Namely, since F is invertible and $\mathbb{P}(U \leqslant u) = u$, we readily obtain that

$$\mathbb{P}(X \leqslant x) = \mathbb{P}(F^{-1}(U) \leqslant x) = \mathbb{P}(U \leqslant F(x)) = F(x)\,. \tag{1.44}$$

Thus, to generate an outcome, say x, of a random variable X with cdf F, first sample an outcome, say u, from $U \sim \mathsf{U}(0,1)$, compute $F^{-1}(u)$, and set it equal to x. Figure 1.1 illustrates the inverse-transform method summarized in the following algorithm:

Algorithm 1.7.1 (The Inverse-Transform Method)

1. *Generate* U *from* $\mathsf{U}(0,1)$.
2. *Return* $X = F^{-1}(U)$.

Fig. 1.1. The inverse-transform method.

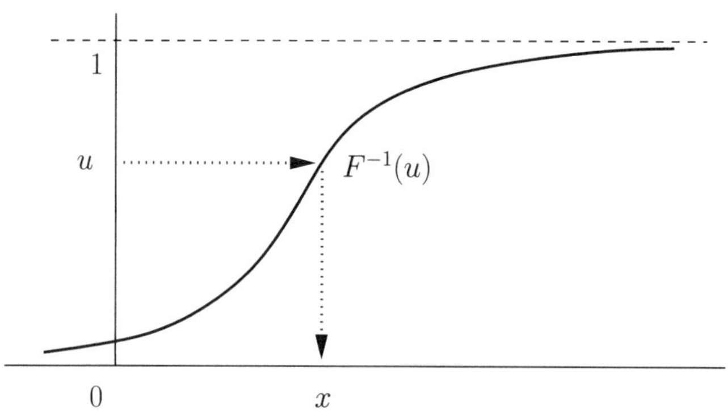

In general, the inverse-transform method requires that the underlying cdf, F, exists in a form for which the corresponding inverse function F^{-1} can be found analytically or algorithmically. Applicable distributions are, for example, the exponential, uniform, Weibull, logistic, and Cauchy distributions.

For a discrete m-point random variable the inverse-transform method can be written as follows:

Algorithm 1.7.2 (The Inverse-Transform Method for a Discrete Distribution)

1. *Generate* $U \sim \mathsf{U}(0,1)$.
2. *Find the smallest positive integer,* k*,* $k = 1, \ldots, m$*, such that* $U \leqslant F(x_k)$ *and return* $X = x_k$.

Much of the execution time in Algorithm 1.7.2 is spent in making the comparisons of Step 2. This time can be reduced by using efficient search techniques; see [47].

The inverse-transform method can be easily extended to random vectors $\mathbf{X} = (X_1, \ldots, X_n)$ of independent random variables from a given joint cdf $F(\mathbf{x})$. In this case we simply apply the method to each component separately. For *dependent* random variables the applicability is quite limited since it requires knowledge of the marginal and conditional distributions of the X_i; see [148].

1.7.2 Alias Method

An alternative to the inverse-transform method for generating discrete random variables, which does not require time consuming search techniques as per Step 2 of Algorithm 1.7.2, is the so-called *alias method* [170]. It is based on the fact that any arbitrary n-point pdf f, with

$$f(x_i) = \mathbb{P}(X = x_i) = p_i \ , \quad i = 1, \ldots, n \ ,$$

can be represented as an *equally weighted mixture* of $n-1$ discrete pdfs, $g^{(k)}$, $k = 1, \ldots, n-1$, each having at most *two* nonzero components. That is, any n-point pdf f can be represented as

$$f = \frac{1}{n-1} \sum_{k=1}^{n-1} g^{(k)} \tag{1.45}$$

for suitably defined 2-point pdfs $g^{(k)}$, $k = 1, \ldots, n-1$; see [170].

The alias method is rather general and efficient, but requires an initial setup and extra storage for the $n-1$ pdfs, $g^{(k)}$. A procedure for computing these two-point pdfs, $g^{(k)}$, $k = 1, \ldots, n-1$ can be found in [47]. Once the representation (1.45) has been established, generation from f is simple and can be written as:

Algorithm 1.7.3

1. *Generate a random random variable U from the discrete uniform pdf* $\mathsf{DU}\{1, n-1\}$. *Let k $(k = 1, \ldots, n-1)$ be the outcome.*
2. *Generate a random random variable X from the two-point pdf $g^{(k)}$, $k = 1, \ldots, n-1$.*

1.7.3 The Composition Method

This method assumes that a cdf, F, can be expressed as a *mixture* of cdfs H_i, that is:

$$F(x) = \sum_{i=1}^{n} p_i \, H_i(x) \ , \tag{1.46}$$

where

$$p_i > 0, \quad \sum_{i=1}^{n} p_i = 1 \ .$$

Let $X_i \sim H_i$ and let Y be a discrete random variable with $\mathbb{P}(Y = i) = p_i$ and independent of X_i, for $1 \leqslant i \leqslant n$. Then the random variable X with cdf F can be represented as

$$X = \sum_{i=1}^{n} X_i \, I_{\{Y=i\}} \ .$$

It follows that in order to generate X from F, we must first generate a discrete random variable Y given above, and then given $Y = i$, generate X_i from H_i.

We thus have

Algorithm 1.7.4 (Composition Method)

1. *Generate the random variable Y according to*

$$\mathbb{P}(Y = i) = p_i \ , \qquad i = 1, \ldots, n \ .$$

2. *Given $Y = i$, return X from the cdf H_i.*

1.7.4 The Acceptance–Rejection Method

The inverse-transform and the composition methods are direct methods in the sense that they deal directly with the cdf of the random variable to be generated. Unfortunately, for many probability distributions it is difficult or impossible to find the inverse transform, that is, to solve

$$F(x) = \int_{-\infty}^{x} f(t) \, dt = u$$

with respect to x. Even if F^{-1} exists in an explicit form, the inverse-transform method may not necessarily be the most efficient variate generation method [47]. The acceptance–rejection method (ARM), which presents an indirect method and is due to John von Neumann, may be appealed to when the above-mentioned direct methods either fail or turn out to be computationally inefficient.

To carry out the ARM we need to specify (1) a pdf h from which it is easy to generate a random variable, and (2) a constant $C \geqslant 1$ such that $C\, h(x) \geqslant f(x)$ for all x. We thus represent $f(x)$ as

$$f(x) = C\, h(x)\, g(x) \ , \tag{1.47}$$

where $0 \leqslant g(x) \leqslant 1$ (see Figure 1.2).

According to the ARM, we generate independently two random variables, U from $\mathsf{U}(0, 1)$ and Y from $h(y)$, and test the inequality $U \leqslant g(Y)$. If the inequality holds, then we accept Y as the required random variable from $f(x)$; otherwise, we reject the pair (U, Y) and try again until a successful pair (U, Y) is obtained. The acceptance–rejection algorithm can be written as follows:

Algorithm 1.7.5 (Acceptance–Rejection Method)

1. *Generate* U *from* $\mathsf{U}(0,1)$.
2. *Generate* Y *from* $h(y)$*, independent of* U.
3. *If* $U \leqslant g(Y)$ *return* $X = Y$. *Otherwise, go to Step 1.*

Fig. 1.2. The acceptance–rejection method.

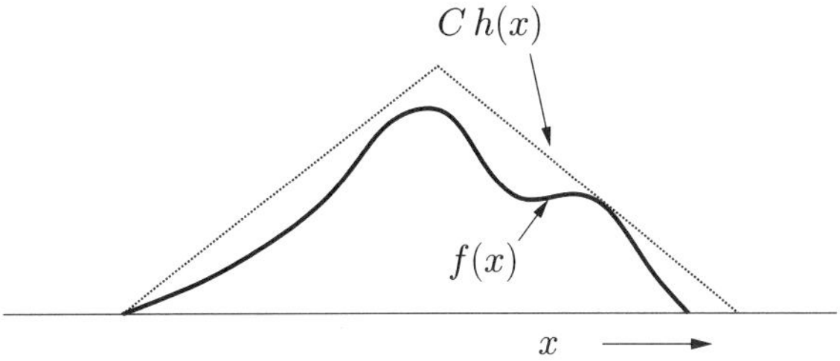

The theoretical justification of the algorithm follows from the application of Bayes' formula to the conditional density $f_Y(x \,|\, U \leqslant g(Y))$, which can be written as

$$f_Y(x \,|\, U \leqslant g(Y)) = \frac{\mathbb{P}(U \leqslant g(Y) \,|\, Y = x)\, h(x)}{\mathbb{P}(U \leqslant g(Y))} \,. \tag{1.48}$$

Direct computations yield

$$\mathbb{P}(U \leqslant g(Y) \,|\, Y = x) = \mathbb{P}(U \leqslant g(x)) = g(x) \tag{1.49}$$

and

$$\begin{aligned} \mathbb{P}(U \leqslant g(Y)) &= \int \mathbb{P}(U \leqslant g(Y) \,|\, Y = x)\, h(x)\, dx \\ &= \int g(x)\, h(x)\, dx = \int \frac{f(x)}{C}\, dx = \frac{1}{C} \,. \end{aligned} \tag{1.50}$$

Upon substitution of (1.49) and (1.50) into (1.48), we obtain

$$f_Y(x \,|\, U \leqslant g(Y)) = C\, h(x)\, g(x) = f(x) \,.$$

The efficiency of the acceptance–rejection method is determined by the acceptance probability $p = \mathbb{P}(U \leqslant g(Y)) = 1/C$ (see (1.50)). Note that the number of trials, say N, before a successful pair (U, Y) occurs has a geometric distribution,

$$\mathbb{P}(N = n) = p(1-p)^{n-1}, \quad n = 1, 2, \ldots, \tag{1.51}$$

with the expected number of trials equal to $1/p = C$.

For this method to be of practical interest, the following criteria must be used in selecting $h(x)$:

1. It should be easy to generate a random variable from $h(x)$.
2. The efficiency, $1/C$, of the procedure should be not too small, that is, $h(x)$ should be "close" to $f(x)$.

Algorithm 1.7.5 is directly applicable to the multidimensional case, using the same reasoning. We need only bear in mind that $\mathbf{Y}$ (see Step 2 of algorithm 1.7.5) becomes an n-dimensional random vector, rather than a one-dimensional random variable. Consequently, we need a convenient way of generating $\mathbf{Y}$ from the multidimensional pdf $h(\mathbf{y})$. However the efficiency of the ARM decreases dramatically with n if $h(\mathbf{x}) \neq f(\mathbf{x})$. In practice it can be used only for $n < 10$; see [148].

1.8 Exercises

Probability Distributions

1. Let $Y = \mathrm{e}^X$, where $X \sim \mathsf{N}(0,1)$. That is, Y has a log-normal distribution. Show that
$$\mathbb{E}\mathrm{e}^{sY} = \infty\,, \qquad \text{for all } s > 0\,.$$
In other words, Y has a heavy-tail distribution.

2. Let $X \sim \mathsf{Beta}(a,b)$. Show that
$$\mathbb{E}X = \frac{a}{a+b} \qquad \text{and} \qquad \mathrm{Var}(X) = \frac{a\,b}{(a+b)^2\,(a+b+1)}\,.$$

3. Let $X_n \sim \mathsf{Beta}(a_n, b_n)$. Suppose $a_n, b_n \to \infty$ and $a_n/b_n \to c$, as $n \to \infty$. Show that X_n converges in probability to $c/(c+1)$, that is,
$$\lim_{n\to\infty} \mathbb{P}\left(\left|X_n - \frac{c}{c+1}\right| > \varepsilon\right) = 0\,,$$
for all $\varepsilon > 0$.

4. Let $U \sim \mathsf{U}[0,1]$. Show that $X = U^{1/v} \sim \mathsf{Beta}(v,1)$ and $1-X \sim \mathsf{Beta}(1,v)$.

5. Show that the cumulant function of the $\mathsf{Beta}(2,1)$ distribution is given by
$$\zeta(s) = \ln\left(\frac{2\,\mathrm{e}^s(s-1)+2}{s^2}\right), \qquad s \in \mathbb{R}\,.$$

6. Consider the collection of pdfs $\{f(\cdot;v), v > 0\}$, with $f(x;v) = v\,x^{v-1}$, $0 \leqslant x \leqslant 1$. Show that this collection (of $\mathsf{Beta}(v,1)$ pdfs) forms an exponential family of the form (1.3), and specify θ, $c(\theta)$, $t(x)$ and $h(x)$ in terms of the parameter v.

Information

7. Random variables X and Y have joint density f given by

$$f(x,y) = \begin{cases} 1/3, & (x,y) = (0,0) \\ 1/3, & (x,y) = (0,1) \\ 0, & (x,y) = (1,0) \\ 1/3, & (x,y) = (1,1) \; . \end{cases}$$

Find the following quantities:
a) $\mathcal{H}(X)$, $\mathcal{H}(Y)$.
b) $\mathcal{H}(X|Y)$, $\mathcal{H}(Y|X)$.
c) $\mathcal{H}(X,Y)$.
d) $\mathcal{H}(X) - \mathcal{H}(X|Y)$.
e) $\mathcal{M}(X,Y)$.

8. Let X, Y and Z be random variables. Prove the following statements:
a) $\mathcal{H}(X \mid Y) \le \mathcal{H}(X)$.
b) $\mathcal{H}(X, Y \mid Z) = \mathcal{H}(X \mid Z) + \mathcal{H}(Y \mid X, Z)$.
c) The mutual information of X and Y satisfies

$$\mathcal{M}(X,Y) = \mathcal{H}(X) - \mathcal{H}(X \mid Y) = \mathcal{H}(Y) - \mathcal{H}(Y \mid X) \; .$$

9. Prove the following statements:
a) $\mathcal{D}(g,h) \geqslant 0$.
b) $\mathcal{D}(g,h)$ is convex in the pair (g,h).
c) If $|\mathcal{X}| = m$ and u is the (discrete) uniform pdf over $\mathcal{X}$, then $\mathcal{D}(g,u) = \ln m - \mathcal{H}(g)$.

10. Let $X_1, X_2, \ldots, X_n$ be a random sample from pdf

$$f(x;\theta) = \theta^{-1} x^{-(1+\theta^{-1})}, \quad x > 1 \, ,$$

for $0 < \theta < 1/2$. The *method of moment estimator* (MME) of θ is obtained by solving, with respect to θ, the equation

$$h(\theta) = N^{-1} \sum_{i=1}^{N} X_i = \overline{X} \, ,$$

where $h(\theta) = \mathbb{E}_\theta X_1 = 1/(1-\theta)$. It follows that the MME is given by

$$\tilde{\theta} = 1 - \frac{1}{\overline{X}} \, .$$

Show that the maximum likelihood estimator (MLE) is given by

$$\widehat{\theta} = \frac{\sum_{i=1}^{N} \ln X_i}{N} \, .$$

Verify the following statements:

a) $\mathbb{E}_\theta \ln X_1 = \theta$,

b) $\mathrm{Var}_\theta(\ln X_1) = \theta^2$,

c) $\mathrm{Var}_\theta(X_1) = \frac{\theta^2}{(1-2\theta)(1-\theta)^2}$,

d) The Fisher information corresponding to X_1 is given by $\mathcal{I}(\theta) = 1/\theta^2$.

Check that, as a consequence of (a)–(c), both $\widehat{\theta}$ and $\tilde{\theta}$ are asymptotically unbiased estimators of θ, that is, $\mathbb{E}_\theta \tilde{\theta} \to \theta$ as $N \to \infty$, and similar for $\widehat{\theta}$. In fact, $\widehat{\theta}$ is unbiased for every N. It is easy to see that by the central limit theorem $\sqrt{N}(\widehat{\theta} - \theta)$ converges in distribution to a $\mathsf{N}(0, \theta^2)$ distribution. Similarly, it can be shown (see Theorem 5.4.1 of [153]) that $\sqrt{N}(\tilde{\theta} - \theta)$ converges in distribution to a $\mathsf{N}(0, \mathrm{Var}_\theta(X_1)/(h'(\theta))^2)$-distribution. Show that for large N the MLE is more efficient than the MME. Finally, note that property (d) shows that the variance of $\widehat{\theta}$ attains the theoretically smallest possible value (Cramér-Rao bound (1.31)) of $1/(\mathcal{I}(\theta)N)$; in other words, the MLE is *(asymptotically) efficient.*

The Score Function Method

11. Calculate the score function $\mathcal{S}$ for the $\mathsf{Beta}(a, b)$ pdf

$$f(x) = \frac{\Gamma(a+b)}{\Gamma(a)\Gamma(b)}\, x^{a-1}\,(1-x)^{b-1},\ x \in [0,1],\ a, b > 0$$

and $\mathsf{Pareto}(a, b)$ pdf

$$f(x) = a\,b\,(1+bx)^{-(a+1)},\ x \in \mathbb{R}_+,\ a, b > 0\,.$$

12. Calculate $\mathbb{E}_v W^2(X; u, v)$ for the $\mathsf{Beta}(v, 1)$ and $\mathsf{Pareto}(v, 1)$ distributions.

13. Show that

$$\nabla_{\mathbf{u}} W(\mathbf{X}; \mathbf{u}, \mathbf{v}) = W(\mathbf{X}; \mathbf{u}, \mathbf{v})\, \mathcal{S}(\mathbf{u}; \mathbf{X})\,.$$

14.* Show that in analogy to (1.40) the covariance matrix of $H(\mathbf{X})\, \nabla_{\boldsymbol{\theta}} W(\mathbf{X}; \boldsymbol{\theta}, \boldsymbol{\eta})$ with $\mathbf{X} \sim f(\cdot; \boldsymbol{\eta})$, is given by

$$\mathbb{E}_{\boldsymbol{\eta}} W^2(\mathbf{X}; \boldsymbol{\theta}, \boldsymbol{\eta})\ \mathbb{E}_{2\boldsymbol{\theta}-\boldsymbol{\eta}}\, H^2(\mathbf{X}) \mathcal{S}(\boldsymbol{\theta}; \mathbf{X})\, \mathcal{S}^T(\boldsymbol{\theta}; \mathbf{X}) - \nabla \ell(\boldsymbol{\theta})[\nabla \ell(\boldsymbol{\theta})]^T\,.$$

15.* Consider the exponential pdf $f(x; \theta) = \theta\, \exp(-\theta\, x)$. Show that if $S(x)$ is a monotonically increasing function, then the expected performance $\ell = \mathbb{E}_\theta S(X)$ — which is assumed to be finite — is a monotonically decreasing, convex function of $\theta \in (0, \infty)$.

Generating Random Variables

16. Apply the inverse-transform method to generate random variables from a Laplace distribution (shifted two-sided exponential distribution), with

$$f(x) = \frac{1}{2\beta} \exp\left(\frac{-|x-\theta|}{\beta}\right), \quad x \in \mathbb{R}, \; \beta > 0 .$$

17. Apply the inverse-transform algorithm to generate random variables from the piecewise-constant pdf

$$f(x) = \begin{cases} C_i , & x_{i-1} \leqslant x \leqslant x_i , \quad i = 1, 2, \ldots, n , \\ 0 , & \text{otherwise} , \end{cases}$$

where $x_0 < x_1 < \cdots < x_{n-1} < x_n$ and $C_i \geqslant 0$, $i = 1, \ldots, n$.

18.* Let

$$f(x) = \begin{cases} C_i \, x , & x_{i-1} \leqslant x < x_i , \quad i = 1, \ldots, n , \\ 0 , & \text{otherwise} , \end{cases}$$

where $0 \leqslant x_0 < x_1 < \cdots < x_{n-1} < x_n$ and $C_i \geqslant 0$, $i = 1, \ldots, n$. Using the inverse-transform method, show that if $X \sim f$, then X can be written as

$$X = \left[x_{\tau-1}^2 + \frac{2(U - F_{\tau-1})}{C_\tau} \right]^{1/2} ,$$

where $U \sim \mathsf{U}(0,1)$, $F_i = \sum_{k=1}^{i} \int_{x_{k-1}}^{x_k} C_k \, y \, dy$, $i = 1, 2, \ldots, n$, and $\tau = \inf\{i : F_i \geqslant U\}$. Describe an algorithm for generating random variables from $f(x)$.

19. Apply the inverse-transform method to generate random variables from the following (discrete) pdf:

$$f(x) = \begin{cases} \dfrac{1}{n+1} , & x = 0, 1, \ldots, n , \\ 0 , & \text{otherwise} . \end{cases}$$

20. Using the acceptance–rejection algorithm generate a random variable from the pdf

$$f(x) = \begin{cases} C_i \, x , & x_{i-1} \leqslant x < x_i , \quad i = 1, \ldots, n , \\ 0 , & \text{otherwise} , \end{cases}$$

where $0 = x_0 < x_1 < \cdots < x_{n-1} < x_n = 1$ and $C_i \geqslant 0$, $i = 1, \ldots, n$. Represent $f(x)$ as

$$f(x) = C\, h(x)\, g(x)\,,$$

where

$$h(x) = 2\,x\,, \quad 0 < x < 1\,.$$

Calculate the efficiency of the algorithm for the case

$$C_1 = C_2 = \cdots = C_n\,, \quad x_i = \frac{i}{n}\,, \quad i = 1, \ldots, n\,.$$

21. Apply the acceptance–rejection algorithm for generating a random variable from the pdf

$$f(x) = \sqrt{\frac{2}{\pi}}\, \mathrm{e}^{-x^2/2}, \quad x \geqslant 0\,,$$

using the representation $f(x) = C\, h(x)\, g(x)$, where

$$h(x) = \frac{1}{\beta}\, \mathrm{e}^{-x/\beta}, \quad x > 0\,,\ \beta > 0\,,$$

for fixed β.

22. Generate a random variable from the pdf

$$f(x) = k\, \mathrm{e}^{-x}\,, \quad 0 \leqslant x \leqslant a$$

using
a) the inverse-transform method
b) the acceptance–rejection method with

$$f(x) = C\, h(x)\, g(x)\,, \text{ where } h(x) = \lambda\, \mathrm{e}^{-\lambda x},\ x > 0\,.$$

Find the efficiency of the acceptance–rejection method for the cases $a = 1$, $a \to 0$ and $a \to \infty$.

23. Let $X_1 \sim \mathsf{Gam}(a, 1)$ and $X_2 \sim \mathsf{Gam}(b, 1)$ be independent. Prove that

$$Y = \frac{X_1}{X_1 + X_2} \sim \mathsf{Beta}(a, b)\,.$$

This implies a general procedure for generating random variables from the $\mathsf{Beta}(a, b)$ distribution. There are many efficient algorithms for generating random variables from a $\mathsf{Gam}(a, 1)$ distribution; see for example [142].

2

A Tutorial Introduction to the Cross-Entropy Method

2.1 Introduction

The aim of this chapter is to provide a gentle and self-contained introduction to the cross-entropy (CE) method. We refer to Section 1.1 for additional background information on the CE method, including many references.

We wish to show that

1. the CE method presents a simple, efficient, and general method for solving a great variety of *estimation and optimization* problems, especially NP-hard combinatorial deterministic and stochastic (noisy) problems,
2. the CE method is a valuable tool for *Monte-Carlo simulation*, in particular when very small probabilities need to be accurately estimated (so-called rare-event simulation).

The CE method has its origins in an adaptive algorithm for rare-event simulation, based on variance minimization [144]. This procedure was soon modified [145] to a randomized optimization technique, where the original variance minimization program was replaced by an associated cross-entropy minimization problem; see Section 1.1.

In the field of rare-event simulation, the CE method is used in conjunction with *importance sampling* (IS), a well-known variance reduction technique in which the system is simulated under a different set of parameters, called the *reference parameters* — or, more generally, a different probability distribution — so as to make the occurrence of the rare event more likely. A major drawback of the conventional IS technique is that the optimal reference parameters to be used in IS are usually very difficult to obtain. Traditional techniques for estimating the optimal reference parameters [148] typically involve time-consuming *variance minimization* (VM) programs. The advantage of the CE method is that it provides a simple and fast adaptive procedure for estimating the optimal reference parameters in the IS. Moreover, the CE method also enjoys asymptotic convergence properties. For example, it is shown in [85] that for *static* models — cf. Remark 2.3 — under mild regularity conditions the CE method terminates with probability 1 in a finite number of iterations, and delivers a consistent and asymptotically normal estimator for the optimal reference parameters. Recently the CE method has been successfully applied to the estimation of rare-event probabilities in dynamic models, in particular queueing models involving both *light-* and *heavy*-tail input distributions; see [46, 15] and Chapter 3.

In the field of optimization problems — combinatorial or otherwise — the CE method can be readily applied by first translating the underlying optimization problem into an associated estimation problem, the so-called *associated stochastic problem* (ASP), which typically involves rare-event estimation. Estimating the rare-event probability and the associated optimal reference parameter for the ASP via the CE method translates effectively back into solving the original optimization problem. Many combinatorial optimization problems (COPs) can be formulated as optimization problems concerning a weighted graph. Depending on the particular problem, the ASP introduces randomness in either

(a) the *nodes* of the graph, in which case we speak of a *stochastic node network* (SNN), or

(b) the *edges* of the graph, in which case we speak of a *stochastic edge network* (SEN).

Examples of SNN problems are the maximal cut (max-cut) problem, the buffer allocation problem and clustering problems. Examples of SEN problems are the travelling salesman problem (TSP), the quadratic assignment problem, the clique problem, and optimal policy search in Markovian decision problems (MDPs). We should emphasize that the CE method may be applied to both deterministic and stochastic COPs. In the latter the objective function itself is random or needs to be estimated via simulation. Stochastic COPs typically occur in stochastic scheduling, flow control, and routing of data networks [24] and in various simulation-based optimization models [148], such as optimal buffer allocation [9]. Chapter 6 deals with noisy optimization problems, for which the CE method is ideally suited.

Recently it was found that the CE method has a strong connection with the fields of neural computation and reinforcement learning. Here CE has been successfully applied to clustering and vector quantization and several MDPs under uncertainty. Indeed, the CE algorithm can be viewed as a stochastic learning algorithm involving the following two iterative phases:

1. Generation of a sample of random data (trajectories, vectors, etc.) according to a specified random mechanism.
2. Updating the parameters of the random mechanism, typically parameters of pdfs, on the basis of the data, to produce a "better" sample in the next iteration.

The significance of the cross-entropy concept is that it defines a precise mathematical framework for deriving fast and "good" updating/learning rules.

The rest of the chapter is organized as follows. In Section 2.2 we present two toy examples that illustrate the basic methodology behind the CE method. The general theory and algorithms are detailed in Section 2.3, for rare-event simulation, and Section 2.4, for Combinatorial Optimization. Finally, in Section 2.5 we discuss the application of the CE method to the max-cut and the TSP, and provide numerical examples of the performance of the algorithm.

Our intention is not to compare the CE method with other heuristics, but demonstrate its beauty and simplicity and promote CE for further applications to optimization and rare-event simulation. This chapter is based partly on [44].

2.2 Methodology: Two Examples

In this section we illustrate the methodology of the CE method via two toy examples, one dealing with rare-event simulation, and the other with combinatorial optimization.

2.2.1 A Rare-Event Simulation Example

Consider the weighted graph of Figure 2.1, with random weights $X_1, \ldots, X_5$. Suppose the weights are independent and exponentially distributed random variables with means $u_1, \ldots, u_5$, respectively. Denote the probability density function (pdf) of $\mathbf{X}$ by $f(\cdot; \mathbf{u})$; thus,

$$f(\mathbf{x}; \mathbf{u}) = \exp\left(-\sum_{j=1}^{5} \frac{x_j}{u_j} \right) \prod_{j=1}^{5} \frac{1}{u_j} \, . \tag{2.1}$$

Let $S(\mathbf{X})$ be the total length of the shortest path from node A to node B.

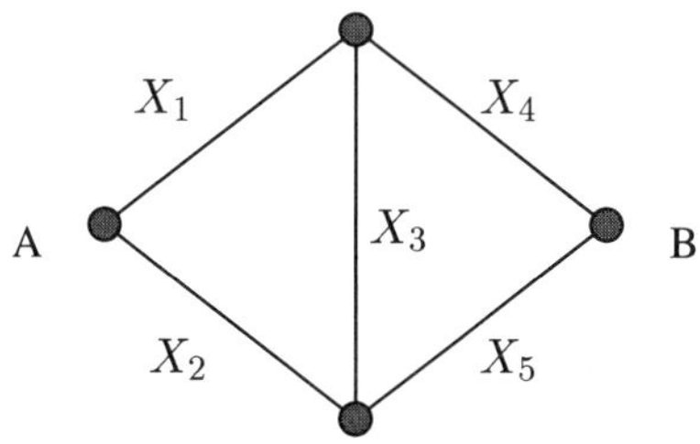

Fig. 2.1. Shortest path from A to B.

We wish to estimate from simulation

$$\ell = \mathbb{P}(S(\mathbf{X}) \geqslant \gamma) = \mathbb{E} I_{\{S(\mathbf{X}) \geqslant \gamma\}} , \tag{2.2}$$

that is, the probability that the length of the shortest path $S(\mathbf{X})$ will exceed some fixed γ. A straightforward way to estimate ℓ in (2.2) is to use *crude Monte Carlo* (CMC) simulation. That is, we draw a random sample $\mathbf{X}_1, \dots, \mathbf{X}_N$ from the distribution of $\mathbf{X}$ and use

$$\frac{1}{N} \sum_{i=1}^{N} I_{\{S(\mathbf{X}_i) \geqslant \gamma\}} \tag{2.3}$$

as the unbiased estimator of ℓ. However, for large γ the probability ℓ will be very small and CMC requires a very large simulation effort. Namely, N needs to be very large in order to estimate ℓ accurately — that is, to obtain a small relative error of 0.01, say. A better way to perform the simulation is to use *importance sampling* (IS). That is, let g be another probability density such that $g(\mathbf{x}) = 0 \Rightarrow I_{\{S(\mathbf{x}) \geqslant \gamma\}} f(\mathbf{x}) = 0$. Using the density g we can represent ℓ as

$$\ell = \int I_{\{S(\mathbf{x}) \geqslant \gamma\}} \frac{f(\mathbf{x})}{g(\mathbf{x})} \, g(\mathbf{x}) \, \mathrm{d}\mathbf{x} = \mathbb{E}_g I_{\{S(\mathbf{X}) \geqslant \gamma\}} \frac{f(\mathbf{X})}{g(\mathbf{X})} , \tag{2.4}$$

where the subscript g means that the expectation is taken with respect to g, which is called the *importance sampling* (IS) density. An unbiased estimator of ℓ is

$$\widehat{\ell} = \frac{1}{N} \sum_{i=1}^{N} I_{\{S(\mathbf{X}_i) \geqslant \gamma\}} \, W(\mathbf{X}_i) , \tag{2.5}$$

where $\widehat{\ell}$ is called the *importance sampling* (IS) or the *likelihood ratio* (LR) estimator,

$$W(\mathbf{x}) = f(\mathbf{x}) / g(\mathbf{x}) \tag{2.6}$$

is called the *likelihood ratio* (LR), and $\mathbf{X}_1, \dots, \mathbf{X}_N$ is a *random sample* from g, that is, $\mathbf{X}_1, \dots, \mathbf{X}_n$ are i.i.d. random vectors with density g. In the particular case where there is no "change of measure," that is, $g = f$, we have $W = 1$, and the LR estimator in (2.6) reduces to the CMC estimator (2.3).

Let us restrict ourselves to g such that $X_1, \dots, X_5$ are independent and exponentially distributed with means $v_1, \dots, v_5$. Then

$$W(\mathbf{x}; \mathbf{u}, \mathbf{v}) = \frac{f(\mathbf{x}; \mathbf{u})}{f(\mathbf{x}; \mathbf{v})} = \exp\left(-\sum_{j=1}^{5} x_j \left(\frac{1}{u_j} - \frac{1}{v_j} \right) \right) \prod_{j=1}^{5} \frac{v_j}{u_j} . \tag{2.7}$$

In this case the "change of measure" is determined by the parameter vector $\mathbf{v} = (v_1, \dots, v_5)$. The main problem now is how to select a $\mathbf{v}$ which gives the most accurate estimate of ℓ for a given simulation effort. As we shall see soon one of the strengths of the CE method for rare-event simulation is that it provides a fast way to determine/estimate the optimal parameters. To this end, without going into the details, a quite general CE algorithm for rare-event estimation is outlined next.

Algorithm

1. Define $\widehat{\mathbf{v}}_0 = \mathbf{u}$. Set $t = 1$ (iteration counter).
2. Generate a random sample $\mathbf{X}_1, \ldots, \mathbf{X}_N$ according to the pdf $f(\cdot; \widehat{\mathbf{v}}_{t-1})$. Calculate the performances $S(\mathbf{X}_i)$ for all i, and order them from smallest to biggest, $S_{(1)} \leqslant \ldots \leqslant S_{(N)}$. Let $\widehat{\gamma}_t$ be the sample $(1-\varrho)$-quantile of performances: $\widehat{\gamma}_t = S_{(\lceil (1-\varrho)N \rceil)}$, provided this is less than γ. Otherwise, put $\widehat{\gamma}_t = \gamma$.
3. Use the **same** sample to calculate, for $j = 1, \ldots, n(= 5)$,

$$\widehat{v}_{t,j} = \frac{\sum_{i=1}^{N} I_{\{S(\mathbf{X}_i) \geqslant \widehat{\gamma}_t\}} W(\mathbf{X}_i; \mathbf{u}, \widehat{\mathbf{v}}_{t-1})\, X_{ij}}{\sum_{i=1}^{N} I_{\{S(\mathbf{X}_i) \geqslant \widehat{\gamma}_t\}} W(\mathbf{X}_i; \mathbf{u}, \widehat{\mathbf{v}}_{t-1})} \,. \tag{2.8}$$

4. If $\widehat{\gamma}_t = \gamma$ then proceed to Step 5; otherwise set $t = t + 1$ and reiterate from Step 2.
5. Let T be the final iteration. Generate a sample $\mathbf{X}_1, \ldots, \mathbf{X}_{N_1}$ according to the pdf $f(\cdot; \widehat{\mathbf{v}}_T)$ and estimate ℓ via the IS estimator

$$\widehat{\ell} = \frac{1}{N_1} \sum_{i=1}^{N_1} I_{\{S(\mathbf{X}_i) \geqslant \gamma\}} W(\mathbf{X}_i; \mathbf{u}, \widehat{\mathbf{v}}_T) \,. \tag{2.9}$$

Note that in Steps 2–4 the optimal IS parameter is estimated. In the final step (Step 5) this parameter is used to estimate the probability of interest. Note also that the algorithm assumes availability of the parameters ϱ (typically between 0.01 and 0.1), N and N_1 in advance.

As an example, consider the case where the *nominal* parameter vector $\mathbf{u}$ is given by $(0.25, 0.4, 0.1, 0.3, 0.2)$. Suppose we wish to estimate the probability that the minimum path is greater than $\gamma = 2$. Crude Monte Carlo with 10^7 samples gave an estimate $1.65 \cdot 10^{-5}$ with an estimated *relative error*, RE, (that is, $\sqrt{\mathrm{Var}(\widehat{\ell})}/\ell$) of 0.165. With 10^8 samples we got the estimate $1.30 \cdot 10^{-5}$ with RE 0.03.

Table 2.1 displays the results of the CE method, using $N = 1000$ and $\varrho = 0.1$. This table was computed in less than half a second.

Table 2.1. Evolution of the sequence $\{(\widehat{\gamma}_t, \widehat{\mathbf{v}}_t)\}$.

t	$\widehat{\gamma}_t$	$\widehat{\mathbf{v}}_t$				
0		0.250	0.400	0.100	0.300	0.200
1	0.575	0.513	0.718	0.122	0.474	0.335
2	1.032	0.873	1.057	0.120	0.550	0.436
3	1.502	1.221	1.419	0.121	0.707	0.533
4	1.917	1.681	1.803	0.132	0.638	0.523
5	2.000	1.692	1.901	0.129	0.712	0.564

Using the estimated optimal parameter vector of $\widehat{\mathbf{v}}_5 = (1.692, 1.901, 0.129, 0.712, 0.564)$, the final step with $N_1 = 10^5$ now gave an estimate of $1.34 \cdot 10^{-5}$ with an estimated RE of 0.03. The simulation time was only 3 seconds, using a Matlab implementation on a Pentium III 500 MHz processor. In contrast, the CPU time required for the CMC method with 10^7 samples is approximately 630 seconds, and with 10^8 samples approximately 6350. We see that with a minimal amount of work we have reduced our simulation effort (CPU time) by roughly a factor of 625.

2.2.2 A Combinatorial Optimization Example

Consider a binary vector $\mathbf{y} = (y_1, \ldots, y_n)$. Suppose that we do not know which components of $\mathbf{y}$ are 0 and which are 1. However, we have an "oracle" which for each binary *input* vector $\mathbf{x} = (x_1, \ldots, x_n)$ returns the performance or response,

$$S(\mathbf{x}) = n - \sum_{j=1}^{n} |x_j - y_j| \,,$$

representing the number of *matches* between the elements of $\mathbf{x}$ and $\mathbf{y}$. Our goal is to present a random search algorithm which reconstructs* (decodes) the unknown vector $\mathbf{y}$ by maximizing the function $S(\mathbf{x})$ on the space of n-dimensional binary vectors.

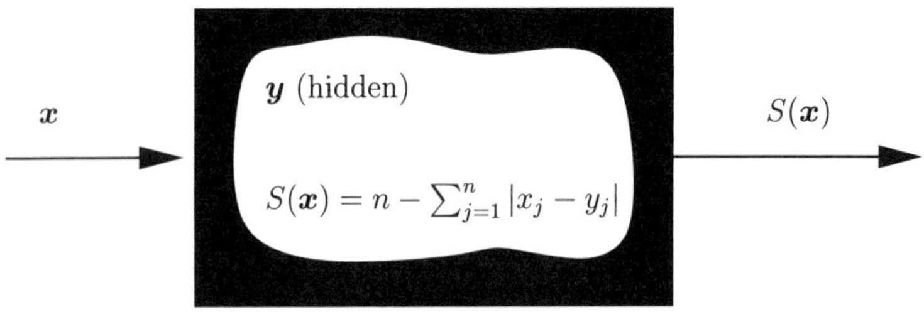

Fig. 2.2. A "device" for reconstructing vector $\mathbf{y}$.

A naive way is to repeatedly generate binary vectors $\mathbf{X} = (X_1, \ldots, X_n)$ such that $X_1, \ldots, X_n$ are independent Bernoulli random variables with success probabilities $p_1, \ldots, p_n$. We write $\mathbf{X} \sim \mathsf{Ber}(\mathbf{p})$, where $\mathbf{p} = (p_1, \ldots, p_n)$. Note that if $\mathbf{p} = \mathbf{y}$, which corresponds to the degenerate case of the Bernoulli distribution, we have $S(\mathbf{X}) = n$, $\mathbf{X} = \mathbf{y}$, and the naive search algorithm yields the optimal solution with probability 1. The CE method for combinatorial optimization consists of creating a sequence of parameter vectors $\widehat{\mathbf{p}}_0, \widehat{\mathbf{p}}_1, \ldots$ and

* Of course, in this toy example the vector $\mathbf{y}$ can be easily reconstructed from the input vectors $(0, 0, \ldots, 0)$, $(1, 0, \ldots, 0)$, $(0, 1, 0, \ldots, 0)$, $\ldots$, $(0, \ldots, 0, 1)$ only.

levels $\widehat{\gamma}_1, \widehat{\gamma}_2, \ldots$, such that $\widehat{\gamma}_1, \widehat{\gamma}_2, \ldots$, converges to the optimal performance (n here) and $\widehat{\mathbf{p}}_0, \widehat{\mathbf{p}}_1, \ldots$ converges to the optimal degenerated parameter vector that coincides with $\mathbf{y}$. Again, the CE procedure — which is similar to the rare-event procedure described in the CE algorithm in Section 2.2.1 — is outlined below, without detail.

Algorithm

1. Start with some $\widehat{\mathbf{p}}_0$. Let $t = 1$.
2. Draw a sample $\mathbf{X}_1, \ldots, \mathbf{X}_N$ of Bernoulli vectors with success probability vector $\widehat{\mathbf{p}}_{t-1}$. Calculate the performances $S(\mathbf{X}_i)$ for all i, and order them from smallest to biggest, $S_{(1)} \leqslant \ldots \leqslant S_{(N)}$. Let $\widehat{\gamma}_t$ be sample $(1 - \varrho)$-quantile of the performances: $\widehat{\gamma}_t = S_{(\lceil (1-\varrho)N \rceil)}$.
3. Use the **same** sample to calculate $\widehat{\mathbf{p}}_t = (\widehat{p}_{t,1}, \ldots, \widehat{p}_{t,n})$ via
$$\widehat{p}_{t,j} = \frac{\sum_{i=1}^{N} I_{\{S(\mathbf{X}_i) \geqslant \widehat{\gamma}_t\}} I_{\{X_{ij}=1\}}}{\sum_{i=1}^{N} I_{\{S(\mathbf{X}_i) \geqslant \widehat{\gamma}_t\}}}, \tag{2.10}$$
$j = 1, \ldots, n$, where $\mathbf{X}_i = (X_{i1}, \ldots, X_{in})$.
4. If the stopping criterion is met, then **stop**; otherwise set $t = t + 1$ and reiterate from Step 2.

A possible stopping criterion is to stop when $\widehat{\gamma}_t$ does not change for a number of subsequent iterations. Another possible stopping criterion is to stop when the vector $\widehat{\mathbf{p}}_t$ has converged to a degenerate — that is, binary — vector. Note that the interpretation of (2.10) is very simple: to *update* the j-th success probability we count how many vectors of the last sample $\mathbf{X}_1, \ldots, \mathbf{X}_N$ have a performance greater than or equal to $\widehat{\gamma}_t$ *and* have the j-th coordinate equal to 1, and we divide this by the number of vectors that have a performance greater than or equal to $\widehat{\gamma}_t$.

As an example, consider the case where $\mathbf{y} = (1, 1, 1, 1, 1, 0, 0, 0, 0, 0)$. Using the initial parameter vector $\widehat{\mathbf{p}}_0 = (1/2, 1/2, \ldots, 1/2)$, and taking $N = 50$ and $\varrho = 0.1$, the algorithm above yields the results given in Table 2.2. We see that the $\widehat{\mathbf{p}}_t$ and $\widehat{\gamma}_t$ converge very quickly to the optimal parameter vector $\mathbf{p}^* = \mathbf{y}$ and optimal performance $\gamma^* = n$, respectively.

Table 2.2. Evolution of the sequence $\{(\widehat{\gamma}_t, \widehat{\mathbf{p}}_t)\}$.

t	$\widehat{\gamma}_t$	$\widehat{\mathbf{p}}_t$									
0		0.50	0.50	0.50	0.50	0.50	0.50	0.50	0.50	0.50	0.50
1	7	0.60	0.40	0.80	0.40	1.00	0.00	0.20	0.40	0.00	0.00
2	9	0.80	0.80	1.00	0.80	1.00	0.00	0.00	0.40	0.00	0.00
3	10	1.00	1.00	1.00	1.00	1.00	0.00	0.00	0.00	0.00	0.00
4	10	1.00	1.00	1.00	1.00	1.00	0.00	0.00	0.00	0.00	0.00

Remark 2.1 (Likelihood ratio term). Note that the randomized optimization algorithm above is almost the *same* as the rare-event simulation algorithm in the previous Section 2.2.1. The most important difference is the absence of the likelihood ratio term W in Step 3. The reason is that for the optimization algorithm the choice of the initial parameter $\mathbf{u}$ is *quite arbitrary*, so using W would be meaningless, while in rare-event simulation it is an essential part of the estimation problem. For more details see Remark 2.5.

2.3 The CE Method for Rare-Event Simulation

In this section we discuss the main ideas behind the CE algorithm for rare-event simulation. When reading this section, the reader is encouraged to refer back to the toy example presented in Section 2.2.1.

Let $\mathbf{X} = (X_1, \ldots, X_n)$ be a random vector taking values in some space $\mathscr{X}$. Let $\{f(\cdot; \mathbf{v})\}$ be a family of probability density functions (pdfs) on $\mathscr{X}$, with respect to some base measure μ, where $\mathbf{v}$ is a real-valued parameter (vector). Thus,

$$\mathbb{E}H(\mathbf{X}) = \int_{\mathscr{X}} H(\mathbf{x})\, f(\mathbf{x}; \mathbf{v})\, \mu(d\mathbf{x}) ,$$

for any (measurable) function H. In most (or all) applications μ is either a counting measure or the Lebesgue measure. For simplicity, for the rest of this section we take $\mu(d\mathbf{x}) = d\mathbf{x}$.

Let S be some real function on $\mathscr{X}$. Suppose we are interested in the probability that $S(\mathbf{X})$ is greater than or equal to some real number γ — which we will refer to as *level* — under $f(\cdot; \mathbf{u})$. This probability can be expressed as

$$\ell = \mathbb{P}_{\mathbf{u}}(S(\mathbf{X}) \geqslant \gamma) = \mathbb{E}_{\mathbf{u}}\, I_{\{S(\mathbf{X}) \geqslant \gamma\}} \,.$$

If this probability is very small, say smaller than 10^{-5}, we call $\{S(\mathbf{X}) \geqslant \gamma\}$ a *rare event.*

A straightforward way to estimate ℓ is to use crude Monte-Carlo simulation: Draw a random sample $\mathbf{X}_1, \ldots, \mathbf{X}_N$ from $f(\cdot\,; \mathbf{u})$; then

$$\frac{1}{N} \sum_{i=1}^{N} I_{\{S(\mathbf{X}_i) \geqslant \gamma\}}$$

is an unbiased estimator of ℓ. However this poses serious problems when $\{S(\mathbf{X}) \geqslant \gamma\}$ is a rare event since a large simulation effort is required to estimate ℓ accurately, that is, with a small relative error or a narrow confidence interval.

An alternative is based on importance sampling: take a random sample $\mathbf{X}_1, \ldots, \mathbf{X}_N$ from an *importance sampling* (different) density g on $\mathscr{X}$, and estimate ℓ using the LR estimator (see (2.5))

$$\widehat{\ell} = \frac{1}{N} \sum_{i=1}^{N} I_{\{S(\mathbf{X}_i) \geqslant \gamma\}} \frac{f(\mathbf{X}_i; \mathbf{u})}{g(\mathbf{X}_i)} . \tag{2.11}$$

The best way to estimate ℓ is to use the change of measure with density

$$g^*(\mathbf{x}) = \frac{I_{\{S(\mathbf{x}) \geqslant \gamma\}} f(\mathbf{x}; \mathbf{u})}{\ell} . \tag{2.12}$$

Namely, by using this change of measure we have in (2.11)

$$I_{\{S(\mathbf{X}_i) \geqslant \gamma\}} \frac{f(\mathbf{X}_i; \mathbf{u})}{g^*(\mathbf{X}_i)} = \ell ,$$

for all i; see[148]. Since ℓ is a constant, the estimator (2.11) has zero variance, and we need to produce only $N = 1$ sample.

The obvious difficulty is of course that this g^* depends on the unknown parameter ℓ. Moreover, it is often convenient to choose a g in the family of densities $\{f(\cdot; \mathbf{v})\}$. The idea now is to choose the *reference parameter* (sometimes called tilting parameter) $\mathbf{v}$ such that the *distance* between the density g^* above and $f(\cdot; \mathbf{v})$ is minimal. A particularly convenient measure of "distance" between two densities g and h is the *Kullback-Leibler distance*, which is also termed the *cross-entropy* between g and h. The Kullback-Leibler distance is defined as:

$$\mathcal{D}(g, h) = \mathbb{E}_g \ln \frac{g(\mathbf{X})}{h(\mathbf{X})} = \int g(\mathbf{x}) \ln g(\mathbf{x}) \, d\mathbf{x} - \int g(\mathbf{x}) \ln h(\mathbf{x}) \, d\mathbf{x} . \tag{2.13}$$

We note that $\mathcal{D}$ is not a "distance" in the formal sense; for example, it is not symmetric.

Minimizing the Kullback-Leibler distance between g^* in (2.12) and $f(\cdot; \mathbf{v})$ is equivalent to choosing $\mathbf{v}$ such that $-\int g^*(\mathbf{x}) \ln f(\mathbf{x}; \mathbf{v}) \, d\mathbf{x}$ is minimized, which is equivalent to solving the maximization problem

$$\max_{\mathbf{v}} \int g^*(\mathbf{x}) \ln f(\mathbf{x}; \mathbf{v}) \, d\mathbf{x} . \tag{2.14}$$

Substituting g^* from (2.12) into (2.14) we obtain the maximization program

$$\max_{\mathbf{v}} \int \frac{I_{\{S(\mathbf{x}) \geqslant \gamma\}} f(\mathbf{x}; \mathbf{u})}{\ell} \ln f(\mathbf{x}; \mathbf{v}) \, d\mathbf{x} , \tag{2.15}$$

which is equivalent to the program

$$\max_{\mathbf{v}} D(\mathbf{v}) = \max_{\mathbf{v}} \mathbb{E}_{\mathbf{u}} \, I_{\{S(\mathbf{X}) \geqslant \gamma\}} \ln f(\mathbf{X}; \mathbf{v}) , \tag{2.16}$$

where D is implicitly defined above. Again using importance sampling, with a change of measure $f(\cdot; \mathbf{w})$ we can rewrite (2.16) as

$$\max_{\mathbf{v}} D(\mathbf{v}) = \max_{\mathbf{v}} \mathbb{E}_{\mathbf{w}} I_{\{S(\mathbf{X})\geqslant\gamma\}} W(\mathbf{X};\mathbf{u},\mathbf{w}) \ln f(\mathbf{X};\mathbf{v}) , \tag{2.17}$$

for *any* reference parameter $\mathbf{w}$, where

$$W(\mathbf{x};\mathbf{u},\mathbf{w}) = \frac{f(\mathbf{x};\mathbf{u})}{f(\mathbf{x};\mathbf{w})}$$

is the *likelihood ratio*, at $\mathbf{x}$, between $f(\cdot;\mathbf{u})$ and $f(\cdot;\mathbf{w})$. The optimal solution of (2.17) can be written as

$$\mathbf{v}^* = \operatorname*{argmax}_{\mathbf{v}} \mathbb{E}_{\mathbf{w}} I_{\{S(\mathbf{X})\geqslant\gamma\}} W(\mathbf{X};\mathbf{u},\mathbf{w}) \ln f(\mathbf{X};\mathbf{v}) . \tag{2.18}$$

We may *estimate* $\mathbf{v}^*$ by solving the following stochastic program (also called *stochastic counterpart* of (2.17))

$$\max_{\mathbf{v}} \widehat{D}(\mathbf{v}) = \max_{\mathbf{v}} \frac{1}{N} \sum_{i=1}^{N} I_{\{S(\mathbf{X}_i)\geqslant\gamma\}} W(\mathbf{X}_i;\mathbf{u},\mathbf{w}) \ln f(\mathbf{X}_i;\mathbf{v}) , \tag{2.19}$$

where $\mathbf{X}_1,\ldots,\mathbf{X}_N$ is a random sample from $f(\cdot;\mathbf{w})$. In typical applications the function $\widehat{D}$ in (3.28) is convex and differentiable with respect to $\mathbf{v}$ [149], in which case the solution of (3.28) may be readily obtained by solving (with respect to $\mathbf{v}$) the following system of equations:

$$\frac{1}{N} \sum_{i=1}^{N} I_{\{S(\mathbf{X}_i)\geqslant\gamma\}} W(\mathbf{X}_i;\mathbf{u},\mathbf{w}) \nabla \ln f(\mathbf{X}_i;\mathbf{v}) = \mathbf{0} , \tag{2.20}$$

The advantage of this approach is that the solution of (2.20) can often be calculated *analytically*. In particular, this happens if the distributions of the random variables belong to a *natural exponential family* (NEF).

It is important to note that the CE program (2.19) is useful only if the probability of the "target event" $\{S(\mathbf{X}) \geqslant \gamma\}$ is not too small under $\mathbf{w}$, say greater than 10^{-5}. For rare-event probabilities, however, the program (2.19) is difficult to carry out. Namely, due to the rareness of the events $\{S(\mathbf{X}_i) \geqslant \gamma\}$, most of the indicator random variables $I_{\{S(\mathbf{X}_i)\geqslant\gamma\}}$, $i = 1,\ldots,N$ will be zero, for moderate N. The same holds for the derivatives of $\widehat{D}(\mathbf{v})$ as given in the left-hand side of (2.20). A *multilevel* algorithm can be used to overcome this difficulty. The idea is to construct a sequence of reference parameters $\{\mathbf{v}_t,\, t \geqslant 0\}$ and a sequence of levels $\{\gamma_t,\ t \geq 1\}$, and iterate in both γ_t and $\mathbf{v}_t$ (see Algorithm 2.3.1 below).

We initialize by choosing a not very small ϱ, say $\varrho = 10^{-2}$ and by defining $\mathbf{v}_0 = \mathbf{u}$. Next, we let γ_1 ($\gamma_1 < \gamma$) be such that, under the original density $f(\mathbf{x};\mathbf{u})$, the probability $\ell_1 = \mathbb{E}_{\mathbf{u}} I_{\{S(\mathbf{X})\geqslant\gamma_1\}}$ is at least ϱ. We then let $\mathbf{v}_1$ be the optimal CE reference parameter for estimating ℓ_1, and repeat the last two steps iteratively with the goal of estimating the pair $\{\ell, \mathbf{v}^*\}$. In other words, each iteration of the algorithm consists of two main *phases*. In the first phase γ_t is updated, in the second $\mathbf{v}_t$ is updated. Specifically, starting with $\mathbf{v}_0 = \mathbf{u}$ we obtain the subsequent γ_t and $\mathbf{v}_t$ as follows:

1. **Adaptive updating of γ_t.** For a fixed $\mathbf{v}_{t-1}$, let γ_t be a $(1-\varrho)$-*quantile* of $S(\mathbf{X})$ under $\mathbf{v}_{t-1}$. That is, γ_t satisfies

$$\mathbb{P}_{\mathbf{v}_{t-1}}(S(\mathbf{X}) \geqslant \gamma_t) \geqslant \varrho\,, \tag{2.21}$$

$$\mathbb{P}_{\mathbf{v}_{t-1}}(S(\mathbf{X}) \leqslant \gamma_t) \geqslant 1-\varrho\,, \tag{2.22}$$

where $\mathbf{X} \sim f(\cdot;\mathbf{v}_{t-1})$.
A simple estimator $\widehat{\gamma}_t$ of γ_t can be obtained by drawing a random sample $\mathbf{X}_1,\ldots,\mathbf{X}_N$ from $f(\cdot;\mathbf{v}_{t-1})$, calculating the performances $S(\mathbf{X}_i)$ for all i, ordering them from smallest to biggest: $S_{(1)} \leqslant \ldots \leqslant S_{(N)}$ and finally, evaluating the sample $(1-\varrho)$-quantile as

$$\widehat{\gamma}_t = S_{(\lceil(1-\varrho)N\rceil)}\,. \tag{2.23}$$

Note that $S_{(j)}$ is called the j-th *order-statistic* of the sequence $S(\mathbf{X}_1),\ldots,S(\mathbf{X}_N)$. Note also that $\widehat{\gamma}_t$ is chosen such that the event $\{S(\mathbf{X}) \geqslant \widehat{\gamma}_t\}$ is not too rare (it has a probability of around ϱ), and therefore updating the reference parameter via a procedure such as (2.23) is not void of meaning.
2. **Adaptive updating of $\mathbf{v}_t$.** For fixed γ_t and $\mathbf{v}_{t-1}$, derive $\mathbf{v}_t$ from the solution of the following CE program

$$\max_{\mathbf{v}} D(\mathbf{v}) = \max_{\mathbf{v}} \mathbb{E}_{\mathbf{v}_{t-1}} I_{\{S(\mathbf{X})\geqslant\gamma_t\}} W(\mathbf{X};\mathbf{u},\mathbf{v}_{t-1}) \ln f(\mathbf{X};\mathbf{v})\,. \tag{2.24}$$

The stochastic counterpart of (2.24) is as follows: for fixed $\widehat{\gamma}_t$ and $\widehat{\mathbf{v}}_{t-1}$, derive $\widehat{\mathbf{v}}_t$ from the solution of following program

$$\max_{\mathbf{v}} \widehat{D}(\mathbf{v}) = \max_{\mathbf{v}} \frac{1}{N}\sum_{i=1}^{N} I_{\{S(\mathbf{X}_i)\geqslant\widehat{\gamma}_t\}} W(\mathbf{X}_i;\mathbf{u},\widehat{\mathbf{v}}_{t-1}) \ln f(\mathbf{X}_i;\mathbf{v})\,. \tag{2.25}$$

Thus, at the first iteration, starting with $\widehat{\mathbf{v}}_0 = \mathbf{u}$, to get a good estimate for $\widehat{\mathbf{v}}_1$, the target event is artificially made less rare by (temporarily) using a level $\widehat{\gamma}_1$ which is chosen smaller than γ. The value of $\widehat{\mathbf{v}}_1$ obtained in this way will (hopefully) make the event $\{S(\mathbf{X}) \geqslant \gamma\}$ less rare in the next iteration, so in the next iteration a value $\widehat{\gamma}_2$ can be used which is closer to γ itself. The algorithm terminates when at some iteration t a level is reached which is at least γ and thus the original value of γ can be used without getting too few samples.

As mentioned before, the optimal solutions of (2.24) and (2.25) can often be obtained *analytically*, in particular when $f(\mathbf{x};\mathbf{v})$ belongs to a NEF.

The above rationale results in the following algorithm.

Algorithm 2.3.1 (Main CE Algorithm for Rare-Event Simulation)

1. *Define* $\widehat{\mathbf{v}}_0 = \mathbf{u}$. *Set* $t = 1$ *(iteration = level counter).*
2. *Generate a sample* $\mathbf{X}_1, \ldots, \mathbf{X}_N$ *from the density* $f(\cdot; \mathbf{v}_{t-1})$ *and compute the sample* $(1-\varrho)$*-quantile* $\widehat{\gamma}_t$ *of the performances according to (2.23), provided* $\widehat{\gamma}_t$ *is less than* γ. *Otherwise set* $\widehat{\gamma}_t = \gamma$.
3. *Use the* **same** *sample* $\mathbf{X}_1, \ldots, \mathbf{X}_N$ *to solve the stochastic program (2.25). Denote the solution by* $\widehat{\mathbf{v}}_t$.
4. *If* $\widehat{\gamma}_t < \gamma$, *set* $t = t + 1$ *and reiterate from Step 2. Else proceed with Step 5.*
5. *Estimate the rare-event probability* ℓ *using the LR estimate*

$$\widehat{\ell} = \frac{1}{N} \sum_{i=1}^{N} I_{\{S(\mathbf{X}_i) \geqslant \gamma\}} W(\mathbf{X}_i; \mathbf{u}, \widehat{\mathbf{v}}_T) , \tag{2.26}$$

where T *denotes the final number of iterations (= number of levels used).*

Example 2.2. We return to the example in Section 2.2.1. In this case, from (2.1) we have

$$\frac{\partial}{\partial v_j} \ln f(\mathbf{x}; \mathbf{v}) = \frac{x_j}{v_j^2} - \frac{1}{v_j} ,$$

so that the j-th equation of (2.20) becomes

$$\sum_{i=1}^{N} I_{\{S(\mathbf{X}_i) \geqslant \gamma\}} W(\mathbf{X}_i; \mathbf{u}, \mathbf{w}) \left(\frac{X_{ij}}{v_j^2} - \frac{1}{v_j} \right) = 0 , \quad j = 1, \ldots, 5 ;$$

therefore

$$v_j = \frac{\sum_{i=1}^{N} I_{\{S(\mathbf{X}_i) \geqslant \gamma\}} W(\mathbf{X}_i; \mathbf{u}, \mathbf{w})\, X_{ij}}{\sum_{i=1}^{N} I_{\{S(\mathbf{X}_i) \geqslant \gamma\}} W(\mathbf{X}_i; \mathbf{u}, \mathbf{w})} , \tag{2.27}$$

which leads to the updating formula (2.8).

Actually, one can show that if the distributions belong to a NEF that is parameterized by the mean, the updating formula always becomes (2.27).

Remark 2.3 (Static Simulation). The above method has been formulated for finite-dimensional random vectors only; this is sometimes referred to as *static* simulation. For infinite-dimensional random vectors or stochastic processes we need a more subtle treatment. We will not go into details here, but rather refer to [46, 15] and Chapter 3. The main point is that Algorithm 2.3.1 holds true without much alteration.

2.4 The CE-Method for Combinatorial Optimization

In this section we discuss the main ideas behind the CE algorithm for combinatorial optimization. When reading this section, the reader is encouraged to refer to the toy example in Section 2.2.2.

Consider the following general maximization problem: Let $\mathscr{X}$ be a finite set of *states*, and let S be a real-valued *performance function* on $\mathscr{X}$. We wish to find the maximum of S over $\mathscr{X}$ and the corresponding state(s) at which this maximum is attained. Let us denote the maximum by γ^*. Thus,

$$S(\mathbf{x}^*) = \gamma^* = \max_{\mathbf{x} \in \mathscr{X}} S(\mathbf{x}) . \tag{2.28}$$

The starting point in the methodology of the CE method is to associate with the optimization problem (2.28) a meaningful *estimation problem.* To this end we define a collection of indicator functions $\{I_{\{S(\mathbf{x}) \geqslant \gamma\}}\}$ on $\mathscr{X}$ for various levels $\gamma \in \mathbb{R}$. Next, let $\{f(\cdot;\mathbf{v}), \mathbf{v} \in \mathscr{V}\}$ be a family of (discrete) probability densities on $\mathscr{X}$, parameterized by a real-valued parameter (vector) $\mathbf{v}$. For a certain $\mathbf{u} \in \mathscr{V}$ we associate with (2.28) the problem of estimating the number

$$\ell(\gamma) = \mathbb{P}_{\mathbf{u}}(S(\mathbf{X}) \geqslant \gamma) = \sum_{\mathbf{x}} I_{\{S(\mathbf{x}) \geqslant \gamma\}} f(\mathbf{x};\mathbf{u}) = \mathbb{E}_{\mathbf{u}} I_{\{S(\mathbf{X}) \geqslant \gamma\}} , \tag{2.29}$$

where $\mathbb{P}_{\mathbf{u}}$ is the probability measure under which the random state $\mathbf{X}$ has pdf $f(\cdot;\mathbf{u})$, and $\mathbb{E}_{\mathbf{u}}$ denotes the corresponding expectation operator. We will call the estimation problem (2.29) the *associated stochastic problem* (ASP). To indicate how (2.29) is associated with (2.28), suppose for example that γ is equal to γ^* and that $f(\cdot;\mathbf{u})$ is the uniform density on $\mathscr{X}$. Note that, typically, $\ell(\gamma^*) = f(\mathbf{x}^*;\mathbf{u}) = 1/|\mathscr{X}|$ — where $|\mathscr{X}|$ denotes the number of elements in $\mathscr{X}$ — is a very small number. Thus, for $\gamma = \gamma^*$ a natural way to estimate $\ell(\gamma)$ would be to use the LR estimator (2.26) with reference parameter $\mathbf{v}^*$ given by

$$\mathbf{v}^* = \operatorname*{argmax}_{\mathbf{v}} \mathbb{E}_{\mathbf{u}}\, I_{\{S(\mathbf{X}) \geqslant \gamma\}} \ln f(\mathbf{X};\mathbf{v}) . \tag{2.30}$$

This parameter could be estimated by

$$\widehat{\mathbf{v}^*} = \operatorname*{argmax}_{\mathbf{v}} \frac{1}{N} \sum_{i=1}^{N} I_{\{S(\mathbf{X}_i) \geqslant \gamma\}} \ln f(\mathbf{X}_i;\mathbf{v}) , \tag{2.31}$$

where the $\mathbf{X}_i$ are generated from pdf $f(\cdot;\mathbf{u})$. It is plausible that, if γ is close to γ^*, that $f(\cdot;\mathbf{v}^*)$ assigns most of its probability mass close to $\mathbf{x}^*$, and thus can be used to generate an approximate solution to (2.28). However, it is important to note that the estimator (2.31) is only of practical use when $I_{\{S(\mathbf{X}) \geqslant \gamma\}} = 1$ for enough samples. This means for example that when γ is close to γ^*, $\mathbf{u}$ needs to be such that $\mathbb{P}_{\mathbf{u}}(S(\mathbf{X}) \geqslant \gamma)$ is not too small. Thus, the choice of $\mathbf{u}$ and γ in (2.28) are closely related. On the one hand we would like to choose γ as close as possible to γ^*, and find (an estimate of) $\mathbf{v}^*$ via the procedure above, which assigns almost all mass to state(s) close to the optimal state. On the other hand, we would like to keep γ relative large in order to obtain an accurate (low RE) estimator for $\mathbf{v}^*$.

The situation is very similar to the rare-event simulation case of Section 2.3. The idea, based essentially on Algorithm 2.3.1, is to adopt a two-phase multilevel approach in which we simultaneously construct a *sequence* of levels $\widehat{\gamma}_1, \widehat{\gamma}_2, \ldots, \widehat{\gamma}_T$ and parameter (vectors) $\widehat{\mathbf{v}}_0, \widehat{\mathbf{v}}_1, \ldots, \widehat{\mathbf{v}}_T$ such that $\widehat{\gamma}_T$ is close to the optimal γ^* and $\widehat{\mathbf{v}}_T$ is such that the corresponding density assigns high probability mass to the collection of states that give a high performance.

This strategy is embodied in the following procedure; see for example [145]:

Algorithm 2.4.1 (Main CE Algorithm for Optimization)

1. *Define* $\widehat{\mathbf{v}}_0 = \mathbf{u}$. *Set* $t = 1$ *(level counter).*
2. *Generate a sample* $\mathbf{X}_1, \ldots, \mathbf{X}_N$ *from the density* $f(\cdot; \mathbf{v}_{t-1})$ *and compute the sample* $(1-\varrho)$*-quantile* $\widehat{\gamma}_t$ *of the performances according to (2.23).*
3. *Use the* **same** *sample* $\mathbf{X}_1, \ldots, \mathbf{X}_N$ *and solve the stochastic program (2.25) with* $W = 1$. *Denote the solution by* $\widehat{\mathbf{v}}_t$.
4. *If for some* $t \geq d$, *say* $d = 5$,

$$\widehat{\gamma}_t = \widehat{\gamma}_{t-1} = \cdots = \widehat{\gamma}_{t-d}\,, \tag{2.32}$$

then **stop** *(let* T *denote the final iteration); otherwise set* $t = t + 1$ *and reiterate from Step 2.*

Note that the initial vector $\widehat{\mathbf{v}}_0$, the sample size N, the stopping parameter d, and the number ϱ have to be specified in advance, but that the rest of the algorithm is "self-tuning."

Remark 2.4 (Smoothed Updating). Instead of updating the parameter vector $\widehat{\mathbf{v}}_{t-1}$ to $\widehat{\mathbf{v}}_t$ directly via (2.31) we use a *smoothed* updating procedure in which

$$\widehat{\mathbf{v}}_t = \alpha\, \widehat{\mathbf{w}}_t + (1-\alpha)\, \widehat{\mathbf{v}}_{t-1}\,, \tag{2.33}$$

where $\widehat{\mathbf{w}}_t$ is the vector derived via (2.25) with $W = 1$. This is especially relevant for optimization problems involving discrete random variables. The main reason why this heuristic smoothed updating procedure performs better is that it prevents the occurrences of 0s and 1s in the parameter vectors; once such an entry is 0 or 1, it often will remain so forever, which is undesirable. We found empirically that a value of α between $0.3 \leq \alpha \leq 0.9$ gives good results. Clearly for $\alpha = 1$ we have the original updating rule in Algorithm 2.4.1.

In many applications we observed numerically that the sequence of pdfs $f(\cdot; \widehat{\mathbf{v}}_0), f(\cdot; \widehat{\mathbf{v}}_1), \ldots$ converges to a degenerate measure (Dirac measure), assigning all probability mass to a single state $\mathbf{x}_T$, for which, by definition, the function value is greater than or equal to $\widehat{\gamma}_T$.

Remark 2.5 (Similarities and differences). Despite the great similarity between Algorithm 2.3.1 and Algorithm 2.4.1, it is important to note that the role of the initial reference parameter $\mathbf{u}$ is significantly different. In Algorithm 2.3.1 $\mathbf{u}$ is the unique *nominal* parameter for estimating $\mathbb{P}_{\mathbf{u}}(S(\mathbf{X}) \geqslant \gamma)$. However, in Algorithm 2.4.1 the choice for the initial parameter $\mathbf{u}$ is fairly arbitrary; it is only used to define the ASP. In contrast to Algorithm 2.3.1 the ASP for Algorithm 2.4.1 is redefined after each iteration. In particular, in Steps 2 and 3 of Algorithm 2.4.1 we determine the optimal reference parameter associated with $\mathbb{P}_{\widehat{\mathbf{v}}_{t-1}}(S(\mathbf{X}) \geqslant \widehat{\gamma}_t)$, instead of $\mathbb{P}_{\mathbf{u}}(S(\mathbf{X}) \geqslant \widehat{\gamma}_t)$. Consequently, the likelihood ratio term W that plays a crucial role in Algorithm 2.3.1 does not appear in Algorithm 2.4.1.

The above procedure can, in principle, be applied to any discrete and continuous optimization problem. For each individual problem two essential ingredients need to be supplied:

1. We need to specify how the samples are generated. In other words, we need to specify the family of densities $\{f(\cdot;\mathbf{v})\}$.
2. We need to calculate the updating rules for the parameters, based on cross-entropy minimization.

In general there are many ways to generate samples from $\mathscr{X}$, and it is not always immediately clear which way of generating the sample will yield better results or easier updating formulas.

Example 2.6. We return to the example from Section 2.2.2. In this case, the random vector $\mathbf{X} = (X_1, \ldots, X_n) \sim \mathsf{Ber}(\mathbf{p})$, and the parameter vector $\mathbf{v}$ is $\mathbf{p}$. Consequently, the pdf is

$$f(\mathbf{X};\mathbf{p}) = \prod_{i=1}^{n} p_i^{X_i}(1-p_i)^{1-X_i} ,$$

and since each X_j can only be 0 or 1,

$$\frac{\partial}{\partial p_j} \ln f(\mathbf{X};\mathbf{p}) = \frac{X_j}{p_j} - \frac{1-X_j}{1-p_j} = \frac{1}{(1-p_j)p_j}(X_j - p_j) .$$

Now we can find the maximum in (2.25) (with $W = 1$) by setting the first derivatives with respect to p_j equal to zero, for $j = 1, \ldots, n$:

$$\frac{\partial}{\partial p_j} \sum_{i=1}^{N} I_{\{S(\mathbf{X}_i) \geqslant \gamma\}} \ln f(\mathbf{X}_i;\mathbf{p}) = \frac{1}{(1-p_j)\,p_j} \sum_{i=1}^{N} I_{\{S(\mathbf{X}_i) \geqslant \gamma\}} \,(X_{ij} - p_j) = 0 .$$

Thus, we get

$$p_j = \frac{\sum_{i=1}^N I_{\{S(\mathbf{X}_i) \geqslant \gamma\}} X_{ij}}{\sum_{i=1}^N I_{\{S(\mathbf{X}_i) \geqslant \gamma\}}} , \tag{2.34}$$

which immediately implies (2.10).

A number of remarks are now in order.

Remark 2.7 (Single-Phase Versus Two-Phase Approach). Algorithm 2.4.1 is a *two-phase* algorithm. That is, at each iteration t both the reference parameter $\mathbf{v}_t$ and the level parameter γ_t are updated. It is not difficult, using the same rationale as before, to formulate a *single-phase* CE algorithm. In particular, consider maximization problem (2.28). Let φ be any increasing "reward"-function of the performances. Let $\{f(\cdot;\mathbf{v})\}$ be a family of densities on $\mathscr{X}$ which contains the Dirac measure at $\mathbf{x}^*$. Then, solving problem (2.28) is equivalent to solving

$$\max_{\mathbf{v}} \mathbb{E}_{\mathbf{v}} \varphi(S(\mathbf{X})) \, ,$$

or solving

$$\max_{\mathbf{v}} \mathbb{E}_{\mathbf{u}} \varphi(S(\mathbf{X})) \ln f(\mathbf{X};\mathbf{v}) \, ,$$

for any $\mathbf{u}$. As before we construct a sequence of parameter vectors $\widehat{\mathbf{v}}_0 = \mathbf{u}, \widehat{\mathbf{v}}_1, \widehat{\mathbf{v}}_2, \ldots$, such that

$$\widehat{\mathbf{v}}_t = \operatorname*{argmax}_{\mathbf{v}} \sum_{i=1}^{N} \varphi(S(\mathbf{X}_i)) \ln f(\mathbf{X}_i;\mathbf{v}) \, ,$$

where $\mathbf{X}_1, \ldots, \mathbf{X}_N$ is a random sample from $f(\cdot;\widehat{\mathbf{v}}_{t-1})$. A reward function without a level parameter γ would simplify Algorithm 2.4.1 substantially. The disadvantage of using this approach is that, typically, it takes too long for Algorithm 2.4.1 to converge, since the large number of "not important" vectors slow down dramatically the convergence of $\{\widehat{\mathbf{v}}_t\}$ to $\mathbf{v}^*$ corresponding to the Dirac measure at $\mathbf{x}^*$. We found numerically that the single-phase CE algorithm is much worse than its two-phase counterpart in Algorithm 2.4.1, in the sense that it is less accurate and more time consuming. Hence it is *crucial* for the CE method to use both phases, that is, follow Algorithm 2.4.1. This is also one of the major differences between CE and ant-based methods, where a single-phase procedure (updating of $\widehat{\mathbf{v}}_t$ alone, no updating of $\widehat{\gamma}_t$) is used [49].

Remark 2.8 (Maximum Likelihood Estimation). It is interesting to note the connection between (2.25) and *maximum likelihood estimation* (MLE). In the MLE problem we are given data $\mathbf{x}_1, \ldots, \mathbf{x}_N$ which are thought to be the outcomes of i.i.d. random variables $\mathbf{X}_1, \ldots, \mathbf{X}_N$ (random sample) each having a distribution $f(\cdot;\mathbf{v})$, where the parameter (vector) $\mathbf{v}$ is an element of some set $\mathscr{V}$. We wish to estimate $\mathbf{v}$ on the basis of the data $\mathbf{x}_1, \ldots, \mathbf{x}_N$. The *maximum likelihood estimate* (MLE) is that parameter $\widehat{\mathbf{v}}$ which maximizes the joint density of $\mathbf{X}_1, \ldots, \mathbf{X}_N$ for the given data $\mathbf{x}_1, \ldots, \mathbf{x}_N$. In other words,

$$\widehat{\mathbf{v}} = \operatorname*{argmax}_{\mathbf{v}} \prod_{i=1}^{N} f(\mathbf{x}_i;\mathbf{v}) \, .$$

The corresponding random variable, obtained by replacing $\mathbf{x}_i$ with $\mathbf{X}_i$ is called the *maximum likelihood estimator* (MLE as well), also denoted by $\widehat{\mathbf{v}}$. Since $\ln(\cdot)$ is an increasing function, we have

$$\widehat{\mathbf{v}} = \operatorname*{argmax}_{\mathbf{v}} \sum_{i=1}^{N} \ln f(\mathbf{X}_i; \mathbf{v}) \,. \tag{2.35}$$

Solving (2.35) is similar to solving (2.25). The only differences are the indicator function $I_{\{S(\mathbf{X}_i) \geqslant \gamma\}}$ and the likelihood ratio W. For $W = 1$ we can write Step 3 in Algorithm 2.4.1 as

$$\widehat{\mathbf{v}}_t = \operatorname*{argmax}_{\mathbf{v}} \sum_{\mathbf{X}_i : S(\mathbf{X}_i) \geqslant \widehat{\gamma}_t} \ln f(\mathbf{X}_i; \mathbf{v}) \,.$$

In other words, $\widehat{\mathbf{v}}_t$ is equal to the MLE of $\widehat{\mathbf{v}}_{t-1}$ based only on the vectors $\mathbf{X}_i$ in the random sample for which the performance is greater than or equal to $\widehat{\gamma}_t$. For example, in Example 2.6 the MLE of p_j based on a random sample $\mathbf{X}_1, \ldots, \mathbf{X}_N$ is

$$\widehat{p}_j = \frac{\sum_{i=1}^{N} X_{ij}}{N} \,.$$

Thus, if we base the MLE only on those vectors that have performance greater than or equal to γ, we obtain (2.34) immediately.

Remark 2.9 (Parameters). The choice for the sample size N and the parameter ϱ depends on the size of the problem and the number of parameters in the ASP. In particular, for a SNN-type problem it is suggested to take the sample size as $N = c\,n$, where n is the number of *nodes* and c a constant ($c > 1$), say $5 \leq c \leq 10$. In contrast, for a SEN-type problem it is suggested to take $N = c\,n^2$, where n^2 is the number of *edges* in the network. It is crucial to realize that the sample sizes $N = c\,n$ and $N = c\,n^2$ (with $c > 1$) are associated with the number of ASP parameters (n and n^2) that one needs to estimate for the SNN and SEN problems, respectively (see also the max-cut and the TSP examples below). Clearly, in order to estimate k parameters reliably, one needs to take *at least* a sample $N = c\,k$ for some constant $c > 1$. Regarding ϱ, it is suggested to take ϱ around 0.01, provided n is reasonably large, say $n \geq 100$; and it is suggested to take a larger ϱ, say $\varrho \approx \ln(n)/n$, for $n < 100$.

Alternatively, the choice of N and ϱ can be determined *adaptively*. For example, in [85] an adaptive algorithm is proposed that adjusts N automatically. The FACE algorithm discussed in Chapter 5 is another option.

2.5 Two Applications

In this section we discuss the application of the CE method to two combinatorial optimization problems, namely the max-cut problem, which is a typical SNN-type problem; and the travelling salesman problem, which is a typical SEN-type problem. We demonstrate the usefulness of the CE method and its fast convergence in a number of numerical examples. We further illustrate the dynamics of the CE method and show how fast the reference parameters converge. For a more in-depth study of the max-cut problem and the related *partition problem* we refer to Sections 4.5 and 4.6. Similarly, the TSP is discussed in more detail in Section 4.7.

2.5.1 The Max-Cut Problem

The max-cut problem in a graph can be formulated as follows. Given a weighted graph $G(V, E)$ with node set $V = \{1, \ldots, n\}$ and edge set E, partition the nodes of the graph into two subsets V_1 and V_2 such that the sum of the weights of the edges going from one subset to the other is maximized. We assume the weights are nonnegative. We note that the max-cut problem is an NP-hard problem. Without loss of generality, we assume that the graph is complete. For simplicity we assume the graph is not directed. We can represent the possibly zero edge weights via a nonnegative, symmetric *cost* matrix $C = (c_{ij})$ where c_{ij} denotes the weight of the edge from i to j.

Formally, a *cut* is a partition $\{V_1, V_2\}$ of V. For example if $V = \{1, \ldots, 6\}$, then $\{\{1, 3, 4\}, \{2, 5, 6\}\}$ is a possible cut. The *cost* of a cut is sum of the weights across the cut. As an example, consider the following 6×6 cost matrix

$$C = \begin{pmatrix} 0 & c_{12} & c_{13} & 0 & 0 & 0 \\ c_{21} & 0 & c_{23} & c_{24} & 0 & 0 \\ c_{31} & c_{32} & 0 & c_{34} & c_{35} & 0 \\ 0 & c_{42} & c_{43} & 0 & c_{45} & c_{46} \\ 0 & 0 & c_{53} & c_{54} & 0 & c_{56} \\ 0 & 0 & 0 & c_{64} & c_{65} & 0 \end{pmatrix} . \tag{2.36}$$

For example the cost of the cut $\{\{1, 3, 4\}, \{2, 5, 6\}\}$ is

$$c_{12} + c_{32} + c_{42} + c_{45} + c_{53} + c_{46} .$$

It will be convenient to represent a cut via a *cut vector* $\mathbf{x} = (x_1, \ldots, x_n)$, where $x_i = 1$ if node i belongs to same partition as 1, and 0 otherwise. By definition $x_1 = 1$. For example, the cut $\{\{1, 3, 4\}, \{2, 5, 6\}\}$ can be represented via the cut vector $(1, 0, 1, 1, 0, 0)$.

Let $\mathscr{X}$ be the set of all cut vectors $\mathbf{x} = (1, x_2, \ldots, x_n)$ and let $S(\mathbf{x})$ be the corresponding cost of the cut. We wish to maximize S via the CE method. Thus, (a) we need to specify how the random cut vectors are generated, and (b) calculate the corresponding updating formulas. The most natural and easiest way to generate the cut vectors is to let $X_2, \ldots, X_n$ be independent Bernoulli random variables with success probabilities $p_2, \ldots, p_n$, exactly as in the second toy example; see Section 2.2.2.

It immediately follows (see Example 2.6) that the updating formula in Algorithm 2.4.1 at the t-th iteration is given by

$$\widehat{p}_{t,i} = \frac{\sum_{k=1}^{N} I_{\{S(\mathbf{X}_k) \geqslant \widehat{\gamma}_t\}} \, I_{\{X_{ki}=1\}}}{\sum_{k=1}^{N} I_{\{S(\mathbf{X}_k) \geqslant \widehat{\gamma}_t\}}} , \tag{2.37}$$

$i = 2, \ldots, n$.

A Synthetic Max-Cut Problem

Since the max-cut problem is NP hard [57, 125], no efficient method for solving the max-cut problem exists. The naive total enumeration routine is only feasible for small graphs, say for those with $n \leq 30$ nodes. Although the branch-and-bound heuristic can solve medium size problems exactly, it too will run into problems when n becomes large.

In order to verify the accuracy of the CE method we construct an artificial graph such that the solution is available in advance. In particular, for $m \in \{1, \ldots, n\}$ consider the following symmetric cost matrix:

$$C = \begin{pmatrix} Z_{11} & B_{12} \\ B_{21} & Z_{22} \end{pmatrix} , \tag{2.38}$$

where Z_{11} is an $m \times m$ symmetric matrix in which all the upper-diagonal elements are generated from a $\mathsf{U}(a, b)$ distribution (and all lower-diagonal elements follow by symmetry), Z_{22} is a $(n - m) \times (n - m)$ symmetric matrix which is generated in a similar way as Z_{11}, and all the other elements are c, apart from the diagonal elements, which are of course 0.

It is not difficult to see that for $c > b(n - m)/m$ the optimal cut is given, by construction, by $V^* = \{V_1^*, V_2^*\}$, with

$$V_1^* = \{1, \ldots, m\} \quad \text{and} \quad V_2^* = \{m + 1, \ldots, n\} , \tag{2.39}$$

and the optimal value of the cut is

$$\gamma^* = c\, m\, (n - m) . \tag{2.40}$$

Of course a similar optimal solution and optimal value can be found for the general case where the elements in Z_{11} and Z_{22} are generated via an arbitrary bounded support distribution with the maximal value of the support less than c.

Table 2.3 lists a typical output of Algorithm 2.4.1 applied to the synthetic max-cut problem, for a network with $n = 400$ nodes. The convergence of the reference vectors $\{\mathbf{p}_t\}$ to the optimal $\mathbf{p}^*$ is further illustrated in Figures 2.3 and 2.4.

In this table besides the $(1-\varrho)$-quantile of the performances $\widehat{\gamma}_t$, we also list the *best* of the performances in each iteration, denoted by $S_{t,(N)}$, and the Euclidean distance

$$||\widehat{\mathbf{p}}_t - \mathbf{p}^*|| = \sqrt{(\widehat{p}_{t,i} - p_i^*)^2} \ ,$$

as a measure of how close the reference vector is to the optimal reference vector $\mathbf{p}^* = (1, 1, \ldots, 1, 0, 0, \ldots, 0)$.

In this particular example, we took $m = 200$ and generated Z_{11} and Z_{22} from $\mathsf{U}(0,1)$ distribution; and the elements in B_{12} (and B_{21}) are constant $c = 1$. The CE parameters were chosen as follows: rarity parameter $\varrho = 0.1$; smoothing parameter $\alpha = 1.0$ (no smoothing); stopping constant $d = 3$; and number of samples per iteration $N = 1000$.

The CPU time was only 100 seconds, using a Matlab implementation on a Pentium III, 500 Mhz processor. We see that the CE algorithm converges quickly, yielding the exact optimal solution 40000 in less than 23 iterations.

Table 2.3. A typical evolution of Algorithm 2.4.1 for the synthetic max-cut problem with $n = 400$, $d = 3$, $\varrho = 0.1$, $\alpha = 1.0$, $N = 1000$.

t	$\widehat{\gamma}_t$	$S_{t,(N)}$	$\lVert\widehat{\mathbf{p}}_t - \mathbf{p}^*\rVert$
1	30085.3	30320.9	9.98
2	30091.3	30369.4	10.00
3	30113.7	30569.8	9.98
4	30159.2	30569.8	9.71
5	30350.8	30652.9	9.08
6	30693.6	31244.7	8.37
7	31145.1	31954.8	7.65
8	31711.8	32361.5	6.94
9	32366.4	33050.3	6.27
10	33057.8	33939.9	5.58
11	33898.6	34897.9	4.93
12	34718.9	35876.4	4.23
13	35597.1	36733.0	3.60
14	36368.9	37431.7	3.02
15	37210.5	38051.2	2.48
16	37996.7	38654.5	1.96
17	38658.8	39221.9	1.42
18	39217.1	39707.8	1.01
19	39618.3	40000.0	0.62
20	39904.5	40000.0	0.29
21	40000.0	40000.0	0.14
22	40000.0	40000.0	0.00
23	40000.0	40000.0	0.00

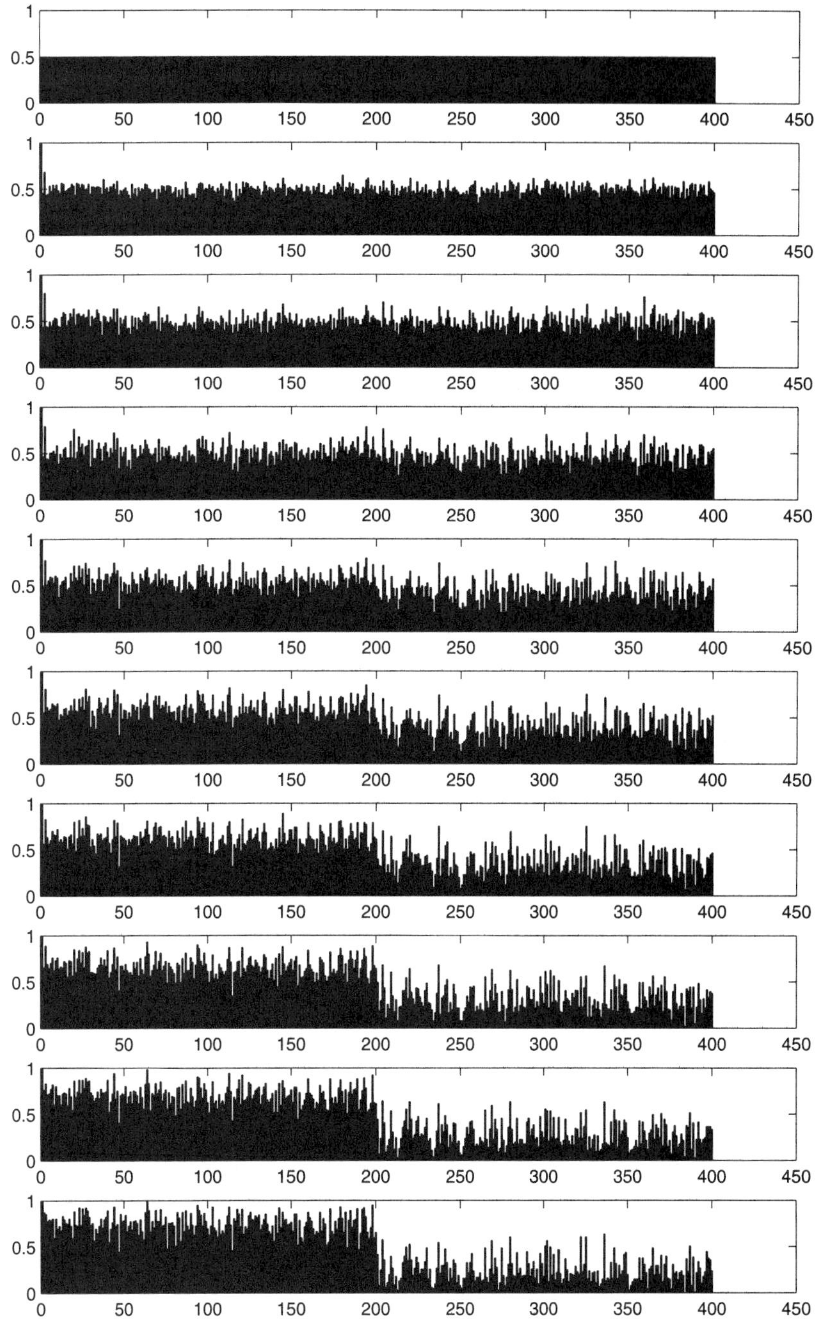

Fig. 2.3. Sequence of reference vectors for the synthetic max-cut problem with 400 nodes. Iterations 0,1,...,9.

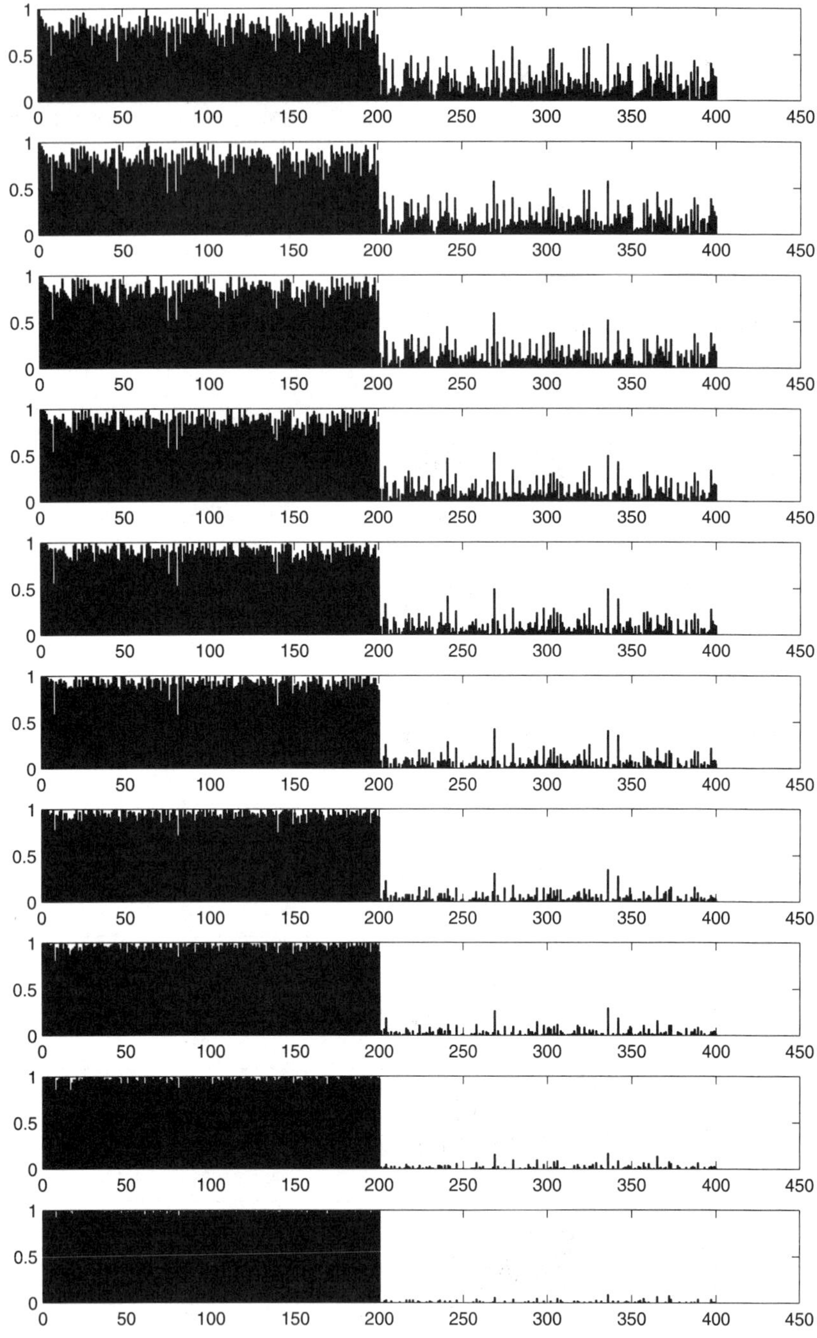

Fig. 2.4. Sequence of reference vectors for the synthetic max-cut problem with 400 nodes. Iterations 10,...,19.

2.5.2 The Travelling Salesman Problem

The travelling salesman problem (TSP) can be formulated as follows: Consider a weighted graph G with n nodes, labeled $1, 2, \ldots, n$. The nodes represent cities, and the edges represent the roads between the cities. Each edge from i to j has weight or cost c_{ij}, representing the length of the road. The problem is to find the shortest *tour* that visits all the cities exactly once[†] (except the starting city, which is also the terminating city); see Figure 2.5.

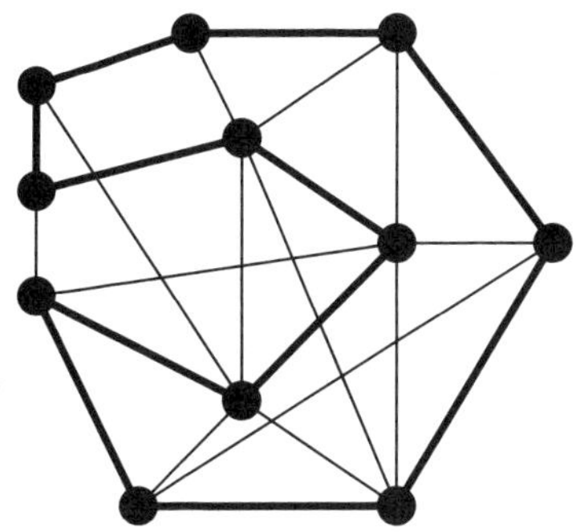

Fig. 2.5. Find the shortest tour $\mathbf{x}$ visiting all nodes.

Without loss of generality, let us assume that the graph is *complete*, and that some of the weights may be $+\infty$. Let $\mathcal{X}$ be the set of all possible tours and let $S(\mathbf{x})$ the total length of tour $\mathbf{x} \in \mathcal{X}$. We can represent each tour via a *permutation* of $(1, \ldots, n)$. For example for $n = 4$, the permutation $(1, 3, 2, 4)$ represents the tour $1 \to 3 \to 2 \to 4 \to 1$. In fact, we may as well represent a tour via a permutation $\mathbf{x} = (x_1, \ldots, x_n)$ with $x_1 = 1$. From now on we identify a tour with its corresponding permutation, where $x_1 = 1$. We may now formulate the TSP as follows.

$$\min_{\mathbf{x} \in \mathcal{X}} S(\mathbf{x}) = \min_{\mathbf{x} \in \mathcal{X}} \left\{ \sum_{i=1}^{n-1} c_{x_i, x_{i+1}} + c_{x_n, 1} \right\} . \tag{2.41}$$

Note that the number of elements in $\mathcal{X}$ is typically very large:

$$|\mathcal{X}| = (n-1)! \tag{2.42}$$

This is exactly the setting of Section 2.4, so we can use the CE method to solve (2.41). Note however that we need to modify Algorithm 2.4.1 since we have here a *minimization* problem.

[†] In some versions of the TSP, cities can be visited more than once.

In order to apply the CE algorithm we need to specify (a) how to generate the random tours, and (b) how to update the parameters at each iteration.

The easiest way to explain how the tours are generated and how the parameters are updated is to relate (2.41) to an *equivalent* minimization problem. Let

$$\tilde{\mathscr{X}} = \{(x_1, \ldots, x_n) \,:\, x_1 = 1, \quad x_i \in \{1, \ldots, n\}, \quad i = 2, \ldots, n\} ,$$

be the set of vectors that correspond to paths of length n that start in 1 and can visit the same city more than once. Note that $|\tilde{\mathscr{X}}| = n^{n-1}$, and $\mathscr{X} \subset \tilde{\mathscr{X}}$. When $n = 4$, we have for example $\mathbf{x} = (1, 3, 1, 3) \in \tilde{\mathscr{X}}$, corresponding to the path $1 \to 3 \to 1 \to 3 \to 1$. Define the function $\tilde{S}$ on $\tilde{\mathscr{X}}$ by $\tilde{S}(\mathbf{x}) = S(\mathbf{x})$, if $\mathbf{x} \in \mathscr{X}$ and $\tilde{S}(\mathbf{x}) = \infty$, otherwise. Then, obviously (2.41) is equivalent to the minimization problem

$$\text{minimize } \tilde{S}(\mathbf{x}) \text{ over } \mathbf{x} \in \tilde{\mathscr{X}} . \tag{2.43}$$

A simple method to generate a random path $\mathbf{X} = (X_1, \ldots, X_n)$ in $\tilde{\mathscr{X}}$ is to use a Markov chain on the graph G, starting at node 1, and stopping after n steps. Let $P = (p_{ij})$ denote the one-step transition matrix of this Markov chain. We assume that the diagonal elements of P are 0, and that all other elements of P are strictly positive, but otherwise P is a general $n \times n$ stochastic matrix.

The pdf $f(\cdot; P)$ of $\mathbf{X}$ is thus parameterized by the matrix P and its logarithm is given by

$$\ln f(\mathbf{x}; P) = \sum_{r=1}^{n} \sum_{i,j} I_{\{\mathbf{x} \in \tilde{\mathscr{X}}_{ij}(r)\}} \ln p_{ij} ,$$

where $\tilde{\mathscr{X}}_{ij}(r)$ is the set of all paths in $\tilde{\mathscr{X}}$ for which the r-th transition is from node i to j. The updating rules for this modified optimization problem follow from (2.25) ($W = 1$), with $\{S(\mathbf{X}_i) \geqslant \gamma\}$ replaced with $\{\tilde{S}(\mathbf{X}_i) \leqslant \gamma\}$, under the condition that the rows of P sum up to 1. Using Lagrange multipliers $u_1, \ldots, u_n$ we obtain the maximization problem

$$\max_{P} \min_{u_1,\ldots,u_n} \left[\mathbb{E}_P I_{\{\tilde{S}(\mathbf{X}) \leqslant \gamma\}} \ln f(\mathbf{X}; P) + \sum_{i=1}^{n} u_i \left(\sum_{j=1}^{n} p_{ij} - 1 \right) \right] . \tag{2.44}$$

Differentiating the expression within square brackets above with respect to p_{ij}, yields, for all $j = 1, \ldots, n$,

$$\frac{\mathbb{E}_P I_{\{\tilde{S}(\mathbf{X}) \leqslant \gamma\}} \sum_{r=1}^{n} I_{\{\mathbf{X} \in \tilde{\mathscr{X}}_{ij}(r)\}}}{p_{ij}} + u_i = 0 . \tag{2.45}$$

Summing over $j = 1, \ldots, n$ gives $\mathbb{E}_P I_{\{\tilde{S}(\mathbf{X}) \leqslant \gamma\}} \sum_{r=1}^{n} I_{\{\mathbf{X} \in \tilde{\mathscr{X}}_i(r)\}} = -u_i$, where $\tilde{\mathscr{X}}_i(r)$ is the set of paths for which the r-th transition starts from node i. It follows that the optimal p_{ij} is given by

$$p_{ij} = \frac{\mathbb{E}_P I_{\{\tilde{S}(\mathbf{X}) \leqslant \gamma\}} \sum_{r=1}^{n} I_{\{\mathbf{X} \in \tilde{\mathcal{X}}_{ij}(r)\}}}{\mathbb{E}_P I_{\{\tilde{S}(\mathbf{X}) \leqslant \gamma\}} \sum_{r=1}^{n} I_{\{\mathbf{X} \in \tilde{\mathcal{X}}_{i}(r)\}}} . \tag{2.46}$$

The corresponding estimator is

$$\widehat{p}_{ij} = \frac{\sum_{k=1}^{N} I_{\{\tilde{S}(\mathbf{X}_k) \leqslant \gamma\}} \sum_{r=1}^{n} I_{\{\mathbf{X}_k \in \tilde{\mathcal{X}}_{ij}(r)\}}}{\sum_{k=1}^{N} I_{\{\tilde{S}(\mathbf{X}_k) \leqslant \gamma\}} \sum_{r=1}^{n} I_{\{\mathbf{X}_k \in \tilde{\mathcal{X}}_{i}(r)\}}} . \tag{2.47}$$

This has a very simple interpretation. To update p_{ij} we simply take the fraction of times that the transitions from i to j occurred, taking into account only those paths that have a total length less than or equal to γ.

This is how we could, *in principle*, carry out the sample generation and parameter updating for problem (2.43). We generate the path via a Markov process with transition matrix P, and use updating formula (2.47). However, *in practice*, we would never generate the paths in this way, since the majority of these tours would be irrelevant since they would not constitute a tour, and therefore their $\tilde{S}$ values would be ∞. In order to avoid the generation of irrelevant tours, we proceed as follows.

Algorithm 2.5.1 (Generation of permutations (tours) in the TSP)

1. *Define* $P^{(1)} = P$ *and* $X_1 = 1$. *Let* $k = 1$.
2. *Obtain* $P^{(k+1)}$ *from* $P^{(k)}$ *by first setting the* X_k*-th column of* $P^{(k)}$ *to 0 and then normalizing the rows to sum up to 1. Generate* X_{k+1} *from the distribution formed by the* X_k*-th row of* $P^{(k+1)}$.
3. *If* $k = n - 1$ *then* **stop**; *otherwise set* $k = k + 1$ *and reiterate from Step 2.*

It is important to realize that the updating formula remains the same; by using Algorithm 2.5.1 we are merely *speeding up* our naive way of generating the paths. Moreover, since we now only generate *tours*, the updated value for p_{ij} can be estimated as

$$\widehat{p}_{ij} = \frac{\sum_{k=1}^{N} I_{\{S(\mathbf{X}_k) \leqslant \gamma\}} I_{\{\mathbf{X}_k \in \mathcal{X}_{ij}\}}}{\sum_{k=1}^{N} I_{\{S(\mathbf{X}_k) \leqslant \gamma\}}} , \tag{2.48}$$

where $\mathcal{X}_{ij}$ is the set of tours in which the transition from i to j is made. This has the same "natural" interpretation as discussed for (2.47).

To complete the algorithm, we need to specify the initialization conditions and the stopping criterion. For the initial matrix $\widehat{P}_0$ we could simply take all off-diagonal elements equal to $1/(n-1)$ and for the stopping criterion use formula (2.32).

Numerical Examples

To demonstrate the usefulness of the CE algorithm and its fast and accurate convergence we provide a number of numerical examples. The first example concerns the benchmark TSP `ft53` taken from the URL

`http://www.iwr.uni-heidelberg.de/groups/comopt/software/TSPLIB95/atsp/`

Table 2.4 presents a typical evolution of the CE Algorithm for the problem `ft53`, which defines an asymmetric fully connected graph of size 53, where the cost of each edge c_{ij} is given. The CE parameters were: stopping parameter $d = 5$, rarity parameter $\varrho = 0.01$, sample size $N = 10\,n^2 = 28090$, and smoothing parameter $\alpha = 0.7$. The *relative experimental error* of the solution is

$$\varepsilon = \frac{\widehat{\gamma}_T - \widehat{\gamma}^*}{\widehat{\gamma}^*} = 0.015\,, \tag{2.49}$$

where $\widehat{\gamma}^* = 6905$ is the best known solution. The CPU time was approximately 6 minutes. In Table 2.4, $S_{t,(1)}$ denotes the length of smallest tour in iteration t. We also included the quantity P_t^{mm}, which is the minimum of the maximum elements in each row of matrix $\widehat{P}_t$.

Table 2.4. A typical evolution of Algorithm 2.4.1 for the TSP `ft53` with $n = 53$ nodes, $d = 5$, $\varrho = 0.01$, $\alpha = 0.7$, $N = 10\,n^2 = 28090$.

t	$\widehat{\gamma}_t$	$S_{t,(1)}$	P_t^{mm}
1	23234.00	21111.00	0.0354
2	20611.00	18586.00	0.0409
3	18686.00	16819.00	0.0514
4	17101.00	14890.00	0.0465
5	15509.00	13459.00	0.0698
6	14449.00	12756.00	0.0901
7	13491.00	11963.00	0.0895
8	12773.00	11326.00	0.1065
9	12120.00	10357.00	0.0965
10	11480.00	10216.00	0.1034
11	11347.00	9952.00	0.1310
12	10791.00	9525.00	0.1319
13	10293.00	9246.00	0.1623
14	10688.00	9176.00	0.1507
15	9727.00	8457.00	0.1346
16	9263.00	8424.00	0.1436

t	$\widehat{\gamma}_t$	$S_{t,(1)}$	P_t^{mm}
17	9422.00	8614.00	0.1582
18	9155.00	8528.00	0.1666
19	8661.00	7970.00	0.1352
20	8273.00	7619.00	0.1597
21	8096.00	7485.00	0.1573
22	7868.00	7216.00	0.1859
23	7677.00	7184.00	0.2301
24	7519.00	7108.00	0.2421
25	7420.00	7163.00	0.2861
26	7535.00	7064.00	0.3341
27	7506.00	7072.00	0.3286
28	7199.00	7008.00	0.3667
29	7189.00	7024.00	0.3487
30	7077.00	7008.00	0.4101
31	7068.00	7008.00	0.4680

Similar performances were found for other TSPs in the benchmark library above. Table 2.5 presents the performance of Algorithm 2.4.1 for a selection of case studies from this library. In all numerical results we use the same CE parameters as for the ft53 problem, that is $\varrho = 10^{-2}$, $N = 10\,n^2, \alpha = 0.7$ (smoothing parameter in (2.33)) and $d = 5$ (in (2.32)). To study the variability in the solutions, each problem was repeated 10 times. In Table 2.5, n denotes the number of nodes of the graph, $\bar{T}$ denotes the average total number of iterations needed before stopping, $\bar{\gamma}_1$ and $\bar{\gamma}_T$ denote the average initial and final estimates of the optimal solution, γ^* denotes the best known solution, $\bar{\varepsilon}$ denotes the average relative experimental error based on 10 replications, ε_* and ε^* denote the smallest and the largest relative error among the 10 generated shortest paths, and finally CPU denotes the average CPU time in seconds. We found that decreasing the sample size N from $N = 10\,n^2$ to $N = 5\,n^2$ all relative experimental errors ε in Table 2.5 increase at most by a factor of 1.5.

Table 2.5. Case studies for the TSP.

file	n	$\bar{T}$	$\bar{\gamma}_1$	$\bar{\gamma}_T$	γ^*	$\bar{\varepsilon}$	ε_*	ε^*	CPU
br17	17	23.8	68.2	39.0	39	0.000	0.000	0.000	9
ftv33	34	31.2	3294.0	1312.2	1286	0.020	0.000	0.062	73
ftv35	36	31.5	3714.0	1490.0	1473	0.012	0.004	0.018	77
ftv38	39	33.8	4010.8	1549.8	1530	0.013	0.004	0.032	132
p43	43	44.5	9235.5	5624.5	5620	0.010	0.000	0.001	378
ftv44	45	35.5	4808.2	1655.8	1613	0.027	0.013	0.033	219
ftv47	48	40.2	5317.8	1814.0	1776	0.021	0.006	0.041	317
ry48p	48	40.8	40192.0	14845.5	14422	0.029	0.019	0.050	345
ft53	53	39.5	20889.5	7103.2	6905	0.029	0.025	0.035	373
ftv55	56	40.0	5835.8	1640.0	1608	0.020	0.002	0.043	408
ftv64	65	43.2	6974.2	1850.0	1839	0.006	0.000	0.014	854
ftv70	71	47.0	7856.8	1974.8	1950	0.013	0.004	0.037	1068
ft70	70	42.8	64199.5	39114.8	38673	0.011	0.003	0.019	948

Dynamics

Finally, as an illustration of the dynamics of the CE algorithm, we display below the sequence of matrices $\widehat{P}_0, \widehat{P}_1, \ldots$ for a TSP with n=10 cities, where the optimal tour is $(1, 2, 3, \ldots, 10, 1)$. A graphical illustration of the convergence is given in Figure 2.6, where we omitted $\widehat{P}_0$ whose off-diagonal elements are all equal to 1/9 and diagonal elements equal to 0.

$$\widehat{P}_1 = \begin{pmatrix} 0.00 & 0.31 & 0.04 & 0.08 & 0.04 & 0.19 & 0.08 & 0.08 & 0.12 & 0.08 \\ 0.04 & 0.00 & 0.33 & 0.08 & 0.17 & 0.08 & 0.08 & 0.04 & 0.04 & 0.12 \\ 0.08 & 0.08 & 0.00 & 0.23 & 0.04 & 0.04 & 0.12 & 0.19 & 0.08 & 0.15 \\ 0.12 & 0.19 & 0.08 & 0.00 & 0.12 & 0.08 & 0.08 & 0.08 & 0.19 & 0.08 \\ 0.08 & 0.08 & 0.19 & 0.08 & 0.00 & 0.23 & 0.08 & 0.04 & 0.15 & 0.08 \\ 0.04 & 0.04 & 0.08 & 0.04 & 0.12 & 0.00 & 0.50 & 0.08 & 0.08 & 0.04 \\ 0.23 & 0.08 & 0.08 & 0.04 & 0.08 & 0.04 & 0.00 & 0.27 & 0.08 & 0.12 \\ 0.08 & 0.15 & 0.04 & 0.04 & 0.19 & 0.08 & 0.08 & 0.00 & 0.27 & 0.08 \\ 0.08 & 0.08 & 0.04 & 0.12 & 0.08 & 0.15 & 0.08 & 0.04 & 0.00 & 0.35 \\ 0.21 & 0.08 & 0.17 & 0.08 & 0.04 & 0.12 & 0.08 & 0.12 & 0.08 & 0.00 \end{pmatrix}.$$

$$\widehat{P}_2 = \begin{pmatrix} 0.00 & 0.64 & 0.03 & 0.06 & 0.04 & 0.04 & 0.06 & 0.04 & 0.04 & 0.06 \\ 0.03 & 0.00 & 0.58 & 0.07 & 0.07 & 0.05 & 0.05 & 0.03 & 0.03 & 0.08 \\ 0.05 & 0.05 & 0.00 & 0.52 & 0.04 & 0.03 & 0.08 & 0.04 & 0.05 & 0.15 \\ 0.04 & 0.13 & 0.05 & 0.00 & 0.22 & 0.18 & 0.05 & 0.04 & 0.25 & 0.05 \\ 0.06 & 0.04 & 0.09 & 0.04 & 0.00 & 0.60 & 0.04 & 0.03 & 0.04 & 0.06 \\ 0.03 & 0.03 & 0.05 & 0.03 & 0.04 & 0.00 & 0.71 & 0.05 & 0.05 & 0.03 \\ 0.20 & 0.04 & 0.05 & 0.03 & 0.05 & 0.03 & 0.00 & 0.51 & 0.05 & 0.04 \\ 0.05 & 0.08 & 0.03 & 0.04 & 0.23 & 0.05 & 0.05 & 0.00 & 0.42 & 0.05 \\ 0.05 & 0.05 & 0.04 & 0.07 & 0.07 & 0.10 & 0.05 & 0.03 & 0.00 & 0.54 \\ 0.50 & 0.05 & 0.04 & 0.05 & 0.04 & 0.08 & 0.05 & 0.14 & 0.05 & 0.00 \end{pmatrix}.$$

$$\widehat{P}_3 = \begin{pmatrix} 0.00 & 0.76 & 0.02 & 0.04 & 0.03 & 0.03 & 0.04 & 0.03 & 0.03 & 0.04 \\ 0.02 & 0.00 & 0.73 & 0.05 & 0.05 & 0.04 & 0.04 & 0.02 & 0.02 & 0.05 \\ 0.03 & 0.03 & 0.00 & 0.70 & 0.02 & 0.02 & 0.05 & 0.02 & 0.03 & 0.09 \\ 0.02 & 0.07 & 0.03 & 0.00 & 0.59 & 0.10 & 0.03 & 0.02 & 0.13 & 0.03 \\ 0.04 & 0.03 & 0.06 & 0.03 & 0.00 & 0.73 & 0.03 & 0.02 & 0.03 & 0.04 \\ 0.02 & 0.02 & 0.04 & 0.02 & 0.03 & 0.00 & 0.79 & 0.04 & 0.04 & 0.02 \\ 0.12 & 0.02 & 0.03 & 0.02 & 0.03 & 0.02 & 0.00 & 0.69 & 0.03 & 0.02 \\ 0.03 & 0.05 & 0.02 & 0.02 & 0.14 & 0.03 & 0.03 & 0.00 & 0.66 & 0.03 \\ 0.03 & 0.03 & 0.02 & 0.05 & 0.05 & 0.06 & 0.03 & 0.02 & 0.00 & 0.71 \\ 0.69 & 0.03 & 0.02 & 0.03 & 0.02 & 0.05 & 0.03 & 0.09 & 0.03 & 0.00 \end{pmatrix}.$$

$$\widehat{P}_4 = \begin{pmatrix} 0.00 & 0.82 & 0.01 & 0.03 & 0.02 & 0.02 & 0.03 & 0.02 & 0.02 & 0.03 \\ 0.01 & 0.00 & 0.80 & 0.03 & 0.03 & 0.03 & 0.03 & 0.01 & 0.01 & 0.04 \\ 0.02 & 0.02 & 0.00 & 0.79 & 0.02 & 0.01 & 0.03 & 0.02 & 0.02 & 0.07 \\ 0.01 & 0.04 & 0.02 & 0.00 & 0.73 & 0.06 & 0.02 & 0.01 & 0.09 & 0.02 \\ 0.03 & 0.02 & 0.04 & 0.02 & 0.00 & 0.81 & 0.02 & 0.01 & 0.02 & 0.03 \\ 0.01 & 0.01 & 0.03 & 0.01 & 0.02 & 0.00 & 0.84 & 0.03 & 0.03 & 0.01 \\ 0.09 & 0.02 & 0.02 & 0.01 & 0.02 & 0.01 & 0.00 & 0.78 & 0.02 & 0.02 \\ 0.02 & 0.03 & 0.01 & 0.02 & 0.09 & 0.02 & 0.02 & 0.00 & 0.76 & 0.02 \\ 0.02 & 0.02 & 0.02 & 0.03 & 0.03 & 0.05 & 0.02 & 0.01 & 0.00 & 0.79 \\ 0.78 & 0.02 & 0.02 & 0.02 & 0.02 & 0.03 & 0.02 & 0.06 & 0.02 & 0.00 \end{pmatrix}.$$

$$\widehat{P}_5 = \begin{pmatrix} 0.00 & 0.86 & 0.01 & 0.02 & 0.02 & 0.02 & 0.02 & 0.02 & 0.02 & 0.02 \\ 0.01 & 0.00 & 0.85 & 0.03 & 0.03 & 0.02 & 0.02 & 0.01 & 0.01 & 0.03 \\ 0.02 & 0.02 & 0.00 & 0.84 & 0.01 & 0.01 & 0.03 & 0.01 & 0.02 & 0.05 \\ 0.01 & 0.03 & 0.01 & 0.00 & 0.80 & 0.05 & 0.01 & 0.01 & 0.06 & 0.01 \\ 0.02 & 0.02 & 0.03 & 0.02 & 0.00 & 0.85 & 0.02 & 0.01 & 0.02 & 0.02 \\ 0.01 & 0.01 & 0.02 & 0.01 & 0.02 & 0.00 & 0.88 & 0.02 & 0.02 & 0.01 \\ 0.06 & 0.01 & 0.02 & 0.01 & 0.02 & 0.01 & 0.00 & 0.84 & 0.02 & 0.01 \\ 0.02 & 0.02 & 0.01 & 0.01 & 0.07 & 0.02 & 0.02 & 0.00 & 0.82 & 0.02 \\ 0.02 & 0.02 & 0.01 & 0.02 & 0.02 & 0.03 & 0.02 & 0.01 & 0.00 & 0.84 \\ 0.84 & 0.02 & 0.01 & 0.02 & 0.01 & 0.03 & 0.02 & 0.05 & 0.02 & 0.00 \end{pmatrix}.$$

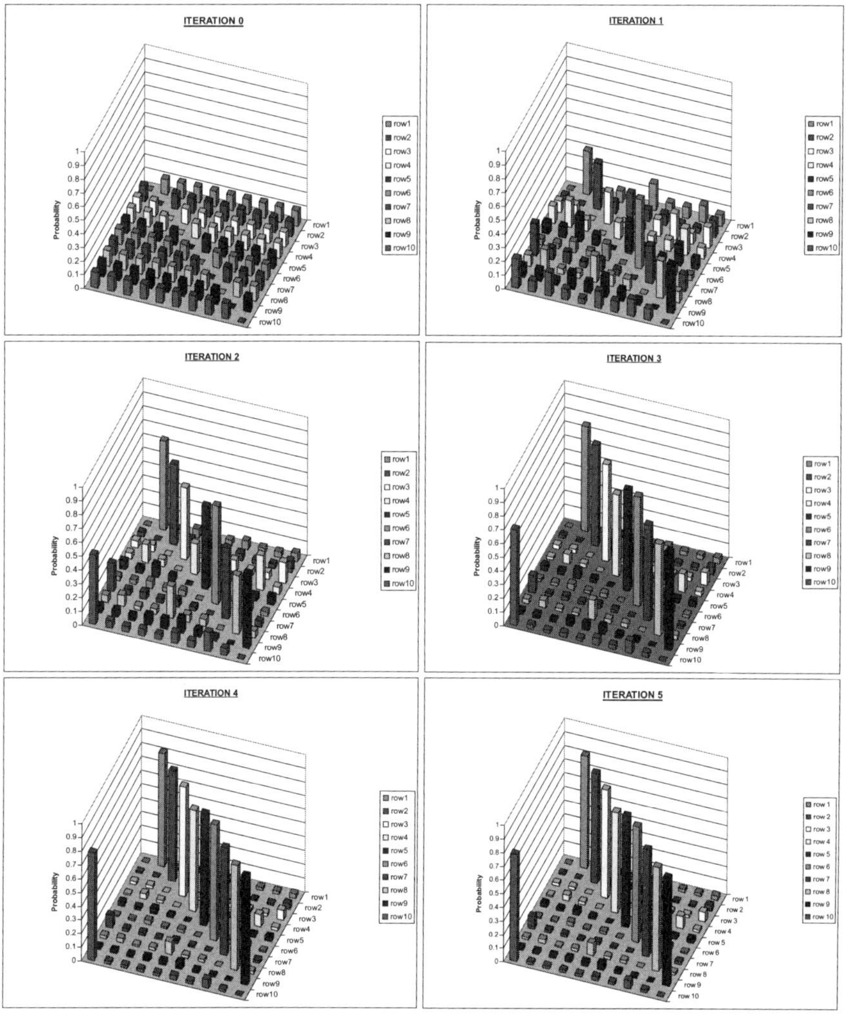

Fig. 2.6. Convergence of the reference parameter (matrix) for a 10 node TSP.

2.6 Exercises

1. Implement and repeat the rare-event simulation toy example corresponding to Table 2.1.

2. Implement and repeat the combinatorial optimization toy example corresponding to Table 2.2.

3. Extend the program used in Exercise 2 to include smoothed updating, and apply the program to a larger example, say $n = 50$. Observe if and how the choice of parameters affects the accuracy and speed of the algorithm.

4. Consider the one-phased analog of the two-phased CE program in Exercise 3; see Remark 2.7. Use the reward function $\varphi(s) = s$.
   a) Show that the updating formulas for the p_j's become:
$$\widehat{p}_{t,j} = \frac{\sum_{i=1}^N S(\mathbf{X}_i)\, X_{ij}}{\sum_{i=1}^N S(\mathbf{X}_i)} .$$
   b) Compare numerically the performances of the one- and two-phased algorithms.

5. Verify (2.40) and show that $c > b(n-m)m$ is a sufficient condition for V^* in (2.39) to be the optimal cut.

6. In the famous *n-queens problem*, introduced by Gauss in 1850, the objective is to position n queens on an $n \times n$ chess board such that no one queen can be taken by any other. Write a computer program that solves the 8-queens problem using the CE method. We may assume that the queens occupy different rows. The positions of the queens may thus be represented as a vector $\mathbf{x} \in \{1,\ldots,8\}^8$, where x_i indicates the column that the queen of row i occupies. For example, the configuration of Figure 2.7 corresponds to $\mathbf{x} = (2,3,7,4,8,5,1,6))$. A straightforward way to generate random vectors $\mathbf{X}$ is to draw each X_i independently from a probability vector $(p_{i1},\ldots,p_{i8})$, $i = 1,\ldots,8$. The performance function S could be chosen such that it represent that number of times the queens can attack each other. That is, the sum of the number of queens minus one, in each row, column and diagonal. In Figure 2.7, $S(\mathbf{x}) = 1$. The updating formulas for the p_{ij} are easy. Excluding symmetry, there are 12 different solutions to the problem. Find them all, by running your algorithm several times. Take $N = 500$, $\alpha = 0.7$ and $\varrho = 0.1$.

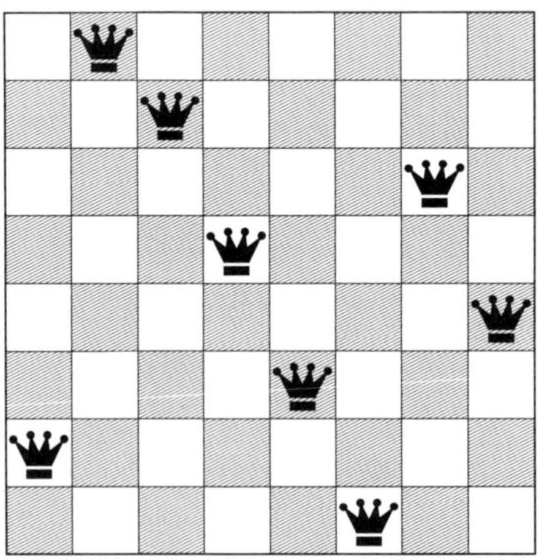

Fig. 2.7. Position the 8 queens such that no queen can attack another.

3

Efficient Simulation via Cross-Entropy

3.1 Introduction

The performance of modern systems, such as coherent reliability systems, inventory systems, insurance risk, storage systems, computer networks and telecommunication networks is often characterized by probabilities of *rare events* and is frequently studied through simulation. Analytical solutions or accurate approximations for such rare-event probabilities are only available for a very restricted class of systems. Consequently, one often has to resort to simulation. However, estimation of rare-event probabilities with crude Monte Carlo techniques requires a prohibitively large numbers of trials, as illustrated in Example 1.3. Thus, new techniques are required for this type of problem. Two methods, called *splitting/RESTART* and *importance sampling* (IS) have been extensively investigated by the simulation community in the last decade.

The basic idea of splitting proposed by Kahn and Harris [93] is to partition the state space of the system into a series of nested subsets and to consider the rare event as the intersection of a nested sequence of events. When a given subset is entered by a sample trajectory during the simulation, numerous random retrials are generated with the initial state for each retrial being the state of the system at the entry point. By doing so, the system trajectory is split into a number of new subtrajectories, hence the name "splitting." A similar idea has been developed by Villén-Altamarino and Villén-Altamarino [166, 167] into a refined simulation technique under the name *RESTART* which has been extended by different authors [58, 59, 63, 64, 65, 72, 71, 79, 151, 152] to the multiple threshold case.

In this chapter, however, we focus on importance sampling techniques. The main idea of IS [68, 160] is to make the occurrence of rare events more frequent by carrying out the simulation under a different probability distribution — the so-called *change of measure* (CoM) — and to estimate the probability of interest via a corresponding *likelihood ratio* (LR) estimator. The aim is to select a CoM that minimizes the variance of the LR estimator. It is well known that, in theory, there exists a CoM that yields a *zero-variance* LR estimator. However, in practice such an optimal CoM cannot be computed since it depends on the underlying quantity/quantities being estimated.

Prominent among the CoMs is the *exponential change of measure* (ECM). Recall, see (1.6), that here instead of the original pdf $f(x)$, the simulation is carried out under an "exponentially twisted" pdf $f_\theta(x) = c\,e^{\theta x} f(x)$, where θ is the twisting parameter. ECM often yields efficient IS estimates; see for example Sadowsky [150] and Asmussen and Rubinstein [18], but is usually feasible only for relative simple models; see also [82, 101, 154]. For such models the (asymptotic) optimal twisting parameter θ^* is often derived via the theory of *large deviations* [32].

An alternative approach to ECM is to use an IS pdf, say $f(\mathbf{x}; \mathbf{v})$, which belongs to the *same parametric family* as the original distribution (also called the *nominal* distribution), say $f(\mathbf{x}; \mathbf{u})$. We shall call such an approach the *standard likelihood ratio* (SLR) approach. Similar to ECM, the SLR approach typically does not lead to the optimal zero-variance estimator, but yields significant variance reduction; see for instance [148] and below. The advantage of such an approach is that (a) it can be applied to rather general models, and (b) the optimal reference parameter $\mathbf{v}^*$ of the IS density $f(\mathbf{x}; \mathbf{v})$ can be derived with standard optimization techniques.

In this chapter, which is partly based on [85] and [103], we give a detailed description of the CE method, and explain how it can be used efficiently in importance sampling simulation, especially with regard to rare events. The CE method was proposed in [144] as an *adaptive* IS algorithm for rare-event simulation, in which the reference parameter $\mathbf{v}^*$ is estimated by minimizing the sample variance of the SLR estimator. The proposed algorithm is called the *variance minimization* (VM) algorithm. In [145] this IS algorithm was further modified to minimize, instead of the sample variance, the sample Kullback-Leibler distance, or *cross-entropy* (CE) distance, between the theoretical zero-variance change of measure and the importance sampling distribution. The estimation method thus obtained is called the *simulated cross-entropy* or just the *cross-entropy* (CE) method. One of the most useful aspects of the CE method is that the optimal reference parameter for an SLR estimator can be effectively estimated without requiring a detailed pre-analysis of the model (for example involving large deviations). We will mostly limit the exposition to *static* simulation models, that is, models in which time does not play a role. Many problems in operations research and manufacturing fall into that category. Various problems in finance — an area that has received much recent attention from the simulation community — also have a static nature [31]. However, an application of the CE method to a *dynamic* model, a queueing model, will be given in Section 3.10. As mentioned, an attractive feature of the CE method is that it can be readily modified for efficient estimation of the optimal solution of NP-hard combinatorial optimization problems and for machine learning, which is the subject of the next chapters.

We show that the CE method is readily applicable to both light- *and* heavy-tailed distributions; see Section 1.2. Because by definition the exponential moments do not exist for heavy-tailed distributions, the exponential change of measure is intrinsically impossible for heavy-tailed distributions when a positive twisting parameter is required. So an alternative method must be used. In their landmark paper, Asmussen, Binswanger, and Højgaard [14] consider various estimators for rare events of the form $\{S_N > x\}$, where S_N is the random or deterministic sum of i.i.d. positive random variables with subexponential pdf, $f(x)$ say. Two logarithmically efficient estimators are given. The first one, based on Asmussen and Binswanger [13] uses conditional Monte Carlo [148] in combination with order statistics. The second estimator uses importance sampling, where the IS density, $h(x)$ say, consists of two parts: for small values of x, $g(x)$ is proportional to $f(x)$ and for large values of x, $g(x)$ is much larger than $f(x)$, decreasing slightly faster than $1/x$. Juneja and Shahabuddin [91] consider a similar problem to the one in [14] and their approach is to estimate $\{S_N > x\}$ via IS using a density $h(x)$ which is obtained from the original $f(x)$ by "twisting" the *hazard rate*. Several variations of this idea are considered. Note that all the above heavy-tail methods have limited application since they deal basically only with the estimation of probabilities of the above events $\{S_N > x\}$.

We address the selection of a proper class of IS distributions for both light- and heavy-tailed distributions via the *transform likelihood ratio* (TLR) method [103]. The idea is to transform the random variables and to apply a change of measure to the distribution of the transformed random variables. This simple "change of variable" technique allows us to transform an original rare-event probability with heavy-tail distributions to an equivalent (auxiliary) one with an arbitrary tail distribution, such as the uniform or exponential distribution, and then we apply a change of measure to the new (auxiliary) distribution. We typically transform to light-tailed distributions, and then apply the ECM or the SLR method to obtain a convenient class of IS distributions. Recall that in the latter case, the IS distribution belongs to the *same parametric family* as the original auxiliary one. As mentioned before we shall use the CE method to estimate the optimal parameter vector of the (parametric) IS distribution.

The remainder of this chapter is organized as follows: In Section 3.2 we describe the main ideas behind importance sampling and the SLR approach. In Section 3.3 the CE method is discussed. In Section 3.4 we present an adaptive CE algorithm for the estimation of the optimal parameters in the importance sampling density for rare-event simulation. The procedure, which involves a sequence of stochastic optimization subproblems, can be applied quite generally. It works particularly well if the underlying distributions have finite support or if they belong to a natural exponential family, since in those cases there are *analytical* solutions to those optimization subproblems. We provide various examples in Section 3.5. Convergence issues of the CE method are illustrated and discussed in Section 3.6. In Section 3.7 we show how CE can be used for finding the root of certain nonlinear equations. Section 3.8 deals with TLR method, and its application to heavy-tail distributions. In

Section 3.10 we show how CE can be applied to efficiently estimate the tail probabilities of waiting time in a queueing system. Section 3.11 discusses the "big-step" modifications of the CE method. Various numerical examples are given in Section 3.12. A direct proof of polynomial complexity pertaining to Weibull random variables with heavy tails is given in the appendix of this chapter.

3.2 Importance Sampling

Here we review some background material from [148]. Let ℓ be the expected performance of a stochastic system given in the form

$$\ell = \mathbb{E}_f H(\mathbf{X}) = \mathbb{E}_f \, \varphi(S(\mathbf{X});\gamma) = \int \varphi(S(\mathbf{x});\gamma)\, f(\mathbf{x})\, \mu(\mathrm{d}\mathbf{x}) \,. \tag{3.1}$$

S here is the *sample performance function*, $\varphi(\cdot;\gamma)$ is a real-valued function of the sample performance, which depends on a parameter γ, f is the density of $\mathbf{X}$ with respect to some measure μ; see Section 1.2. The subscript f in $\mathbb{E}_f H(\mathbf{X})$ means that the expectation is taken with respect to the density f.

Examples of $\varphi(S(\mathbf{X});\gamma)$ are indicator functions

$$\varphi(S(\mathbf{X});\gamma) = I_{\{S(\mathbf{X})\geqslant\gamma\}} \tag{3.2}$$

and Boltzmann functions

$$\varphi(S(\mathbf{X});\gamma) = \exp(-S(\mathbf{X})/\gamma) \,. \tag{3.3}$$

An example of the above framework is the *stochastic shortest path* problem illustrated in Section 2.2.1. The shortest path in a stochastic network can be defined as

$$S(\mathbf{X}) = \min_{j=1,\ldots,p} \sum_{i\in\mathcal{B}_j} X_i \,. \tag{3.4}$$

Here, $\mathcal{B}_j$ is the j-th complete path from a source to a sink, p is the number of complete paths, and $\mathbf{X} = (X_1,\ldots,X_n)$ is the random vector whose components X_i, $i = 1,\ldots,n$, represent the durations (weights) of the links. For the longest path we replace min with max in (3.4).

Let g be another probability density such that Hf is *dominated* by g. That is, $g(\mathbf{x}) = 0 \Rightarrow H(\mathbf{x})f(\mathbf{x}) = 0$. Using the density g we can represent ℓ as

$$\ell = \int H(\mathbf{x})\,\frac{f(\mathbf{x})}{g(\mathbf{x})}\,g(\mathbf{x})\,\mu(\mathrm{d}\mathbf{x}) = \mathbb{E}_g H(\mathbf{X})\,\frac{f(\mathbf{X})}{g(\mathbf{X})} \,, \tag{3.5}$$

where the subscript g means that the expectation is taken with respect to g, which is called the *importance sampling* (IS) density.

An unbiased estimator of ℓ is

$$\widehat{\ell} = \frac{1}{N} \sum_{i=1}^{N} H(\mathbf{X}_i)\, W(\mathbf{X}_i) \,, \tag{3.6}$$

where $\widehat{\ell}$ is called the *importance sampling* (IS) or the *likelihood ratio* (LR) estimator,

$$W(\mathbf{x}) = f(\mathbf{x})/g(\mathbf{x}) \tag{3.7}$$

is called the *likelihood ratio* (LR), and $\mathbf{X}_1, \ldots, \mathbf{X}_N$ is a *random sample* from g, that is, $\mathbf{X}_1, \ldots, \mathbf{X}_N$ are i.i.d. random vectors with density g. In the particular case where there is no "change of measure" that is, $(g = f)$, we have $W = 1$, and the LR estimator in (3.6) reduces to the following *crude Monte Carlo* (CMC) estimator:

$$\frac{1}{N} \sum_{i=1}^{N} H(\mathbf{X}_i) \,, \tag{3.8}$$

where $\mathbf{X}_1, \ldots, \mathbf{X}_N$ is a random sample from the density f.

The choice of the IS density g is crucial for the variance of the estimator $\widehat{\ell}$ in (3.6), and we consider next the problem of minimizing the variance of $\widehat{\ell}$ with respect to g, that is

$$\min_{g} \mathrm{Var}_g \left\{ H(\mathbf{X}) \, \frac{f(\mathbf{X})}{g(\mathbf{X})} \right\} . \tag{3.9}$$

It is well known (see for example [148]) that the solution of the problem (3.9) is

$$g^*(\mathbf{x}) = \frac{|H(\mathbf{x})| \, f(\mathbf{x})}{\int |H(\mathbf{x})| \, f(\mathbf{x}) \, \mu(\mathrm{d}\mathbf{x})} . \tag{3.10}$$

Note that if $H(\mathbf{x}) \geqslant 0$, then

$$g^*(\mathbf{x}) = \frac{H(\mathbf{x}) \, f(\mathbf{x})}{\ell} \tag{3.11}$$

and

$$\mathrm{Var}_{g^*}(\widehat{\ell}) = \mathrm{Var}_{g^*}(H(\mathbf{X})W(\mathbf{X})) = \mathrm{Var}_{g^*}(\ell) = 0 .$$

The density g^* as per (3.10) and (3.11) is called the *optimal importance sampling density.*

Example 3.1. Let $X \sim \mathsf{Exp}(u^{-1})$, and $H(X) = I_{\{X \geqslant \gamma\}}$, for some $\gamma > 0$. Let f denote the pdf of X. Consider the estimation of

$$\ell = \mathbb{E}_u H(X) = \int_{\gamma}^{\infty} u^{-1} \, \mathrm{e}^{-x\,u^{-1}} \, dx = \mathrm{e}^{-\gamma\, u^{-1}} .$$

We have

$$g^*(x) = I_{\{x \geqslant \gamma\}} \, u^{-1} \, \mathrm{e}^{-x\,u^{-1}} \, \mathrm{e}^{\gamma u^{-1}} = I_{\{x \geqslant \gamma\}} \, u^{-1} \, \mathrm{e}^{-(x-\gamma)\,u^{-1}} \; .$$

Thus, the optimal importance sampling density of X is the shifted exponential distribution. Note that Hf is dominated by g^*, but f itself is not dominated by g^*. Since g^* is optimal, the LR estimator $\widehat{\ell}$ is constant, that is

$$\widehat{\ell} = H(X)W(X) = \frac{H(X)f(X)}{H(X)f(X)/\ell} = \ell \, .$$

In general, implementation of the optimal importance sampling density g^* as per (3.10) and (3.11) is problematic. The main difficulty lies in the fact that to derive $g^*(\mathbf{x})$ one needs to know ℓ. But ℓ is precisely the quantity we want to estimate from the simulation! In most simulation studies the situation is even worse, since the analytical expression for the sample performance H is unknown in advance. To overcome this difficulty one can make a pilot run with the underlying model, obtain a sample $H(\mathbf{X}_1), \ldots, H(\mathbf{X}_N)$, and then use it to estimate g^*. It is important to note that sampling from such an artificially constructed density might be a very complicated and time consuming task, especially when g is a high-dimensional density.

Since sampling from the optimal importance sampling density g^* is problematic we consider next the case when the underlying density f belongs to some parametric family $\mathcal{F} = \{f(\cdot; \mathbf{v}), \;\mathbf{v} \in \mathcal{V}\}$. From now on, we will assume that this is the case. Let $f(\cdot; \mathbf{u})$ denote the density of the random vector $\mathbf{X}$ in (3.1), for some fixed "nominal" parameter $\mathbf{u} \in \mathcal{V}$. We now restrict the choice of IS densities g to those from the same parametric family $\mathcal{F}$; so g differs from the original density $f(\cdot; \mathbf{u})$ by a single parameter (vector) $\mathbf{v}$, which we will call the *reference parameter*. We will write the likelihood ratio in (3.7), with $g(\mathbf{x}) = f(\mathbf{x}; \mathbf{v})$, as

$$W(\mathbf{X}; \mathbf{u}, \mathbf{v}) = \frac{f(\mathbf{X}; \mathbf{u})}{f(\mathbf{X}; \mathbf{v})} \, . \tag{3.12}$$

In this case the LR estimator $\widehat{\ell}$ in (3.6) becomes

$$\widehat{\ell} = \frac{1}{N} \sum_{i=1}^{N} H(\mathbf{X}_i) \, W(\mathbf{X}_i; \mathbf{u}, \mathbf{v}) \, , \tag{3.13}$$

where $\mathbf{X}_1, \ldots, \mathbf{X}_N$ is a random sample from $f(\cdot; \mathbf{v})$. We will call (3.13) the *standard likelihood ratio (SLR) estimator*, in contrast to the (nonparametric) LR estimator (2.11). To find an optimal $\mathbf{v}$ in the SLR estimator $\widehat{\ell}$ we need to consider the program (3.9), which reduces to

$$\min_{\mathbf{v}} \, \mathrm{Var}_{\mathbf{v}} H(\mathbf{X}) \, W(\mathbf{X}; \mathbf{u}, \mathbf{v}) \, . \tag{3.14}$$

Since under $f(\cdot; \mathbf{v})$ the expectation $\ell = \mathbb{E}_{\mathbf{v}} H(\mathbf{X}) \, W(\mathbf{X}; \mathbf{u}, \mathbf{v})$ is constant, the optimal solution of (3.14) coincides with that of

$$\min_{\mathbf{v}} V(\mathbf{v}) = \min_{\mathbf{v}} \mathbb{E}_{\mathbf{v}} H^2(\mathbf{X})\, W^2(\mathbf{X};\mathbf{u},\mathbf{v}) \,. \tag{3.15}$$

The above optimization problem can still be difficult to solve, since the density with respect to which the expectation is computed depends on the decision variable $\mathbf{v}$. To overcome this obstacle, we rewrite (3.15) as

$$\min_{\mathbf{v}} V(\mathbf{v}) = \min_{\mathbf{v}} \mathbb{E}_{\mathbf{w}} H^2(\mathbf{X})\, W(\mathbf{X};\mathbf{u},\mathbf{v})\, W(\mathbf{X};\mathbf{u},\mathbf{w}) \,. \tag{3.16}$$

Note that (3.16) is obtained from (3.15) by multiplying and dividing the integrand by $f(\mathbf{x};\mathbf{w})$ where $\mathbf{w}$ is an *arbitrary* reference parameter. Note also that in (3.15) and (3.16) the expectation is taken with respect to the densities $f(\cdot;\mathbf{v})$ and $f(\cdot;\mathbf{w})$, respectively. Moreover, $W(\mathbf{X};\mathbf{u},\mathbf{w}) = f(\mathbf{X};\mathbf{u})/f(\mathbf{X};\mathbf{w})$, and $\mathbf{X} \sim f(\mathbf{x};\mathbf{w})$. Note finally that for the particular case $\mathbf{w} = \mathbf{u}$ we obtain from (3.16)

$$\min_{\mathbf{v}} V(\mathbf{v}) = \min_{\mathbf{v}} \mathbb{E}_{\mathbf{u}} H^2(\mathbf{X})\, W(\mathbf{X};\mathbf{u},\mathbf{v}) \,. \tag{3.17}$$

We shall call each of the equivalent problems (3.14)–(3.17), the *variance minimization* (VM) problem; and we call the parameter vector ${}_*\mathbf{v}$, that minimizes the programs (3.14)–(3.17) the *optimal VM reference parameter vector.*

Example 3.2 (Example 3.1 continued). Consider again estimating $\ell = \mathbb{P}_u(X \geqslant \gamma) = \exp(-\gamma u^{-1})$. In this case, using the family $\{f(x;v), v > 0\}$ defined by $f(x;v) = v^{-1}\exp(xv^{-1}), x \geqslant 0$, the program (3.17) reduces to:

$$\min_{v} V(v) = \min_{v} \frac{v}{u^2} \int_{\gamma}^{\infty} \mathrm{e}^{-(2u^{-1}-v^{-1})x}\, dx = \min_{v \geqslant u/2} \frac{v^2}{u} \frac{\mathrm{e}^{-\gamma(2u^{-1}-v^{-1})}}{(2v-u)} \,.$$

Note that this follows directly from (1.40) and Table 1.5. The optimal reference parameter ${}_*v$ is given by

$${}_*v = \frac{1}{2}\left\{\gamma + u + \sqrt{\gamma^2 + u^2}\right\} = \gamma + \frac{u}{2} + \mathcal{O}((u/\gamma)^2) \,,$$

where $\mathcal{O}(x^2)$ is a function of x such that

$$\lim_{x \to 0} \frac{\mathcal{O}(x^2)}{x^2} = \text{constant.}$$

We see that for $\gamma \gg u$, ${}_*v$ is approximately equal to γ.

Typically, the optimal reference parameter ${}_*\mathbf{v}$ that minimizes the programs (3.14)–(3.17) is not available analytically, since the expected performance $\mathbb{E}_{\mathbf{v}} H(\mathbf{X})$ cannot be evaluated exactly. To overcome this difficulty, we can replace the expected values in (3.14)–(3.17) by their sample (Monte Carlo) counterparts and then take the optimal solutions of the associated Monte Carlo programs as estimators of ${}_*\mathbf{v}$. For example, the Monte Carlo counterpart of (3.17) is

$$\min_{\mathbf{v}} \frac{1}{N} \sum_{i=1}^{N} H^2(\mathbf{X}_i)\, W(\mathbf{X}_i\,; \mathbf{u}, \mathbf{v})\;, \tag{3.18}$$

where $\mathbf{X}_1, \ldots, \mathbf{X}_N$ is an i.i.d. sample generated from the original density $f(\cdot; \mathbf{u})$. This type of procedure has been well studied in the literature, appearing under various names such as *stochastic counterpart method, stochastic optimization*, or the *simulated VM program.* Under proper assumptions, it is possible to show that the optimal solutions of (3.18) converge to ${}_*\mathbf{v}$ as N goes to infinity and it has a normal asymptotic distribution. See [148] for more details.

We now borrow from [148] a simple recursive algorithm for estimating the optimal VM reference parameter ${}_*\mathbf{v}$ (see Algorithm 8.4.1 in [148]), which is based on the stochastic counterpart of (3.16), that is, on

$$\min_{\mathbf{v}} \widehat{V}(\mathbf{v}) = \min_{\mathbf{v}} \frac{1}{N} \sum_{i=1}^{N} H^2(\mathbf{X}_i)\, W(\mathbf{X}_i\,; \mathbf{u}, \mathbf{v})\, W(\mathbf{X}_i\,; \mathbf{u}, \mathbf{w})\;, \tag{3.19}$$

where $\mathbf{X}_1, \ldots, \mathbf{X}_N$ is an i.i.d. sample from $f(\cdot; \mathbf{w})$, for any $\mathbf{w}$.

Algorithm 3.2.1 (VM Algorithm)

1. *Choose an initial tilting vector* $\mathbf{v}_{(0)}$*, for example* $\mathbf{v}_{(0)} = \mathbf{u}$*. Set* $k = 1$*.*
2. *Generate a random sample* $\mathbf{X}_1, \ldots, \mathbf{X}_N$ *from* $f(\cdot; \mathbf{v}_{(k-1)})$*.*
3. *Solve the stochastic program (3.19). Denote the solution by* $\mathbf{v}_{(k)}$*, thus*

$$\mathbf{v}_{(k)} = \operatorname*{argmin}_{\mathbf{v}} \frac{1}{N} \sum_{i=1}^{N} H^2(\mathbf{X}_i)\, W(\mathbf{X}_i\,; \mathbf{u}, \mathbf{v})\, W(\mathbf{X}_i\,; \mathbf{u}, \mathbf{v}_{(k-1)})\;. \tag{3.20}$$

4. *If convergence is reached, say at* $k = K$*, then* **stop***; otherwise increase* k *by 1 and reiterate from Step 2.*

Remark 3.3 (Stopping Criterion). A possible stopping criterion is to stop when all the parameters have ceased to increase or decrease monotonously. More precisely, for the j-th component of $\mathbf{v}$ let K_j be the iteration at which the sequence $v_{(0),j}, v_{(1),j}, v_{(2),j}, \ldots$ starts "fluctuating," that is, K_j is such that either

$$v_{(0),j} \leqslant v_{(1),j} \leqslant v_{(2),j} \cdots \leqslant v_{(K_j-1),j} > v_{(K_j),j}\;,$$

or

$$v_{(0),j} \geqslant v_{(1),j} \geqslant v_{(2),j} \cdots \geqslant v_{(K_j-1),j} < v_{(K_j),j}\;.$$

Now stop at the iteration

$$K = \max_j K_j\;.$$

3.3 Kullback-Leibler Cross-Entropy

An alternative to variance minimization for estimating the optimal reference parameter vector in (3.13) is based on the Kullback-Leibler *cross-entropy*, which defines a "distance" between the two pdfs g and h and can be written (see also Section 1.5) as

$$\begin{aligned}\mathcal{D}(g,h) &= \int g(\mathbf{x}) \ln \frac{g(\mathbf{x})}{h(\mathbf{x})}\, \mu(d\mathbf{x}) \\ &= \int g(\mathbf{x}) \ln g(\mathbf{x})\, \mu(d\mathbf{x}) - \int g(\mathbf{x}) \ln h(\mathbf{x})\, \mu(d\mathbf{x}) \,.\end{aligned} \tag{3.21}$$

Recall from Section 1.5 that $\mathcal{D}(g,h) \geqslant 0$, with equality if and only if $g = h$ (on a set of μ-measure 1). The idea behind the CE method is to choose the IS density h such that the Kullback-Leibler distance between the optimal importance sampling density g^* in (3.10) and h is *minimal*; that is, the CE optimal IS density h^* is the solution to the *functional* optimization program

$$\min_h \mathcal{D}\left(g^*, h\right) \,.$$

But, it is immediate from $\mathcal{D}(g^*,h) \geqslant 0$ that $h^* = g^*$. Hence, if we optimize over all densities h, the VM and CE optimal IS densities *coincide.*

However, when using the SLR approach, the class of densities is restricted to some family $\{f(\cdot;\mathbf{v}), \mathbf{v} \in \mathcal{V}\}$ that contains the nominal density $f(\cdot;\mathbf{u})$. The CE method now aims to solve the *parametric* optimization problem

$$\min_{\mathbf{v}} \mathcal{D}\left(g^*, f(\cdot;\mathbf{v})\right) \,,$$

with

$$g^*(\mathbf{x}) = c^{-1} |H(\mathbf{x})| f(\mathbf{x};\mathbf{u}) \,,$$

where $c = \int |H(\mathbf{x})| f(\mathbf{x};\mathbf{u})\, \mu(d\mathbf{x})$. Since the first term in the right-hand side of (3.21) does not depend on $\mathbf{v}$, minimizing the Kullback-Leibler distance between g^* above and $f(\cdot;\mathbf{v})$ is equivalent to *maximizing*, with respect to $\mathbf{v}$,

$$\int |H(\mathbf{x})|\, f(\mathbf{x};\mathbf{u}) \ln f(\mathbf{x};\mathbf{v})\, \mu(d\mathbf{x}) = \mathbb{E}_{\mathbf{u}} |H(\mathbf{X})| \ln f(\mathbf{X};\mathbf{v}) \,.$$

Let us now assume for simplicity that $H(\mathbf{X}) \geqslant 0$, thereby dropping the absolute signs in the formulas above. We find that the optimal reference parameter (with respect to Kullback-Leibler distance) is the solution to

$$\max_{\mathbf{v}} D(\mathbf{v}) = \max_{\mathbf{v}} \mathbb{E}_{\mathbf{u}}\, H(\mathbf{X}) \ln f(\mathbf{X};\mathbf{v}) \,. \tag{3.22}$$

This is equivalent to the program

$$\max_{\mathbf{v}} D(\mathbf{v}) = \max_{\mathbf{v}} \mathbb{E}_{\mathbf{w}}\, H(\mathbf{X})\, W(\mathbf{X};\mathbf{u},\mathbf{w}) \ln f(\mathbf{X};\mathbf{v}) \,, \tag{3.23}$$

for *any* tilting parameter $\mathbf{w}$, where $W(\mathbf{X};\mathbf{u},\mathbf{w})$ is again the likelihood ratio given in (3.12).

In analogy to the VM program (3.17) we call (3.23) the *cross-entropy* (CE) program; and we call the parameter vector $\mathbf{v}^*$ that minimizes the program (3.23) the *optimal CE reference parameter vector.*

Similar to (3.19), we can estimate the optimal solution $\mathbf{v}^*$ by the optimal solution of the program

$$\max_{\mathbf{v}} \widehat{D}(\mathbf{v}) = \max_{\mathbf{v}} \frac{1}{N} \sum_{i=1}^{N} H(\mathbf{X}_i)\, W(\mathbf{X}_i;\mathbf{u},\mathbf{w}) \ln f(\mathbf{X}_i;\mathbf{v}) \,, \tag{3.24}$$

where $\mathbf{X}_1,\ldots,\mathbf{X}_N$ is a random sample from $f(\cdot;\mathbf{w})$. We shall call the program (3.24), the *stochastic counterpart* of the CE program (3.23) or simply the *simulated CE* program to distinguish it from the VM counterpart (3.19). In typical applications the function $\widehat{D}$ in (3.24) is concave and differentiable with respect to $\mathbf{v}$ [149] and, thus, the solution of (3.24) may be readily obtained by solving (with respect to $\mathbf{v}$) the following system of equations:

$$\frac{1}{N} \sum_{i=1}^{N} H(\mathbf{X}_i)\, W(\mathbf{X}_i;\mathbf{u},\mathbf{w})\, \nabla \ln f(\mathbf{X}_i;\mathbf{v}) = \mathbf{0} \,, \tag{3.25}$$

where the gradient is with respect to $\mathbf{v}$. Similarly, the solution to (3.23) may be obtained by solving

$$\mathbb{E}_{\mathbf{u}} H(\mathbf{X})\, \nabla \ln f(\mathbf{X};\mathbf{v}) = \mathbf{0} \,, \tag{3.26}$$

provided the expectation and differentiation operators can be interchanged [149]. Note that the function $\mathbf{v} \mapsto \nabla \ln f(\mathbf{x};\mathbf{v})$ is the *score function*, see (1.27). In particular, for exponential families (1.3), with $\boldsymbol{\theta} = \psi(\mathbf{u})$ for some parameterization ψ, the solution to (3.23) satisfies, by (1.28),

$$\mathbb{E}_{\mathbf{u}} H(\mathbf{X}) \left(\frac{\nabla c(\boldsymbol{\eta})}{c(\boldsymbol{\eta})} + \mathbf{t}(\mathbf{x}) \right) = \mathbf{0} \,, \tag{3.27}$$

where $\boldsymbol{\eta} = \psi(\mathbf{v})$. It is interesting to compare formulas (3.25) and (3.26) with similar formulas such as (1.39) and (1.38) of the score function method. We see for example that for $\mathbf{v} = \mathbf{u}$ the left-hand side of (3.26) is simply the gradient of the expected performance: $\nabla \ell(\mathbf{u}) = \nabla \mathbb{E}_{\mathbf{u}} H(\mathbf{X})$. A notable difference in the estimator in (3.25) for the CE method and (1.39) for the score function method is that in the former case $\mathbf{u}$ is fixed and $\mathbf{v}$ variable, whereas in the latter case $\mathbf{u}$ is variable and $\mathbf{v}$ is fixed.

We shall consider the CE programs (3.23) and (3.24) as alternatives to their VM counterparts (3.16) and (3.19), respectively and shall show that their optimal solutions are nearly the same. This leads to the following CE analog of Algorithm 3.2.1:

Algorithm 3.3.1 (CE Algorithm)

1. *Choose an initial tilting vector* $\mathbf{v}^{(0)}$*, for example* $\mathbf{v}^{(0)} = \mathbf{u}$*. Set* $k = 1$*.*
2. *Generate a random sample* $\mathbf{X}_1, \dots, \mathbf{X}_N$ *from* $f(\cdot; \mathbf{v}^{(k-1)})$*.*
3. *Solve the simulated cross-entropy program (3.24). Denote the optimal solution by* $\mathbf{v}^{(k)}$*, thus*

$$\mathbf{v}^{(k)} = \operatorname*{argmax}_{\mathbf{v}} \frac{1}{N} \sum_{i=1}^{N} H(\mathbf{X}_i)\, W(\mathbf{X}_i; \mathbf{u}, \mathbf{v}^{(k-1)}) \ln f(\mathbf{X}_i; \mathbf{v}) \,. \tag{3.28}$$

4. *If convergence is reached, say at* $k = K$*, then* **stop***; otherwise increase* k *by 1 and reiterate from Step 2.*

We could use the same stopping criterion as in Remark 3.3.

We shall show that optimal solutions of the CE programs (3.23) and (3.24) and their VM counterparts (3.16) and (3.19) are nearly the same. The advantage of Algorithm 3.3.1 as compared to Algorithm 3.2.1 is that $\mathbf{v}^{(k)}$ in (3.28) can often be calculated *analytically*. In particular, this happens if the distributions of the random variables belong to the *natural exponential family* (NEF) or if f is a discrete pdf with *finite* support.

Note that for the VM programs there is typically *no analytic solution* even for NEF distributions. Thus, numerical optimization procedures must be used in such cases. This emphasizes a big advantage of the CE approach over its VM counterpart. The following two examples deal with NEF and finite support discrete distributions, respectively.

Example 3.4 (Example 3.2 continued). Consider again estimation of $\ell = \mathbb{P}_u(X \geqslant \gamma) = \exp(-\gamma u^{-1})$. In this case, program (3.22) becomes:

$$\begin{aligned}
\max_v D(v) &= \max_v \int I_{\{x \geqslant \gamma\}} u^{-1} \mathrm{e}^{-x u^{-1}} \ln\left(v^{-1} \mathrm{e}^{-x v^{-1}}\right) dx \\
&= \max_v \int_\gamma^\infty u^{-1} \mathrm{e}^{-x u^{-1}} (-\ln v - x v^{-1})\, dx \\
&= \min_v \frac{\mathrm{e}^{-\gamma u^{-1}} (\gamma + u + v \ln v)}{v} \,.
\end{aligned}$$

It is easily verified that the optimal CE reference parameter v^* is $\gamma + u$. As in the VM case, for $\gamma \gg u$, the optimal reference parameter is approximately γ.

Example 3.5 (NEF distributions). Suppose that the univariate random variable X belongs to a NEF of distributions parameterized by the mean v. Thus (see Section 1.3) the density of X belongs to a class $\{f(x; v)\}$ with

$$f(x; v) = \mathrm{e}^{x\theta(v) - \zeta(\theta(v))}\, h(x) \,,$$

with $\mathbb{E}_v X = \zeta'(\theta(v)) = v$ and $\mathrm{Var}_v(X) = \zeta''(\theta(v))$. Specifically, let $X \sim f(x; u)$ for some nominal reference parameter u. For simplicity assume that

$\mathbb{E}_u H(X) > 0$ and that X is nonconstant. Writing $\theta = \theta(v)$ we have from (3.26) and (1.28) that the optimal reference parameter v^* is given by the solution of

$$\mathbb{E}_u H(X) \left(\frac{c'(\theta)}{c(\theta)} + X \right) = \mathbb{E}_u H(X) \, (-v + X) = 0 \, ,$$

where we have used the fact that $c(\theta) = \mathrm{e}^{-\zeta(\theta)}$. Hence v^* is given by

$$v^* = \frac{\mathbb{E}_u H(X) \, X}{\mathbb{E}_u \, H(X)} = \frac{\mathbb{E}_w W(X; u, w) \, H(X) \, X}{\mathbb{E}_w H(X) \, W(X; u, w)} \, , \tag{3.29}$$

for any reference parameter w. It is not difficult to check that v^* is indeed a unique global maximum of $D(v) = \mathbb{E}_u H(X) \ln f(X; v)$.

The estimate $\widehat{v}$ of v^* in (3.29) can be obtained analytically from the solution of the stochastic program (3.24), that is,

$$\widehat{v} = \frac{\sum_{i=1}^N H(X_i) W(X_i; u, w) X_i}{\sum_{i=1}^N H(X_i) W(X_i; u, w)} \, , \tag{3.30}$$

where $X_1, \ldots, X_N$ is a random sample from the density $f(\cdot; w)$.

A similar explicit formula can be found for the case where $\mathbf{X} = (X_1, \ldots, X_n)$ is a vector of independent random variables such that each component X_j belongs to a NEF parameterized by the mean. In particular, if $\mathbf{u} = (u_1, \ldots, u_n)$ is the nominal reference parameter, then for each $j = 1, \ldots, n$ the density of X_j is given by

$$f_j(x; u_j) = \mathrm{e}^{x\theta(u_j) - \zeta(\theta(u_j))} \, h_j(x) \, .$$

It is easy to see that problem (3.23) under the independence assumption becomes "separable," that is, it reduces to n subproblems of the form above. Thus, we find that the optimal reference parameter vector $\mathbf{v}^* = (v_1^*, \ldots, v_n^*)$ is given as

$$v_j^* = \frac{\mathbb{E}_{\mathbf{u}} H(\mathbf{X}) \, X_j}{\mathbb{E}_{\mathbf{u}} \, H(\mathbf{X})} = \frac{\mathbb{E}_{\mathbf{w}} H(\mathbf{X}) \, W(\mathbf{X}; \mathbf{u}, \mathbf{w}) \, X_j}{\mathbb{E}_{\mathbf{w}} H(\mathbf{X}) \, W(\mathbf{X}; \mathbf{u}, \mathbf{w})} \, . \tag{3.31}$$

Moreover, we can estimate the j-th component of $\mathbf{v}^*$ as

$$\widehat{v}_j = \frac{\sum_{i=1}^N H(\mathbf{X}_i) W(\mathbf{X}_i; \mathbf{u}, \mathbf{w}) X_{ij}}{\sum_{i=1}^N H(\mathbf{X}_i) W(\mathbf{X}_i; \mathbf{u}, \mathbf{w})} \, , \tag{3.32}$$

where $\mathbf{X}_1, \ldots, \mathbf{X}_N$ is a random sample from the density $f(\cdot; \mathbf{w})$, and X_{ij} is the jth component of $\mathbf{X}_i$.

Example 3.6 (Finite support discrete distributions). Let X be a discrete random variable with finite support, that is, X can only take a finite number of values, say $\{a_1, \ldots, a_m\}$. Let $u_j = \mathbb{P}(X = a_j), j = 1, \ldots, m$ and define

$\mathbf{u} = (u_1, \ldots, u_m)$. The distribution of X is thus trivially parameterized by the vector $\mathbf{u}$. We can write the density of X as

$$f(x; \mathbf{u}) = \sum_{j=1}^{m} u_j \, I_{\{x = a_j\}} \, .$$

From the discussion at the beginning of this section we know that the optimal CE and VM parameter *coincide*, since we optimize over *all* densities on $\{a_1, \ldots, a_m\}$. By (3.10) the VM (and CE) optimal density is given by

$$\begin{aligned} f(x; \mathbf{v}^*) &= \frac{H(x) f(x; \mathbf{u})}{\sum_x H(x) f(x; \mathbf{u})} \\ &= \frac{\sum_{j=1}^{m} H(a_j) u_j I_{\{x=a_j\}}}{\mathbb{E}_{\mathbf{u}} H(X)} \\ &= \sum_{j=1}^{m} \frac{H(a_j) u_j}{\mathbb{E}_{\mathbf{u}} H(X)} I_{\{x=a_j\}} \\ &= \sum_{j=1}^{m} \left(\frac{\mathbb{E}_{\mathbf{u}} H(X) I_{\{X=a_j\}}}{\mathbb{E}_{\mathbf{u}} H(X)} \right) I_{\{x=a_j\}} \, , \end{aligned}$$

so that

$$v_j^* = \frac{\mathbb{E}_{\mathbf{u}} H(X) \, I_{\{X=a_j\}}}{\mathbb{E}_{\mathbf{u}} H(X)} = \frac{\mathbb{E}_{\mathbf{w}} H(X) \, W(X; \mathbf{u}, \mathbf{w}) I_{\{X=a_j\}}}{\mathbb{E}_{\mathbf{w}} H(X) \, W(X; \mathbf{u}, \mathbf{w})} \, , \tag{3.33}$$

for any reference parameter $\mathbf{w}$, provided that $\mathbb{E}_{\mathbf{w}} H(X) W(X; \mathbf{u}, \mathbf{w}) > 0$.

The vector $\mathbf{v}^*$ can be estimated from the stochastic counterpart of (3.33), that is as

$$\widehat{v}_j = \frac{\sum_{i=1}^{N} H(X_i) \, W(X_i; \mathbf{u}, \mathbf{w}) I_{\{X_i = a_j\}}}{\sum_{i=1}^{N} H(X_i) \, W(X_i; \mathbf{u}, \mathbf{w})} \, , \tag{3.34}$$

where $X_1, \ldots, X_N$ is an i.i.d. sample from the density $f(\cdot; \mathbf{w})$.

A similar result holds for a random vector $\mathbf{X} = (X_1, \ldots, X_n)$ where $X_1, \ldots, X_n$ are independent discrete random variables with finite support, characterized by the parameter vectors $\mathbf{u}_1, \ldots, \mathbf{u}_n$. Because of the independence assumption, the CE problem (3.23) separates into n subproblems of the form above, and all the components of the optimal CE reference parameter $\mathbf{v}^* = (\mathbf{v}_1^*, \ldots, \mathbf{v}_n^*)$, which is now a vector of vectors, follow from (3.34). Note that in this case the optimal VM and CE reference parameters are usually not equal, since we are not optimizing the cross entropy over all densities. See, however, Proposition 4.2 for an important case where they *do* coincide and yield a zero-variance LR estimator.

The updating rule (3.34), which involves discrete finite support distributions and, in particular, the Bernoulli distribution, will be extensively used for combinatorial optimization problems in the rest of the book.

Example 3.7. Let X be a random variable having the following discrete distribution

$$u_i = \mathbb{P}(X = i), \; i = 1, ..., m; \quad \sum_{i=1}^{m} u_i = 1 \; .$$

Our goal is to estimate $\ell = \mathbb{P}(X \geqslant \gamma)$ using the LR estimator (3.6), with $H(X) = I_{\{X \geqslant \gamma\}}$. Assume for simplicity that $\gamma \in \{1, \ldots, m\}$. From Example 3.6 we see immediately that both VM and CE optimal reference parameters are given by

$$_*v_i = v_i^* = \begin{cases} 0 & \text{if } 1 \leqslant i \leqslant \gamma - 1 \\ \dfrac{u_i}{\sum_{i=\gamma}^{m} u_i} & \text{if } \gamma \leqslant i \leqslant m \; . \end{cases} \tag{3.35}$$

Using this change of measure yields obviously a *zero-variance* LR estimator (3.6).

3.4 Estimation of Rare-Event Probabilities

This section we shift our attention to the CE method in the specific context of rare-event simulation. Note that the results of the previous sections have a more general scope.

Let $S(\mathbf{X})$ again denote the sample performance, where $\mathbf{X} \sim f(\cdot; \mathbf{u})$ and suppose that we wish to estimate $\ell = \mathbb{P}_{\mathbf{u}}(S(\mathbf{X}) \geqslant \gamma) = \mathbb{E}_{\mathbf{u}} I_{\{S(\mathbf{X}) \geqslant \gamma\}}$, for some fixed level γ. Note that this estimation problem presents a particular case of (3.1) with $H(\mathbf{X}) = I_{\{S(\mathbf{X}) \geqslant \gamma\}}$. Assume as before that $\mathbf{X}$ has a density $f(\cdot; \mathbf{u})$ in some family $\{f(\cdot; \mathbf{v})\}$. We can estimate the quantity ℓ using the SLR estimator (see also (3.13))

$$\widehat{\ell} = \frac{1}{N} \sum_{i=1}^{N} I_{\{S(\mathbf{X}_i) \geqslant \gamma\}} \, W(\mathbf{X}_i; \mathbf{u}, \mathbf{v}) \; . \tag{3.36}$$

It is worth noticing that the optimal (zero-variance) change of measure for this case has an easy interpretation. Namely, from (3.10) we have, with $\mathcal{A} = \{\mathbf{x} \, : \, S(\mathbf{x}) \geqslant \gamma\}$,

$$g^*(\mathbf{x}) = \begin{cases} f(\mathbf{x}; \mathbf{u}) / \int_{\mathcal{A}} f(\mathbf{x}; \mathbf{u}) \, \mu(d\mathbf{x}) \; , & \text{if } S(\mathbf{x}) \geqslant \gamma \; , \\ 0 \; , & \text{else.} \end{cases}$$

We see that g^* is the conditional density of $\mathbf{X} \sim f(\cdot; \mathbf{u})$ given that the event $\{S(\mathbf{X}) \geqslant \gamma\}$ occurs.

We repeat the main ideas and results of Section 2.3 leading to the main algorithm of this chapter. An important observation to make is that the simulated CE program (3.24) is of little practical use when rare events are involved, because most of the indicators $H(\mathbf{X}_i) = I_{\{S(\mathbf{X}_i) \geqslant \gamma\}}$ will be zero. For these problems a *two-phase* CE procedure is employed in which both the reference parameter $\mathbf{v}$ and the *level* γ are updated, thus avoiding too many indicators that are zero, and creating a sequence of two-tuples $\{(\mathbf{v}_t, \gamma_t)\}$ with the goal of finding/estimating the optimal CE reference parameter $\mathbf{v}^*$.

Starting with $\widehat{\mathbf{v}}_0$ and a ϱ not too small, say $\varrho = 0.01$, the two phases are as follows:

1. **Adaptive updating of** γ_t. For a fixed $\mathbf{v}_{t-1}$, let γ_t be a $(1-\varrho)$-*quantile* of $S(\mathbf{X})$ under $\mathbf{v}_{t-1}$. That is, γ_t satisfies

$$\mathbb{P}_{\mathbf{v}_{t-1}}(S(\mathbf{X}) \geqslant \gamma_t) \geqslant \varrho \,, \tag{3.37}$$

$$\mathbb{P}_{\mathbf{v}_{t-1}}(S(\mathbf{X}) \leqslant \gamma_t) \geqslant 1-\varrho \,, \tag{3.38}$$

where $\mathbf{X} \sim f(\cdot; \mathbf{v}_{t-1})$. A simple estimator $\widehat{\gamma}_t$ of γ_t is the order statistic

$$\widehat{\gamma}_t = S_{(\lceil (1-\varrho)N \rceil)} \,. \tag{3.39}$$

2. **Adaptive updating of** $\mathbf{v}_t$. For fixed γ_t and $\mathbf{v}_{t-1}$, derive $\mathbf{v}_t$ from the solution of the following CE program

$$\max_{\mathbf{v}} D(\mathbf{v}) = \max_{\mathbf{v}} \mathbb{E}_{\mathbf{v}_{t-1}} I_{\{S(\mathbf{X}) \geqslant \gamma_t\}} W(\mathbf{x}; \mathbf{u}, \mathbf{v}_{t-1}) \ln f(\mathbf{X}; \mathbf{v}) \,. \tag{3.40}$$

The stochastic counterpart of (3.40) is as follows: for fixed $\widehat{\gamma}_t$ and $\widehat{\mathbf{v}}_{t-1}$, derive $\widehat{\mathbf{v}}_t$ from the following program

$$\max_{\mathbf{v}} \widehat{D}(\mathbf{v}) = \max_{\mathbf{v}} \frac{1}{N} \sum_{i=1}^{N} I_{\{S(\mathbf{X}_i) \geqslant \widehat{\gamma}_t\}} W(\mathbf{X}_i; \mathbf{u}, \widehat{\mathbf{v}}_{t-1}) \ln f(\mathbf{X}_i; \mathbf{v}) \,. \tag{3.41}$$

As seen before, the optimal solutions of (3.40) and (3.41) can often be obtained *analytically*, in particular when $f(\mathbf{x}; \mathbf{v})$ belongs to a NEF or is a density with finite support; see Examples 3.5 and 3.6.

For ease of reference we repeat Algorithm 2.3.1.

Algorithm 3.4.1 (Main CE Algorithm for Rare-Event Simulation)

1. *Define* $\widehat{\mathbf{v}}_0 = \mathbf{u}$. *Set* $t = 1$ *(iteration = level counter).*
2. *Generate a sample* $\mathbf{X}_1, \ldots, \mathbf{X}_N$ *from the density* $f(\cdot; \mathbf{v}_{t-1})$ *and compute the sample* $(1-\varrho)$*-quantile* $\widehat{\gamma}_t$ *of the performances according to (3.39), provided* $\widehat{\gamma}_t$ *is less than* γ. *Otherwise set* $\widehat{\gamma}_t = \gamma$.
3. *Use the* **same** *sample* $\mathbf{X}_1, \ldots, \mathbf{X}_N$ *to solve the stochastic program (3.41). Denote the solution by* $\widehat{\mathbf{v}}_t$.
4. *If* $\widehat{\gamma}_t < \gamma$, *set* $t = t + 1$ *and reiterate from Step 2. Else proceed with Step 5.*
5. *Estimate the rare-event probability* ℓ *using the SLR estimator* $\widehat{\ell}$ *in* (3.36), *with* $\mathbf{v}$ *replaced by* $\widehat{\mathbf{v}}_T$, *where* T *is the final number of iterations (= number of levels used).*

Remark 3.8. In typical applications a much smaller sample size N in Step 2 can be chosen than the final sample size in Step 5. When we need to distinguish between the two sample sizes, in particular when reporting numerical experiments, we will use the notation N and N_1 for Step 2 and 5, respectively.

Remark 3.9. To obtain a more accurate estimate of $\mathbf{v}^*$ it is sometimes useful, especially when the sample size is relatively small, to repeat Steps 2–4 for a number of additional iterations after level γ has been reached.

Note that Algorithm 3.4.1 breaks down the "hard" problem of estimating the very small probability ℓ into a sequence of "simple" problems, each time generating a sequence of pairs $\{(\widehat{\gamma}_t, \widehat{\mathbf{v}}_t)\}$ depending on ϱ, which is called the *rarity parameter.*

Remark 3.10. It is important to understand the differences between Algorithm 3.3.1 and Algorithm 3.4.1. The major difference is that Algorithm 3.4.1 estimates the optimal reference parameters for *multiple* levels $\{\gamma_t\}$ whereas in Algorithm 3.3.1 only *one* level γ is used. With this in mind, we call Algorithm 3.4.1 a *multilevel* procedure (where the levels are indexed by t) and Algorithm 3.3.1 a *single-level* procedure. At the t-th level in Algorithm 3.4.1 we have to estimate (from the same sample) first γ_t and then $\mathbf{v}_t$.

We also present the *deterministic* version of Algorithm 3.4.1.

Algorithm 3.4.2 (Deterministic Version of the CE Algorithm)

1. *Define* $\mathbf{v}_0 = \mathbf{u}$. *Set* $t = 1$.
2. *Calculate* γ_t *as*

$$\gamma_t = \max\left\{s \,:\, \mathbb{P}_{\mathbf{v}_{t-1}}(S(\mathbf{X}) \geqslant s) \geqslant \varrho\right\}, \tag{3.42}$$

   provided $\gamma_t < \gamma$; *otherwise set* $\gamma_t = \gamma$.
3. *Calculate* $\mathbf{v}_t$ *(see (3.40)) as*

$$\mathbf{v}_t = \operatorname*{argmax}_{\mathbf{v}} \mathbb{E}_{\mathbf{v}_{t-1}} I_{\{S(\mathbf{X}) \geqslant \gamma_t\}} W(\mathbf{x}; \mathbf{u}, \mathbf{v}_{t-1}) \ln f(\mathbf{X}; \mathbf{v}) \,. \tag{3.43}$$

4. *If* $\gamma_t = \gamma$, *then* **stop**; *otherwise set* $t = t + 1$ *and reiterate from Step 2.*

Note that as compared with Algorithm 3.4.1 Step 5 is redundant in Algorithm 3.4.2.

It is not a simple matter to see whether or not Algorithm 3.4.2 and Algorithm 3.4.1 will reach the desired value γ say after a finite number of iterations, provided $\varrho \gg \ell$.

3.5 Examples

For better insight into the single-level Algorithm 3.3.1 (updating only the reference vector) and the multiple-level Algorithm 3.4.1 (updating the reference vector and the level) we now present several examples. Although in some of those examples the quantities of interest can be computed analytically, we present them to illustrate the algorithms. In particular, we show that the optimal reference parameter vectors ${}_*\mathbf{v}$ and $\mathbf{v}^*$ of the VM and CE programs are nearly the same. We start with a most simple example of the deterministic multilevel Algorithm 3.4.2.

Example 3.11. Let $X \sim \mathsf{Exp}(1)$. Suppose we want to estimate the probability $\mathbb{P}(X \geq \gamma)$, $(X \sim \{\mathsf{Exp}(v^{-1})\})$ for large γ, say $\gamma = (16,\ 64,\ 256)$ (corresponding to $\ell \approx (10^{-7},\ 10^{-28},\ 10^{-111})$, respectively. From (3.42) it is obvious that $\gamma_t = \min(\gamma, -v_{t-1} \ln \varrho)$, and from Example 3.4 we find $v_t = \gamma_t + 1$. If we take $\varrho = 0.02$ then it is readily seen that the deterministic Algorithm 3.4.2 terminates at iteration $T = 2, 3$ and 4, respectively. And if we take $\varrho = 0.001$ and the same values of γ, then Algorithm 3.4.2 terminates at iteration $T = 2, 3$ and 3, respectively. Table 3.1 presents the evolution of γ_t and v_t for fixed $\gamma = 64$ and $\varrho = 0.02$ and $\varrho = 0.001$, respectively.

Table 3.1. The evolution of γ_t and v_t for $\gamma = 64$, and $\varrho = 0.02$, and $\varrho = 0.001$.

$\varrho = 0.02$			$\varrho = 0.001$		
t	γ_t	v_t	t	γ_t	v_t
0	–	1	0	–	1
1	3.912	4.912	1	6.907	7.907
2	19.215	20.215	2	54.625	55.625
3	64	65	3	64	65

It follows from Table 3.1 that for $\gamma = 64$ in both cases Algorithm 3.4.2 terminates after three iterations ($\gamma_T = \gamma_3 = 64$). Figure 3.1 illustrates the three-level procedure.

Example 3.12 (Minimum of exponential random variables). Suppose we are interested in estimating $\ell = \ell(\gamma) = \mathbb{P}(S(\mathbf{X}) \geqslant \gamma)$, where

$$S(\mathbf{X}) = \min(X_1, \ldots, X_n) \tag{3.44}$$

and the random variables $X_1, \ldots, X_n$ are exponentially identically distributed with mean u; thus each X_i has density $f(\cdot; u) = u^{-1} \exp(-xu^{-1}), x \geqslant 0$. Obviously,

$$\ell = \prod_{i=1}^{n} \mathbb{P}(X_i \geqslant \gamma) = \mathrm{e}^{-n\gamma u^{-1}} . \tag{3.45}$$

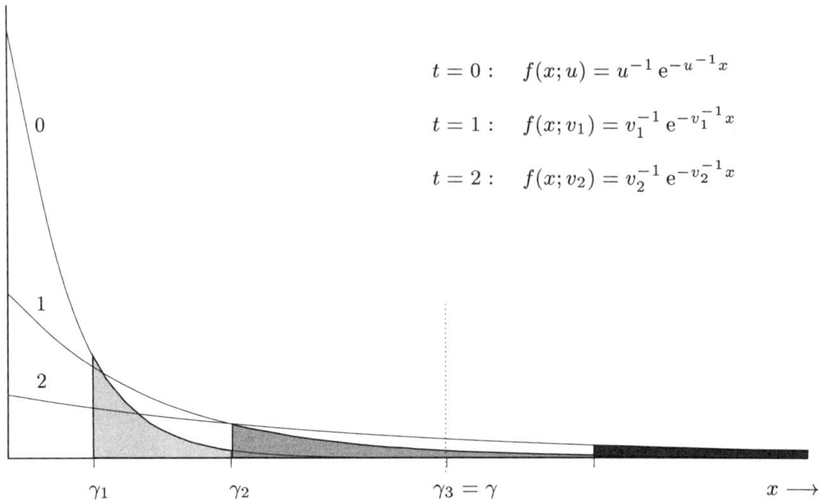

Fig. 3.1. A three-level realization of Algorithm 3.4.2. All shaded regions have area ϱ. The density of $f(\cdot; v_3)$ has been omitted.

For large γ, the squared coefficient of variation (SCV) of the crude Monte Carlo (CMC) estimator (see (1.10)) is

$$\kappa^2(\gamma) \approx \frac{1}{N}\,\mathrm{e}^{n\gamma u^{-1}} .$$

Hence the CMC estimator has *exponential* complexity in γ.

VM Approach. It is easy to verify from (1.40) and Table 1.5 that for i.i.d. and exponentially distributed random variables X_i, we have for any performance function H that

$$\mathbb{E}_u H(\mathbf{X}) W(\mathbf{X}; u, v) = \left(\frac{v^2}{u(2v-u)}\right)^n \mathbb{E}_{\frac{uv}{2v-u}} H(\mathbf{X}) . \tag{3.46}$$

It follows (see also Example 3.2) that the variance of the estimator (3.36) is

$$\begin{aligned}
\mathrm{Var}(\widehat{\ell}) &= \frac{1}{N}\left\{\mathbb{E}_u I_{\{\min(X_1,\ldots,X_n) \geqslant \gamma\}} W(\mathbf{X}; u, v) - \ell^2\right\} \\
&= \frac{1}{N}\left\{\left(\frac{v^2}{u(2v-u)}\right)^n \mathbb{E}_{\frac{uv}{2v-u}} I_{\{\min(X_1,\ldots,X_n) \geqslant \gamma\}} - \ell^2\right\} \\
&= \frac{1}{N}\left\{\left(\frac{v^2}{u(2v-u)}\right)^n \mathrm{e}^{-n\gamma(2u^{-1}-v^{-1})} - \ell^2\right\} \\
&= \frac{1}{N}\left\{\left(\frac{v^2 \mathrm{e}^{\gamma v^{-1}}}{u(2v-u)}\right)^n \ell^2 - \ell^2\right\} .
\end{aligned}$$

Consequently, the SCV of $\widehat{\ell}$ is given by

$$\kappa^2(v, \gamma) = \frac{1}{N} \left\{ \left(\frac{v^2 \mathrm{e}^{\gamma v^{-1}}}{u(2v - u)} \right)^n - 1 \right\} .$$

As in Example 3.2 we can readily obtain that under the variance minimization criterion the optimal reference parameter is given by

$${}_*v = \frac{1}{2} \left\{ \gamma + u + \sqrt{\gamma^2 + u^2} \right\} = \gamma + \frac{u}{2} + \mathcal{O}((u/\gamma)^2) .$$

CE Approach. Since the distribution of $\mathbf{X}$ lies in an exponential family of the form (1.3), with $\theta = u^{-1}, t(\mathbf{x}) = -\sum_{i=1}^n x_i$ and $c(\theta) = \theta^n$, we obtain directly from (3.27) that v^* satisfies

$$\mathbb{E}_u I_{\{S(\mathbf{X}) \geqslant \gamma\}} \left\{ \frac{n}{\eta} - \sum_{i=1}^n X_i \right\} = 0 , \tag{3.47}$$

with $\eta = 1/v^*$. It follows that

$$v^* = \mathbb{E}_u \left[n^{-1} \sum_{i=1}^n X_i \;\middle|\; S(\mathbf{X}) \geqslant \gamma \right] = u + \gamma .$$

For large $\gamma \gg u$ we have ${}_*v \approx v^* \approx \gamma$, so that for both VM and CE approaches

$$\kappa^2(\gamma) \approx \frac{1}{N} \gamma^n \mathrm{e}^n (2u)^{-n}, \tag{3.48}$$

where N is the sample size. That is, for large γ, the SCV $\kappa^2(\gamma)$ of the CMC and of both VM and CE optimal LR estimators increase in γ exponentially and polynomially, respectively. In other words, the CMC and both optimal VM and CE estimators can be viewed as *exponential* and *polynomial* ones.

Example 3.13 (Sum of exponential random variables). Suppose we are interested in estimating

$$\ell(\gamma) = \mathbb{P}(S(\mathbf{X}) \geqslant \gamma) = \mathbb{P}(X_1 + \cdots + X_n \geqslant \gamma) , \tag{3.49}$$

where the random variables $X_1, \ldots, X_n$ are exponentially identically distributed with mean u. Of course, in that case we know that $S(\mathbf{X})$ has a Gamma distribution with parameters n and u^{-1}, so $\mathbb{P}(S(\mathbf{X}) \geqslant \gamma)$ can be computed exactly as

$$\ell(\gamma) = \mathbb{P}_u(S(\mathbf{X}) \geqslant \gamma) \; = \; \sum_{k=0}^{n-1} \frac{\mathrm{e}^{-\gamma u^{-1}} (\gamma u^{-1})^k}{k!} .$$

It is difficult to compute the VM-optimal parameter in this case. However, we can compute the CE-optimal parameter $\mathbf{v}^*$ given by (3.31). Since $X_1, \ldots, X_n$

are i.i.d. it is clear that all components of $\mathbf{v}^*$ are identical. After some algebra, we obtain that

$$v^* = \frac{u \sum_{k=0}^{n} \mathrm{e}^{-\gamma u^{-1}} (\gamma u^{-1})^k / k!}{\sum_{k=0}^{n-1} \mathrm{e}^{-\gamma u^{-1}} (\gamma u^{-1})^k / k!} = \frac{u \mathbb{P}_u (X_1 + \cdots + X_{n+1} \geqslant \gamma)}{\mathbb{P}_u (X_1 + \cdots + X_n \geqslant \gamma)}, \quad (3.50)$$

where X_{n+1} is independent of $X_1, \ldots, X_n$ and has the same distribution as $X_1, \ldots, X_n$. From the above expression, it is easy to see that, when γ/u is large (that is $\gamma \gg u$), we have

$$v^* \approx (u + \gamma)/n \approx \gamma/n . \quad (3.51)$$

Let us consider the asymptotic properties of the SCV of the LR estimator $\widehat{\ell}$ in (3.6), as $\gamma \to \infty$. Similar to the previous example we have

$$N \times \kappa^2(\gamma) = \left(\frac{v^2}{u(2v - u)} \right)^n \frac{\mathbb{P}_{\frac{uv}{2v-u}} (X_1 + \cdots + X_n \geqslant \gamma)}{\ell^2(\gamma)} - 1 , \quad (3.52)$$

where we have used (3.46). For large γ and fixed w the dominating term in $\mathbb{P}_w(X_1 + \cdots + X_n \geqslant \gamma)$ is $(\gamma w^{-1})^{n-1} \exp(-\gamma w^{-1})/(n-1)!$. We can use this to show that for reference parameter $v = \gamma/n$, we have

$$N \times \kappa^2(\gamma) = \frac{\mathrm{e}^n (n-1)!}{n^n} \left\{ \frac{\gamma}{2u} - \frac{2n-3}{4} \right\} - 1 + o(1)$$

as $\gamma \to \infty$. That is, the SCV of the optimal LR estimator grows *linearly* with respect to γ. Notice that, for $v = u$ — which corresponds to the CMC estimator — the SCV increases *exponentially* in γ .

Example 3.14 (Heavy tails). The CE method is not limited to light-tail distributions. We can also apply it to heavy-tail distributions. To illustrate this, we generalize Example 3.12 for $n = 1$ to the Weibull case. Specifically, consider estimation of $\ell = \mathbb{P}(X \geqslant \gamma)$ with $X \sim \mathsf{Weib}(a, u^{-1})$, $0 < a < 1$. That is, X has density

$$f(x; u) = a\, u^{-1} \left(u^{-1} x\right)^{a-1} \mathrm{e}^{-\left(u^{-1} x\right)^a}, \quad x > 0 . \quad (3.53)$$

To estimate ℓ via the CE method we shall use the family of distributions $\{\mathsf{Weib}(a, v^{-1}), v > 0\}$, where a is kept fixed. Note that for $a = 1$ we have the class of exponential distributions.

Using the CE approach, we find the optimal CE reference parameter by solving

$$\max_v D(v) = \max_v \int_\gamma^\infty f(x; u) \ln f(x; v)\, dx ,$$

or, equivalently, by solving

$$\int_\gamma^\infty f(x; u) \frac{d}{dv} \ln f(x; v)\, dx = 0 . \quad (3.54)$$

Substituting (3.53) into (3.54) yields the following simple expression for the optimal CE reference parameter v^*:

$$v^* = (u^a + \gamma^a)^{1/a} . \tag{3.55}$$

This is true for *any* $a > 0$. Note that $\{\mathsf{Weib}(a, u^{-1}), u > 0\}$ is an exponential family of the form (1.3), with $t(x) = -x^a$, $\theta = u^{-a}$, $c(\theta) = \theta$ and $h(x) = ax^{a-1}$. So we can obtain (3.55) similar to (3.47) as the solution to

$$\mathbb{E}_u I_{\{X \geqslant \gamma\}} \left\{ \frac{1}{\eta} - X^a \right\} = 0 , \tag{3.56}$$

with $\eta = v^{-a}$. It follows that the optimal η^* is given by $1/\mathbb{E}_u \left[X^a \,|\, X \geqslant \gamma\right]$, or equivalently

$$v^* = \left(\frac{\mathbb{E}_u I_{\{X \geqslant \gamma\}} X^a}{\mathbb{E}_u I_{\{X \geqslant \gamma\}}} \right)^{1/a} . \tag{3.57}$$

Similar to Example 3.12 the variance of the SLR estimator $\widehat{\ell}$ for any reference parameter $\mathbf{v}$ can be easily found from (1.40) and Table 1.5. Namely, using the exponential family representation above we have

$$\begin{aligned}
\mathrm{Var}(\widehat{\ell}) &= \frac{1}{N} \left\{ \mathbb{E}_u I_{\{X \geqslant \gamma\}} W(X; u, v) - \ell^2 \right\} \\
&= \frac{1}{N} \left\{ \frac{\mathbb{P}_{(2\theta - \eta)^{-1/a}}(X \geqslant \gamma)}{(u/v)^a (2 - (u/v)^a)} - \ell^2 \right\} \\
&= \frac{1}{N} \left\{ \frac{\mathrm{e}^{-\gamma^a (2u^{-a} - v^{-a})}}{(u/v)^a (2 - (u/v)^a)} - \ell^2 \right\} \\
&= \frac{1}{N} \left\{ \frac{\mathrm{e}^{(\gamma/v)^a}}{(u/v)^a (2 - (u/v)^a)} \, \ell^2 - \ell^2 \right\} .
\end{aligned}$$

If we substitute v with v^* and divide by ℓ^2 , we find that the SCV κ^2 of $\widehat{\ell}$ is given by

$$\frac{1}{N} \left\{ \frac{\exp\left(\frac{(\gamma/u)^a}{1 + (\gamma/u)^a} \right) (1 + (\gamma/u)^a)^2}{2(\gamma/u)^a + 1} \right\} .$$

It follows that for large γ/u

$$\kappa^2 \approx \frac{1}{N} \frac{\mathrm{e}}{2} \left(\frac{\gamma}{u} \right)^a .$$

In other words, the SLR estimator $\widehat{\ell}$ has *polynomial* complexity in γ, for any $a > 0$, including the heavy-tail case $0 < a < 1$. It is a common misunderstanding that IS only works for light-tail distributions. In this example we saw that polynomial complexity can be easily obtained by using the CE method. But we can do even better. In Section 3.8 we will see how with the TLR method we can in fact achieve an SLR estimator with *bounded relative error*, meaning that the κ^2 is bounded by c/N for some constant c which does not depend on γ.

Remark 3.15 (Weibull random variables). As a generalization of Example 3.14 consider the estimation of $\ell = \mathbb{P}(S(\mathbf{X}) \geqslant \gamma)$ where the components X_i, $i = 1, \ldots, n$ of $\mathbf{X}$ are independent and $X_i \sim \mathsf{Weib}(a_i, u_i^{-1})$, $i = 1, \ldots, n$. Consider the change of measure where the X_i are still independent but now $\mathsf{Weib}(a_i, v_i^{-1})$ distributed, for some $v_i > 0$. We write this symbolically as

$$X_i \sim \mathsf{Weib}(a_i, u_i^{-1}) \longrightarrow \mathsf{Weib}(a_i, v_i^{-1}) \, .$$

It is readily seen from (3.57) that for fixed a_i, $i = 1, \ldots, n$, programs (3.40) and (3.41) can be solved analytically, and the components of $\mathbf{v}_t = (v_{t,1}, \ldots, v_{t,n})$ and $\widehat{\mathbf{v}}_t = (\widehat{v}_{t,1}, \ldots, \widehat{v}_{t,n})$ in the Weibull pdf are

$$v_{t,i} = \left(\frac{\mathbb{E}_{\mathbf{v}_{t-1}} I_{\{S(\mathbf{X}) \geqslant \gamma_t\}} W(\mathbf{X}; \mathbf{u}, \mathbf{v}_{t-1}) X_i^{a_i}}{\mathbb{E}_{\mathbf{v}_{t-1}} I_{\{S(\mathbf{X}) \geqslant \gamma_t\}} W(\mathbf{X}; \mathbf{u}, \mathbf{v}_{t-1})} \right)^{1/a_i} \tag{3.58}$$

and

$$\widehat{v}_{t,i} = \left(\frac{\sum_{k=1}^N I_{\{S(\mathbf{X}_k) \geqslant \widehat{\gamma}_t\}} W(\mathbf{X}_k; \mathbf{u}, \widehat{\mathbf{v}}_{t-1}) X_{ki}^{a_i}}{\sum_{k=1}^N I_{\{S(\mathbf{X}_k) \geqslant \widehat{\gamma}_t\}} W(\mathbf{X}_k; \mathbf{u}, \widehat{\mathbf{v}}_{t-1})} \right)^{1/a_i} , \tag{3.59}$$

respectively.

A different parameterization of the Weibull distribution gives an even simpler formula. Namely, if we use the change of measure

$$X_i \sim \mathsf{Weib}(a_i, u_i^{-1/a_i}) \longrightarrow \mathsf{Weib}(a_i, v_i^{-1/a_i}) \, ,$$

thus,

$$f(x; v_i) = a_i \, v_i^{-1} \, x^{a_i - 1} \, \mathrm{e}^{-v_i^{-1} x^{a_i}} \, .$$

Then the v-parameters are updated as

$$\widehat{v}_{t,i} = \frac{\sum_{k=1}^N I_k W_k X_{ki}^{a_i}}{\sum_{k=1}^N I_k W_k} \, , \tag{3.60}$$

where we have used the abbreviations $I_k = I_{\{S(\mathbf{X}_k) \geqslant \widehat{\gamma}_t\}}$ and $W_k = W(\mathbf{X}_k; \mathbf{u}, \widehat{\mathbf{v}}_{t-1})$.

Remark 3.16 (Two-parameter update). For the Weibull distribution it is not difficult to formulate a two-parameter updating procedure in which both scale and shape parameter are updated. Specifically, consider the change of measure

$$X_i \sim \mathsf{Weib}(a_i, u_i^{-1/a_i}) \longrightarrow \mathsf{Weib}(b_i, v_i^{-1/b_i}), \quad v_i > 0, \, b_i > 0 \, .$$

The updating formula for the v_i is given in (3.60), but an analytic updating of the parameter vector $\mathbf{b} = (b_1, \ldots, b_n)$ is not available from (3.41). However, it is readily seen that the i-th component of $\nabla_{\mathbf{b}} \ln f(\mathbf{X}; \mathbf{b}, \mathbf{v})$ for the random vector $\mathbf{X}$ with independent components $X_i \sim \mathsf{Weib}(b_i, \widehat{v}_i^{-1/b_i}),\ i = 1, \ldots, n$, equals

$$b_i^{-1} + \ln X_i - \frac{X_i^{b_i}}{\widehat{v}_i} \ln X_i \ . \tag{3.61}$$

Consequently, the i-th component of $\mathbf{b}$ can be obtained from the numerical solution of the following nonlinear equation:

$$\frac{1}{N} \sum_{k=1}^{N} I_k W_k (b_i^{-1} + \ln X_{ki} - \frac{X_{ki}^{b_i}}{\widehat{v}_i} \ln X_{ki}) = 0 \ . \tag{3.62}$$

Substituting $\widehat{v}_i$ from (3.60), into (3.62) we obtain

$$b_i^{-1} + \frac{\sum_{k=1}^{N} I_k W_k \ln X_{ki}}{\sum_{k=1}^{N} I_k W_k} - \frac{\sum_{k=1}^{N} I_k W_k X_{ki}^{b_i} \ln X_{ki}}{\sum_{k=1}^{N} I_k W_k X_{ki}^{b_i}} = 0 \ . \tag{3.63}$$

One might solve (3.63) for example using the bisection method.

Remark 3.17 (Two-parameter families). As noted in Remark 3.16 it is sometimes useful to consider two-parameter families in which both parameters have to be updated. Consider the univariate case where X is a random variable with a density $f(\cdot; u)$. We wish to estimate $\ell = \mathbb{P}(S(X) \geqslant \gamma)$, via IS, using a member of the two-parameter family $\{g(\cdot; \boldsymbol{\eta})\}$, where $\boldsymbol{\eta} = (\eta_1, \eta_2)$. The optimal CE parameters follow, in the usual way, from minimization of the Kullback-Leibler distance to the theoretically optimal IS density for this problem. Specifically, the optimal parameter vector $\boldsymbol{\eta}^*$ is the solution to the maximization problem

$$\max_{\boldsymbol{\eta}} D(\boldsymbol{\eta}) = \max_{\boldsymbol{\eta}} \mathbb{E}_{\boldsymbol{\theta}}\, I_{\{S(X) \geqslant \gamma\}} W(X; u, \boldsymbol{\theta}) \ln g(X; \boldsymbol{\eta}) \ , \tag{3.64}$$

where $X \sim g(\cdot; \boldsymbol{\theta})$ and $W(X; u, \boldsymbol{\theta}) = f(X; u)/g(X; \boldsymbol{\theta})$. As usual it can be estimated by solving the stochastic optimization program

$$\max_{\boldsymbol{\eta}} \widehat{D}(\boldsymbol{\eta}) = \max_{\boldsymbol{\eta}} \frac{1}{N} \sum_{i=1}^{N} I_{\{S(X_i) \geqslant \gamma\}}\, W(X_i; u, \boldsymbol{\theta}) \ln g(X_i; \boldsymbol{\eta}) \ , \tag{3.65}$$

where $X_1, \ldots, X_N$ is a random sample from $g(\cdot; \boldsymbol{\theta})$. This is very similar to the "standard" CE procedure (3.40) and its stochastic equivalent (3.41). Consequently, the main CE Algorithm 3.4.1 and its modifications can be easily adapted to the deployment of the family $\{g(\cdot; \boldsymbol{\eta})\}$ instead of $\{f(\cdot; \mathbf{v})\}$.

We give two examples where the updating formulas are easy, because — as noted in Remark 2.8 — solving (3.65) is very similar to deriving the *maximum likelihood* estimator of $\boldsymbol{\eta}$ on the basis of the random sample $X_1, \ldots, X_n$. The only difference is the presence of the "weighting terms" $I_i W_i$ (using abbreviations similar to those in (3.60)).

Normal Distribution

Consider the density

$$g(x;\boldsymbol{\eta}) = \frac{1}{\sqrt{2\pi\sigma^2}}\,\mathrm{e}^{-\frac{1}{2}\frac{(x-\mu)^2}{\sigma^2}}\,, \quad x \in \mathbb{R}, \quad \boldsymbol{\eta} = (\mu, \sigma^2)\,.$$

The optimal solution of (3.65) follows from minimization of

$$\frac{1}{\sigma^2}\sum_{i=1}^{N} I_i\, W_i\,(X_i-\mu)^2 + \ln(\sigma^2)\sum_{i=1}^{N} I_i\, W_i\,.$$

It is easily seen that this minimum is obtained at $(\widehat{\mu}, \widehat{\sigma^2})$ given by

$$\widehat{\mu} = \frac{\sum_{i=1}^N I_i\, W_i\, X_i}{\sum_{i=1}^N I_i W_i} \tag{3.66}$$

and

$$\widehat{\sigma^2} = \frac{\sum_{i=1}^N I_i\, W_i\,(X_i-\widehat{\mu})^2}{\sum_{i=1}^N I_i\, W_i}\,. \tag{3.67}$$

Shifted Exponential Distribution

Consider the density

$$g(x;\boldsymbol{\eta}) = \eta\,\mathrm{e}^{-\eta\,(x-a)}\,, \quad x \geqslant a, \quad \boldsymbol{\eta} = (\eta, a)\,.$$

Denote the CE optimal parameters by $(\widehat{\eta}, \widehat{a})$. Let $X_{(1)}, \ldots, X_{(N)}$ denote the order statistics of $X_1, \ldots, X_N$, and let $I_{(i)}$ be the indicator in $\{I_1, \ldots, I_N\}$ that corresponds to $X_{(i)}$. Let K be the first index such that $I_{(K)} = 1$. Note that $\widehat{a}$ must be less than or equal to the order statistic $X_{(1)}$, since otherwise $g(X_i; (\widehat{\eta}, \widehat{a})) = 0$, for some i. Thus, we have to maximize

$$\sum_i I_i\, W_i\, \ln\eta - \eta\, I_i\, W_i\,(X_i - a)\,,$$

for $a \leqslant X_{(1)}$ and $\eta \geqslant 0$. It follows that

$$\widehat{a} = X_{(K)}$$

and

$$\widehat{\eta} = \frac{\sum_i I_i\, W_i}{\sum_i I_i\, W_i\,(X_i - \widehat{a})}\,.$$

Remark 3.18 (Hazard rate twisting). It is interesting to note that hazard rate twisting [91] often amounts to SLR. In hazard rate twisting the change of measure for some distribution with pdf f (with support in $\mathbb{R}_+$) and tail distribution function $\bar{F}$ is such that the *hazard rate* (or failure rate) $\lambda(x) = f(x)/\bar{F}(x)$ is changed to $(1-\theta)\lambda(x)$, for some $0 \leqslant \theta < 1$. The pdf of the changed measure is now

$$f_\theta(x) = \lambda(x)(1-\theta)\,\mathrm{e}^{-(1-\theta)\Lambda(x)}\,,$$

where $\Lambda(x) = \int_0^x \lambda(y)\,dy$. In particular, for the $\mathsf{Weib}(a, u^{-1})$ distribution we have $\lambda(x) = au^{-1}(u^{-1}x)^a$ and $\Lambda(x) = (u^{-1}x)^a$, so that

$$f_\theta(x) = (1-\theta)au^{-1}(u^{-1}x)^{a-1}\mathrm{e}^{-(1-\theta)(u^{-1}x)^a}\,,$$

which corresponds to the SLR CoM $\mathsf{Weib}(a, u^{-1}) \longrightarrow \mathsf{Weib}(a, v^{-1})$, with $v^{-1} = (1-\theta)^{1/a}u^{-1}$. Similarly, for the $\mathsf{Pareto}(a, u^{-1})$ distribution, with $\bar{F}(x) = (1 + x/u)^{-a}$, we have $\lambda(x) = au^{-1}(1 + u^{-1}x)^{-1}$ and $\Lambda(x) = a\ln(1 + u^{-1}x)$, so that

$$f_\theta(x) = (1-\theta)au^{-1}(1 + u^{-1}x)^{-((1-\theta)a+1)}\,,$$

so that hazard rate twisting with parameter θ corresponds to the SLR change of measure $\mathsf{Pareto}(a, u^{-1}) \longrightarrow \mathsf{Pareto}(b, u^{-1})$ with $b = (1-\theta)a$. Note that in the Weibull case the scale parameter u^{-1} is changed, whereas in the second case the shape parameter a is changed.

3.6 Convergence Issues

In this section we consider the convergence behavior of Algorithm 3.4.2. In particular, we investigate the conditions under which the algorithm indeed terminates. For this to happen, the sequence $\{\gamma_t\}$ has to cross level γ at some stage $T < \infty$, at which point γ_T is reset to γ and $\mathbf{v}_T$ is determined from (3.43), from which it follows that

$$\mathbf{v}_T = \operatorname*{argmax}_{\mathbf{v}} \mathbb{E}_{\mathbf{u}} I_{\{S(\mathbf{X}) \geqslant \gamma\}} \ln f(\mathbf{X}; \mathbf{v}) = \mathbf{v}^*\,.$$

The next proposition gives a sufficient condition for convergence in the one-dimensional case. Recall that we wish to estimate $\ell = \mathbb{P}(S(X) \geqslant \gamma)$, where $X \in \mathbb{R}$ has a pdf $f(\cdot; u)$ belonging to a family $\{f(\cdot; v)\}$. Let $F(\cdot; v)$ be the cdf corresponding to the pdf $f(\cdot; v)$.

Proposition 3.19 (Convergence). *Let $\{f(\cdot; v)\}$ be a NEF that is parameterized by the mean v, and let S be an unbounded monotone increasing continuous function on $\mathbb{R}$. Suppose $F(x; v)$ is monotone nonincreasing in v. If there exists $c > 0$ such that*

$$F(v + c; v) \leqslant F(u + c; u) \quad \textit{for all } v > u\,, \tag{3.68}$$

then any ϱ satisfying

$$\varrho < 1 - F(u + c; u) \tag{3.69}$$

ensures that Algorithm 3.4.2 terminates in at most γ/c iterations.

Proof. First, observe that

$$v_1 = \frac{\mathbb{E}_u X I_{\{S(X) \geqslant \gamma_1\}}}{\mathbb{E}_u I_{\{S(X) \geqslant \gamma_1\}}} > \mathbb{E}_u X = u = v_0 \; .$$

This inequality follows from the fact that $\mathbb{E}_u X I_{\{S(X) \geqslant \gamma_1\}} - u\,\mathbb{E} I_{\{S(X) \geqslant \gamma_1\}}$ is the covariance between X and $I_{\{S(X) \geqslant \gamma_1\}}$, which must be positive since S is increasing.

Second, let $\Delta v_t = v_t - v_{t-1}$ be the increment of v at iteration t, and similarly let $\Delta\gamma_t$ be the increment in γ. Suppose that for some $t \geqslant 1$ we have $\Delta v_t > 0$. Then, $\Delta\gamma_{t+1} > 0$ as well. To see this, recall that $\mathbb{P}_{v_{t-1}}(S(X) \geqslant \gamma_t) = \varrho$, so that

$$F(x_t; v_{t-1}) = F(x_{t+1}; v_t) \; ,$$

where $x_t = S^{-1}(\gamma_t)$, $t \geqslant 1$. Let us write this as

$$F(x_t; v_{t-1}) = F(x_t + \Delta x_{t+1}; v_{t-1} + \Delta v_t) \; , \tag{3.70}$$

where $\Delta x_{t+1} = x_{t+1} - x_t$. Since $F(x; v)$ is monotone nondecreasing in x, monotone nonincreasing in v, and $\Delta v_t > 0$, we must have that $\Delta x_{t+1} > 0$, which implies $\Delta\gamma_{t+1} > 0$.

Third, suppose, conversely, that $\Delta\gamma_t > 0$ for some $t \geqslant 1$. Then, also $\Delta v_t > 0$. This is proved as follows: Define

$$\psi(\gamma) = \frac{\mathbb{E}_u X I_{\{S(X) \geqslant \gamma\}}}{\mathbb{E}_u I_{\{S(X) \geqslant \gamma\}}} = \frac{\int_{S^{-1}(\gamma)}^{\infty} x f(x; u)\, dx}{\int_{S^{-1}(\gamma)}^{\infty} f(x; u)\, dx} \; .$$

We leave it as an exercise to show that ψ has a strictly positive derivative. Since $v_t = \psi(\gamma_t)$, it follows that $\gamma_t > \gamma_{t-1}$ implies $v_t > v_{t-1}$.

Finally, we show that Algorithm 3.4.2 actually reaches level γ in a finite number of iterations. A sufficient condition is that $\{\gamma_t\}$ increases to ∞, as $t \to \infty$. This in turn is equivalent to $\{v_t\}$ increasing to ∞, because $\psi(\gamma) \geqslant S^{-1}(\gamma)$ is a continuous, monotone increasing and unbounded function. The condition $v_t \to \infty$, or equivalently the divergence of the series $\sum \Delta v_t$, depends on the choice of ϱ. Specifically, ϱ needs to be chosen such that for every $v > u$ the corresponding increment Δv is greater than some fixed constant $c > 0$, with

$$\Delta v = \frac{\int_{F^{-1}(1-\varrho; v)}^{\infty} (x - v) f(x; u)\, dx}{\int_{F^{-1}(1-\varrho; v)}^{\infty} f(x; u)\, dx} \; .$$

To find such a ϱ observe that $\Delta v > F^{-1}(1 - \varrho; v) - v$, so that a sufficient condition for divergence of $\sum \Delta v_t$ is that $F^{-1}(1 - \varrho; v) - v > c$, which is equivalent with (3.69). In particular, if (3.68) holds, then any ϱ satisfying (3.69) guarantees convergence of the CE algorithm with a "step size" of at least c.

Remark 3.20. If (3.68) does not hold then we may still enforce the divergence of $\sum \Delta v_t$ by updating ϱ in each iteration via

$$\varrho_t \leqslant 1 - F(v_t + c; v_t) \, .$$

Example 3.21 (Example 3.12 continued). Recall that in that example we had $S(\mathbf{X}) = \min(X_1, \dots, X_n)$ with $X_i \sim f(x; u) = u^{-1} \mathrm{e}^{-x u^{-1}}$. As seen earlier, the CE-optimal solution is $v^* = \gamma + u$ (see Example 3.4). Consider now γ_t and $\mathbf{v}_t$ defined in (3.42) and (3.43). Since the algorithm stops when $\gamma_t \geqslant \gamma$, and since the distribution of $S(\mathbf{X})$ is continuous, we can write

$$\begin{aligned} \gamma_t &= \max \{x \leqslant \gamma : \exp(-xn/v_{t-1}) \geqslant \varrho\} \\ &= \min \{\gamma, C \, v_{t-1}/n\} \end{aligned} \tag{3.71}$$

where $C = -\ln \varrho > 0$. The parameter v_t defined can then be rewritten as

$$v_t = \gamma_t + u \; = \; \min\{\gamma, C v_{t-1}/n\} + u \, . \tag{3.72}$$

Consider the unidimensional function $g(v) = \min\{\gamma, Cv/n\} + u$. It is easy to see that g has a single fixed point $\overline{v}$. Figure 3.2 illustrates the convergence of the procedure.

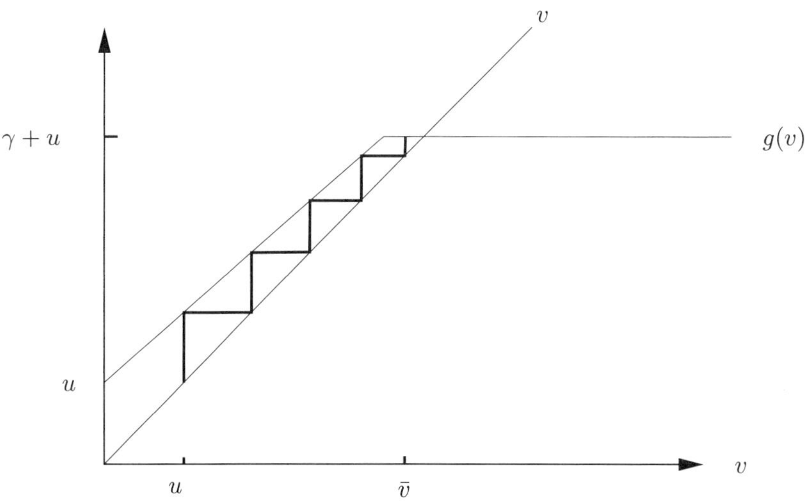

Fig. 3.2. Convergence of Algorithm 3.4.2.

It is easy to see that $\overline{v} = v^*$ if and only if

$$\gamma \; \leqslant \; (C/n)\overline{v} \, . \tag{3.73}$$

Let us compute $\overline{v}$. If $C/n \geqslant 1$, that is, if $C \geqslant n$, then we have $\overline{v} = \gamma + u = v^*$. On the other hand, if $C/n < 1$, that is, if $C < n$, then we have two cases:

1. $\gamma \leqslant (C/n)(\gamma + u)$: in this case, we have that $g(\gamma + u) = \gamma + u$, that is $\overline{v} = \gamma + u$.
2. $\gamma > (C/n)(\gamma + u)$: in this case, we have that $\overline{v} = u/(1 - C/n)$.

In both cases it is easy to check that the optimality condition (3.73) becomes $\gamma \leqslant (C/n)(\gamma + u)$, that is $C \geqslant n\gamma/(\gamma + u)$. Since $C = -\ln(\varrho)$, it follows that Algorithm 3.4.2 will converge to the correct solution if and only if

$$\varrho \leqslant \exp\left(-\frac{n\gamma}{\gamma + u}\right). \tag{3.74}$$

Moreover, if $\varrho \leqslant \exp(-n)$ (which implies (3.74)), that is, if $C/n > 1$, then the differences $v_t - v_{t-1}$ *increase* until the point when γ is hit by γ_t. On the other hand, if $\varrho > \exp(-n)$, then the differences $v_t - v_{t-1}$ *decrease* until the point when γ is hit by γ_t.

At first sight, condition (3.74) seems discouraging, since it requires the parameter ϱ to decrease exponentially with n. Notice however that this example constitutes an intrinsically difficult problem — from (3.45), we see that the probability being estimated goes to zero exponentially in n under *any* parameter. It makes perhaps more sense to consider the behavior of (3.74) for *fixed* n — then we see that $\varrho \leqslant \exp(-n)$ is a sufficient condition for Algorithm 3.4.2 to work, *regardless of the value of* γ. We may also consider what happens when γ is allowed to vary with n; for example, when $\gamma = \Delta/n$ for some $\Delta > 0$, condition (3.74) becomes asymptotically $\varrho \leqslant \exp(-\Delta/u)$.

Example 3.22 (Example 3.13 continued). Recall that in that example we had $S(\mathbf{X}) = X_1 + \cdots + X_n$ with $X_i \sim f(x; u) = u^{-1}e^{-xu^{-1}}$. As seen earlier, the CE-optimal solution is given by (3.50). To study the convergence of the deterministic Algorithm 3.4.2, we define

$$R_n(z) = \sum_{k=0}^{n-1} \frac{\mathrm{e}^{-z}\, z^k}{k!}, \quad z \geqslant 0\,,$$

and let R_n^{-1} denote its inverse on (0,1). To find γ_t for $t = 1, 2, \ldots$ we have to solve γ_t from

$$R_n(\gamma_t/v_{t-1}) = \varrho\,,$$

and check whether this $\gamma_t \geqslant \gamma$, in which case we "reset" γ_t to γ. In other words,

$$\gamma_t = \min\{\gamma,\, v_{t-1}\, R_n^{-1}(\varrho)\}\,.$$

Moreover, for large γ_t we saw in (3.51) that $v_t \approx (u + \gamma_t)/n$. Thus,

$$v_t \approx \min\left\{\frac{\gamma + u}{n},\ \frac{v_{t-1}\, R_n^{-1}(\varrho) + u}{n}\right\}.$$

Consequently, we are in the same situation as in the previous example. In particular, if we define

$$g(v) = \min\left\{\frac{\gamma+u}{n}, \frac{v\,R_n^{-1}(\varrho)+u}{n}\right\} ,$$

then $v_1, v_2, \ldots$ converges to the fixed point $(\gamma+u)/n$ of g if and only if

$$\frac{u}{n - R_n^{-1}(\varrho)} \geqslant \frac{u+\gamma}{n} ,$$

which is equivalent to

$$\varrho \leqslant R_n\left(\frac{n\gamma}{u+\gamma}\right) .$$

Note as γ grows larger, $R_n\left(\frac{n\gamma}{u+\gamma}\right)$ tends to $1/2$, showing that even a ϱ as large as $1/2$ guarantees convergence.

Example 3.23 (Example 3.7 continued). Consider the deterministic Algorithm 3.4.2 for the discrete uniform n-point density, given by

$$u_i = \frac{1}{m},\ i = 1, \ldots, m .$$

It is readily seen that in this case (3.35) becomes

$$v_i^* = \begin{cases} 0, & \text{if } 1 \leqslant i \leqslant \gamma - 1, \\ \dfrac{1}{m-\gamma+1}, & \text{if } \gamma \leqslant i \leqslant m , \end{cases} \tag{3.75}$$

assuming again that $\gamma \in \{1, \ldots, m\}$. Now, consider the sequence of pairs $\{(\gamma_t, \mathbf{v}_t), t = 1, 2, \ldots\}$ for the multilevel approach of Algorithm 3.4.2. First, for a given $\mathbf{v}_t$, $t = 1, 2, \ldots$ (where $\mathbf{v}_0 = \mathbf{u}$, by definition), we have

$$\gamma_{t+1} = \max\{x : \mathbb{P}_{\mathbf{v}_t}(X \geqslant x) \geqslant \varrho\} .$$

From (3.75) we see that $\mathbf{v}_t = (v_{t1}, \ldots, v_{tm})$ satisfies

$$v_{ti} = \begin{cases} 0, & \text{if } 1 \leqslant i \leqslant \gamma_t - 1 , \\ \dfrac{1}{m-\gamma_t+1}, & \text{if } \gamma_t \leqslant i \leqslant m , \end{cases}$$

for a given γ_t. In particular, we have for $t = 1, 2, 3, \ldots$,

$$\mathbb{P}_{\mathbf{v}_t}(X \geqslant \gamma) = \frac{m - \lceil\gamma\rceil + 1}{m - \gamma_t + 1} .$$

It follows that for $t = 1, 2, \ldots$,

$$\gamma_{t+1} = \lfloor (m+1)(1-\varrho) + \varrho\gamma_t \rfloor ,$$

and, similarly, $\gamma_1 = \lfloor 1 + m(1-\varrho) \rfloor$.

Due to the rounding operation, it is difficult to compute a closed form solution for γ_t. We can however use the bounds

$$(m+1)(1-\varrho) + \varrho\gamma_t - 1 \leqslant \gamma_{t+1} \leqslant (m+1)(1-\varrho) + \varrho\gamma_t \,.$$

By applying these bounds recursively, we obtain that

$$\begin{aligned}\gamma_{t+1} &\geqslant ((m+1)(1-\varrho) - 1)\left(1 + \varrho + \cdots + \varrho^{t+1}\right) \\ \gamma_{t+1} &\leqslant ((m+1)(1-\varrho))\left(1 + \varrho + \cdots + \varrho^{t+1}\right),\end{aligned}$$

that is,

$$\left(m + 1 - \frac{1}{1-\varrho}\right)(1 - \varrho^{t+1}) \leqslant \gamma_{t+1} \leqslant (m+1)(1-\varrho^{t+1}) \,.$$

We can infer from the above inequality that, if $m + 1 - \frac{1}{1-\varrho} > \gamma - 1$, that is, if $\varrho < (m - \gamma + 1)/(m - \gamma + 2)$, then for t large enough we have $\gamma_{t+1} \geqslant \gamma$. Therefore, at some stage T of Algorithm 3.4.2, we obtain that $\gamma_T = \gamma$ and $\mathbf{v}_T = \mathbf{v}^*$. In other words, the algorithm finishes after a *finite* number of steps provided

$$\varrho < \frac{m - \gamma + 1}{m - \gamma + 2} \,. \tag{3.76}$$

The examples above illustrate that the parameter ϱ used in the CE algorithm plays a crucial role: we can only expect the CE algorithm to converge to the correct values if ϱ is sufficiently small. To determine, however, *a priori* a desired ϱ can be a difficult task.

A way to overcome this problem is to modify the basic CE algorithm in such a way that either ϱ or N or both are changed *adaptively*. We will consider such a modification, called the *fully adaptive cross entropy* FACE method, in more detail in Chapter 5; see also [145] for related ideas. However, for the rest of this section we briefly consider a modification proposed in [85], which paper provides various proofs of convergence for the CE algorithm.

Let $\mathbf{v}^*$ be a CE-optimal solution, that is,

$$\mathbf{v}^* \in \mathcal{V}^* = \underset{\mathbf{v}}{\operatorname{argmax}}\ \mathbb{E}_{\mathbf{u}} I_{\{S(\mathbf{X}) \geqslant \gamma\}} \ln f(\mathbf{X}; \mathbf{v}) \,, \tag{3.77}$$

where we have not assumed that the maximizer is unique — the argmax set $\mathcal{V}^*$ may contain more than one element.

The basic assumption in [85] is the following:

Assumption A: There exists a set $\mathcal{U}$ such that $\mathcal{U} \cap \mathcal{V}^* \neq \emptyset$ and $\mathbb{P}_{\mathbf{v}}(S(\mathbf{X}) \geqslant \gamma) > 0$ for all $\mathbf{v} \in \mathcal{U}$.

Assumption A simply ensures that the probability being estimated, $\ell = \mathbb{P}(S(\mathbf{X}) \geqslant \gamma)$, does not vanish in a neighborhood of the optimal parameter

$\mathbf{v}^*$. The assumption is trivially satisfied when the distribution of $S(\mathbf{X})$ has infinite tail. For finite support distributions the assumption holds as long as either γ is less than the maximum value of the function $S(\mathbf{x})$, or if there is a positive probability that γ is attained.

We present below from [85] the modified version of Algorithm 3.4.1. The algorithm requires the definition of constants ϱ (typically, $0.01 \leqslant \varrho \leqslant 0.1$), $\beta > 1$ and $\delta > 0$.

Algorithm 3.6.1 (An Adaptive CE Algorithm)

1. *Define* $\varrho_0 = \varrho$, $\widehat{\mathbf{v}}_0 = \mathbf{u}$, $N_0 = N$ *(original sample size). Set* $t = 1$.
2. *Generate a random sample* $\mathbf{X}_1, \dots, \mathbf{X}_{N_{t-1}}$ *from* $f(\cdot; \mathbf{v}_{t-1})$. *Let* $\widehat{\gamma}_t$ *be the sample* $(1 - \varrho_{t-1})$*-quantile of performances* $S(\mathbf{X}_1), \dots, S(\mathbf{X}_{N_{t-1}})$, *provided this is less than* γ. *Otherwise put* $\widehat{\gamma}_t = \gamma$.
3. *Use the* **same** *sample* $\mathbf{X}_1, \dots, \mathbf{X}_{N_{t-1}}$ *to solve the stochastic program (3.41), with* $N = N_{t-1}$. *Denote the solution by* $\widehat{\mathbf{v}}_t$.
4. *If* $\widehat{\gamma}_t = \gamma$ *then proceed with Step 5. Otherwise, check whether there exists a* $\bar{\varrho} \leqslant \varrho$ *such that the sample* $(1 - \bar{\varrho})$*-quantile of* $S(\mathbf{x}_1), \dots, S(\mathbf{x}_{N_{t-1}})$ *is bigger than or equal to* $\min\{\gamma, \widehat{\gamma}_{t-1} + \delta\}$:
   a) *If so, choose* ϱ_{t-1} *the largest such* $\bar{\varrho}$ *and let* $N_t = N_{t-1}$;
   b) *Otherwise set* $\varrho_t = \varrho_{t-1}$ *and* $N = \beta N$.

   Set $t = t + 1$ *and reiterate from Step 2.*
5. *Estimate the rare-event probability* ℓ *using the estimator* (3.36), *with* $\mathbf{v}$ *replaced by* $\widehat{\mathbf{v}}_T$, *where* T *is the final iteration counter.*

It is proved in [85] that, under Assumption A, Algorithm 3.6.1 converges with probability 1 to a solution of (3.24) (with $H(\mathbf{X}_i) = I_{\{S(\mathbf{X}_i) \geqslant \gamma\}}$ and $N = N_T$) after a finite number of iterations T.

We can then compare the approximating solution $\widehat{\mathbf{v}}_T$ and the "true" solution $\mathbf{v}^*$ using the asymptotic analysis for optimal solutions of stochastic optimization problems discussed in [149]. In particular, we obtain the *consistency* result that, as $N \to \infty$, the distance between $\widehat{\mathbf{v}}_T$ and the solution set defined in (3.77) goes to zero (with probability 1) provided that: i) the function $\mathbf{v} \mapsto \ln f(\mathbf{x}; \mathbf{v})$ is continuous, ii) the set $\mathcal{U}$ defined in assumption A is compact, and iii) there exists a function h such that $\mathbb{E}_{\mathbf{u}} h(\mathbf{X}) < \infty$ and $|\ln f(\mathbf{x}; \mathbf{v})| \leqslant h(\mathbf{x})$ for all $\mathbf{x}$ and all $\mathbf{v}$.

Remark 3.24. Instead of using a constant δ in Algorithm 3.6.1 one can also apply a dynamic procedure where δ is changed according to the differences in $\widehat{\gamma}_t$. More specifically, we can set $\delta_t = \widehat{\gamma}_t - \widehat{\gamma}_{t-1}$ between Steps 3 and 4 of the algorithm, and use δ_t in Step 4. The idea behind that is to update ϱ_t as soon as the differences in the $\widehat{\gamma}$'s start decreasing. Such a procedure does not affect theoretical convergence and typically makes $\widehat{\gamma}_t$ reach γ faster; however, in some instances this might cause $\widehat{\gamma}_t$ to increase too fast, which in turn will yield a poor estimate $\widehat{\mathbf{v}}_T$ in Step 2 of the algorithm, due to the (relative) rarity of the event $\{S(\mathbf{X}) \geqslant \widehat{\gamma}_{t-1}\}$ in (3.41). Thus, the more conservative choice δ=constant is recommended unless some pilot studies can be performed.

An even more conservative approach is to take $\delta = 0$ until the sequence $\{\widehat{\gamma}_t\}$ gets "stalled," at which point a positive δ is used again. This approach yields the slowest progression of $\widehat{\gamma}_t$, but in turn the estimate $\widehat{\mathbf{v}}_T$ is more reliable. Notice however that, even if the optimal $\mathbf{v}^*$ could be obtained, some problems might still require a very large sample size in the final estimation Step 5; see the discussion in Section 3.5. Given the limitations of one's computational budget, Algorithm 3.6.1 can be used to detect such a situation — the algorithm can be halted once ϱ_t in Step 4 gets too small (or, equivalently, when N_t gets too large).

3.7 The Root-Finding Problem

We briefly discuss now an application of the CE method to *root finding*. In many practical situations we need to estimate, for given ℓ, the root γ of the nonlinear equation

$$\mathbb{P}_{\mathbf{u}}(S(\mathbf{X}) \geqslant \gamma) = \mathbb{E}_{\mathbf{u}} I_{\{S(\mathbf{X}) \geqslant \gamma\}} = \ell \tag{3.78}$$

rather than estimate ℓ itself. An estimate of γ in (3.78) based on the sample equivalent of $\mathbb{E}_{\mathbf{u}} I_{\{S(\mathbf{X}) \geqslant \gamma\}}$ can be obtained, for example, via stochastic approximation [148].

Alternatively, one can obtain γ using the CE method. In particular, Algorithm 3.4.1 can be modified as follows:

Algorithm 3.7.1 (Root-Finding Algorithm)

1. *Define* $\widehat{\mathbf{v}}_0 = \mathbf{u}$. *Set* $t = 1$ *(iteration = level counter).*
2. *Generate a sample* $\mathbf{X}_1, \ldots, \mathbf{X}_N$ *from the density* $f(\cdot; \mathbf{v}_{t-1})$ *and compute the sample* $(1-\varrho)$*-quantile* $\widehat{\gamma}_t$ *of the performances according to (3.39). Calculate*
$$\widehat{\ell}_t = \frac{1}{N} \sum_{i=1}^{N} I_{\{S(\mathbf{X}_i) \geqslant \widehat{\gamma}_t\}} W(\mathbf{X}_i; \mathbf{u}, \widehat{\mathbf{v}}_{t-1}) \,,$$
provided this is greater than ℓ, *otherwise set* $\widehat{\ell_t} = \ell$.
3. *Use the* **same** *sample* $\mathbf{X}_1, \ldots, \mathbf{X}_N$ *to solve the stochastic program (3.41). Denote the solution by* $\widehat{\mathbf{v}}_t$.
4. *If* $\widehat{\ell_t} = \ell$ *proceed to Step 5; otherwise let* $t = t+1$ *and reiterate from Step 2.*
5. *Let* T *be the final iteration number. Generate a sample* $\mathbf{X}_1, \ldots, \mathbf{X}_{N_1}$ *from the density* $f(\cdot; \widehat{\mathbf{v}}_T)$ *and take as estimator of* γ *the smallest number* $\widehat{\gamma}$ *such that*
$$\frac{1}{N_1} \sum_{i=1}^{N_1} I_{\{S(\mathbf{X}_i) \geqslant \widehat{\gamma}\}} W(\mathbf{X}_i; \mathbf{u}, \widehat{\mathbf{v}}_T) \leqslant \ell \,.$$

3.8 The TLR Method

In this section we present the *transform likelihood ratio* (TLR) method as a simple, convenient and unifying way of constructing efficient IS estimators that are applicable for both light- and heavy-tailed distributions.

Let $\mathbf{X}$ be a random vector. Suppose we wish to estimate

$$\ell = \mathbb{E} I_{\{S(\mathbf{X}) \geqslant \gamma\}} \, .$$

The TLR method comprises two steps. The first is a simple *change of variable* step. That is, we write $\mathbf{X}$ as a function of another random vector $\mathbf{Z}$, for example

$$\mathbf{X} = H(\mathbf{Z}) \, . \tag{3.79}$$

If we define

$$\widetilde{S}(\mathbf{Z}) = S(H(\mathbf{Z})) \, ,$$

then

$$\ell = \mathbb{E} I_{\{\widetilde{S}(\mathbf{Z}) \geqslant \gamma\}} \, .$$

Suppose $\mathbf{Z}$ has density $h(\cdot; \boldsymbol{\theta})$ in some class of densities $\{h(\cdot; \boldsymbol{\eta})\}$. Then we can seek to estimate ℓ efficiently via IS using either the SLR method (staying in the same parametric class) or ECM. The parameter updating can again be done via the CE method. In particular, when using the SLR method we obtain in analogy to (3.36) the estimator

$$\widehat{\ell} = \frac{1}{N} \sum_{i=1}^{N} I_{\{\widetilde{S}(\mathbf{Z}_i) \geqslant \gamma\}} \widetilde{W}(\mathbf{Z}_i; \boldsymbol{\theta}, \boldsymbol{\eta}) \, , \tag{3.80}$$

where

$$\widetilde{W}(\mathbf{Z}_i; \boldsymbol{\theta}, \boldsymbol{\eta}) = \frac{h(\mathbf{Z}_i; \boldsymbol{\theta})}{h(\mathbf{Z}_i; \boldsymbol{\eta})}$$

and $\mathbf{Z}_i \sim h(\mathbf{z}; \boldsymbol{\eta})$. We shall call the SLR estimator (3.80) based on the transformation (3.79), the *transform likelihood ratio* (TLR) estimator.

To find the optimal parameter vector $\boldsymbol{\eta}^*$ of the TLR estimator (3.80) we can solve in analogy to (3.40) the following CE program:

$$\max_{\boldsymbol{\eta}} D(\boldsymbol{\eta}) = \max_{\boldsymbol{\eta}} \mathbb{E}_{\boldsymbol{\eta}_{t-1}} I_{\{\widetilde{S}(\mathbf{Z}) \geqslant \gamma_t\}} \widetilde{W}(\mathbf{Z}; \boldsymbol{\theta}, \boldsymbol{\eta}_{t-1}) \ln h(\mathbf{Z}; \boldsymbol{\eta}) \tag{3.81}$$

and similarly for the stochastic counterpart of (3.81). For example, $h(\mathbf{z}; \boldsymbol{\theta})$ might be any light-tail NEF pdf, (and thus, the optimal reference parameter vector $\boldsymbol{\eta}^*$ could be obtained analytically from the stochastic version (counterpart) of (3.81)), or $h(\mathbf{z}; \boldsymbol{\theta})$ might be a truncated version of the original pdf $f(\mathbf{x})$, denoted as $\widetilde{f}(\mathbf{x}; c)$, where the truncation parameter c could be controllable as well.

It is crucial to understand that in contrast to the SLR estimate (3.36), its TLR counterpart (3.80) involves an additional stage, namely it uses the transformation stage (3.79). As result, the TLR estimate (3.80) presents a *three-stage* procedure rather than on a *two-stage* one (see (3.36)). The three stages of TLR are associated with

1. Transformation from the original pdf f to an auxiliary one h.
2. Updating the parameter vector $\boldsymbol{\eta}$ (at each iteration of Algorithm 3.4.1) using the stochastic counterpart of (3.81).
3. Estimating ℓ according to (3.80) with $\boldsymbol{\eta}$ replaced by $\widehat{\boldsymbol{\eta}}^*$, which presents the solution obtained from Algorithm 3.4.1 at stage two.

Example 3.25 (Inverse-Transform Likelihood Ratio). Consider the single-dimensional case. According to the *inverse transform* (IT) method (see Section 1.7.1) a random variable $X \sim F$ can be written as

$$X = F^{-1}(Z) \,, \tag{3.82}$$

where $Z \sim \mathsf{U}(0,1)$ and F^{-1} is the inverse of the cdf F.

Let $h(\cdot;\nu)$ be another density on $(0,1)$ dominating the uniform density, and parameterized by some reference parameter ν. Thus, $h(x,\nu) > 0$, for all $0 \leqslant x \leqslant 1$. An example is the $\mathsf{Beta}(\nu,1)$-distribution, with density

$$h(z;\nu) = \nu\, z^{\nu-1}, \quad z \in (0,1) \,,$$

with $\nu > 0$ or the $\mathsf{Beta}(1,\nu)$-distribution, with density

$$h(z;\nu) = \nu\,(1-z)^{\nu-1}, \quad z \in (0,1) \,.$$

The TLR estimator is given by

$$\widehat{\ell} = N^{-1} \sum_{i=1}^{N} I_{\{\widetilde{S}(Z_i) \geqslant \gamma\}} \widetilde{W}(Z_i;\nu) \,, \tag{3.83}$$

where $Z_1,\ldots,Z_N$ is a random sample from $h(\cdot;\nu)$ and

$$\widetilde{W}(Z;\nu) = \frac{1}{h(Z;\nu)} \tag{3.84}$$

is the LR. We call (3.83) the *inverse transform likelihood ratio* (ITLR) estimator [144].

Consider next the multivariate case where the components of $\mathbf{X} = (X_1,\ldots, X_n)$ are independent and $X_i \sim F(\cdot;u_i)$ for a fixed parameter vector $\mathbf{u} = (u_1,\ldots,u_n)$. In analogy with the univariate case we wish to estimate, for some performance function S,

$$\ell = \mathbb{E} I_{\{S(\mathbf{X}) \geqslant \gamma\}} = \mathbb{E} I_{\{\widetilde{S}(\mathbf{Z}) \geqslant \gamma\}} \,,$$

where $\widetilde{S}(\mathbf{Z}) = S(F^{-1}(Z_1;u_1),\ldots,F^{-1}(Z_n;u_n))$, $\mathbf{Z} = (Z_1,\ldots,Z_n)$, and Z_j, $j=1,\ldots,n$ are i.i.d. and uniformly distributed on (0,1).

Let $h(\cdot;\boldsymbol{\nu})$ be another density on $(0,1)^n$ dominating the uniform density, and parameterized by some reference parameter vector $\boldsymbol{\nu}$. For example, we could choose h such that the Z_i's are independent with a $\mathsf{Beta}(1,\nu_i)$-distribution, in which case

$$h(\mathbf{z};\boldsymbol{\nu}) = \prod_{i=1}^{n} \nu_i\,(1-z_i)^{\nu_i-1}, \quad \mathbf{z}\in(0,1)^n\,, \tag{3.85}$$

with $\boldsymbol{\nu}=(\nu_1,\ldots,\nu_n)$. As in the univariate case we have the ITLR estimator

$$\widehat{\ell} = N^{-1}\sum_{i=1}^{N} I_{\{\widetilde{S}(\mathbf{Z}_i)\geqslant\gamma\}}\widetilde{W}(\mathbf{Z}_i;\boldsymbol{\nu})\,, \tag{3.86}$$

respectively, where $\mathbf{Z}_1,\ldots,\mathbf{Z}_N$ is a random sample from $h(\cdot;\boldsymbol{\nu})$ and

$$\widetilde{W}(\mathbf{Z};\boldsymbol{\nu}) = \frac{1}{h(\mathbf{Z};\boldsymbol{\nu})}\,. \tag{3.87}$$

Note that Algorithm 3.4.1 remains the same for the ITLR approach, provided the CE programs (3.40) and (3.41) are replaced by

$$\max_{\boldsymbol{\nu}} D(\boldsymbol{\nu}) = \max_{\boldsymbol{\nu}} \mathbb{E}_{\boldsymbol{\nu}_{t-1}} I_{\{\widetilde{S}(\mathbf{Z})\geq\gamma_t\}}\widetilde{W}(\mathbf{Z};\boldsymbol{\nu}_{t-1})\ln h(\mathbf{Z};\boldsymbol{\nu})\,, \tag{3.88}$$

and

$$\max_{\boldsymbol{\nu}} \widehat{D}(\boldsymbol{\nu}) = \max_{\boldsymbol{\nu}} \frac{1}{N}\sum_{i=1}^{N} I_{\{\widetilde{S}(\mathbf{Z}_i)\geq\widehat{\gamma}_t\}}\widetilde{W}(\mathbf{Z}_i;\widehat{\boldsymbol{\nu}}_{t-1})\ln h(\mathbf{Z}_i;\boldsymbol{\nu})\,, \tag{3.89}$$

respectively, where $\mathbf{Z}_i \sim h(\cdot;\widehat{\boldsymbol{\nu}}_{t-1})$.

In particular, for the case (3.85) where the Z_i's are independent and $Z_i \sim \mathsf{Beta}(1,\nu_i)$, $i=1,\ldots,n$ (3.89) can be solved analytically, and it is not difficult to see that the components of $\boldsymbol{\nu}=(\nu_1,\ldots,\nu_n)$ are updated as

$$\widehat{\nu}_{t,j} = -\frac{\displaystyle\sum_{i=1}^{N} I_{\{\widetilde{S}(\mathbf{Z}_i)\geqslant\widehat{\gamma}_t\}}\widetilde{W}(\mathbf{Z}_i;\widehat{\boldsymbol{\nu}}_{t-1})}{\displaystyle\sum_{i=1}^{N} I_{\{\widetilde{S}(\mathbf{Z}_i)\geqslant\widehat{\gamma}_t\}}\widetilde{W}(\mathbf{Z}_i;\widehat{\boldsymbol{\nu}}_{t-1})\ln(1-Z_{ij})}\,, \tag{3.90}$$

where Z_{ij} is the j-th component of $\mathbf{Z}_i$.

The following example shows that (I)TLR can lead to a more efficient estimator than the SLR method.

Example 3.26 (Example 3.12 continued). Suppose, as in Example 3.12, that we are interested in estimating $\ell = \mathbb{P}(S(\mathbf{X})\geqslant\gamma)$, where

$$S(\mathbf{X}) = \min(X_1, \ldots, X_n), \quad X_1, \ldots, X_n \sim \mathsf{Exp}(u^{-1}) . \tag{3.91}$$

In this case we can write

$$X_i = -u \ln(1 - Z_i) , \; i = 1, \ldots, n , \tag{3.92}$$

where $Z_i \sim \mathsf{U}(0,1)$, $i = 1, \ldots, n$ and $Z_1, \ldots, Z_n$ independent. We have

$$\widetilde{S}(\mathbf{Z}) = \min_i(-u \ln(1 - Z_i)) = -u \ln(1 - \min_i Z_i) ,$$

so that

$$\ell = \mathbb{P}(\widetilde{S}(\mathbf{Z}) \geqslant \gamma) = \mathbb{P}(\min_i Z_i \geqslant 1 - \eta) ,$$

with $\eta = \mathrm{e}^{-\gamma u^{-1}}$.

Let $h(\mathbf{z}; \nu) = \prod_{i=1}^n \nu z_i^{\nu-1}$, $\nu > 0$ be the dominating density on $\mathsf{U}^n(0,1)$ for $\mathbf{Z}$. Note that (by symmetry) we choose all component pdfs the *same*, this in contrast to (3.85). To find the optimal ν we need to solve the CE program (3.88), which for this case reduces to

$$\max_{\nu > 0} D(\nu) = \max_{\nu > 0} \mathbb{E} I_{\{\min_i Z_i \geqslant 1-\eta\}} \sum_{i=1}^n (\ln \nu + (\nu - 1) \; \ln Z_i) .$$

Equating the gradient with respect to ν to 0 gives

$$\begin{aligned} \nu^* &= -\frac{n \mathbb{E} I_{\{\min_i Z_i \geqslant 1-\eta\}}}{\mathbb{E} I_{\{\min_i Z_i \geqslant 1-\eta\}} \sum_{i=1}^n \ln Z_i} \\ &= -\frac{n \eta^n}{n \eta^{n-1} \int_{1-\eta}^1 \ln z \, dz} = \frac{\eta}{\ln(1-\eta)(1-\eta) + \eta} . \end{aligned}$$

It follows that for small η we have

$$\nu^* \approx \frac{2}{\eta} . \tag{3.93}$$

To find the asymptotic SCV κ^2 we need to find first the variance of the ITLR estimator $\widehat{\ell}$. Let $V(\nu)$ be the second moment of $I_{\{\widetilde{S}(\mathbf{Z}) \geqslant \gamma\}} \widetilde{W}(\mathbf{Z}; 1, \nu)$. We have

$$\begin{aligned} V(\nu) &= \mathbb{E}_\nu \left\{ \left(\prod_{i=1}^n I_{\{Z_i \geqslant 1-\eta\}} \right)^2 \cdot \left(\frac{1}{h(\mathbf{Z}; \nu)} \right)^2 \right\} \\ &= \left(\mathbb{E}_\nu \left\{ I_{\{Z \geqslant 1-\eta\}} \left(\nu Z^{\nu-1} \right)^{-2} \right\} \right)^n \\ &= \left(\frac{1}{\nu} \int_{1-\eta}^{1} z^{1-\nu} \mathrm{d}z \right)^n \\ &= \left(\frac{(1 - (1-\eta)^{2-\nu})}{\nu(2-\nu)} \right)^n . \end{aligned} \tag{3.94}$$

From (3.94) and (3.93) we have for small η

$$V(\nu^*) \approx \left\{ \frac{\eta 2^{-1}}{2 - 2\eta^{-1}} [1 - (1-\eta)^{2-2\eta^{-1}}] \right\}^n .$$

So that, for small η

$$V(\nu^*) \approx \frac{(e^2-1)^n}{4n} \eta^{2n} .$$

For ℓ we have

$$\ell = \prod_{i=1}^{n} \mathbb{P}(Z_i \geqslant 1 - \eta) = \left(\int_{1-\eta}^{1} 1 dz \right)^n = \eta^n ,$$

Finally,

$$N \times \kappa^2 = \frac{V(\nu^*)}{\ell^2} - 1 \approx \left(\frac{e^2-1}{4} \right)^n - 1 . \tag{3.95}$$

Note that κ^2 in (3.95) does not depend on η and therefore neither does it depend on γ. Consequently, the corresponding estimators have *bounded relative error* in γ. Comparing (3.95) with $N \times \kappa^2 = \gamma^n \mathrm{e}^n (2u)^{-n} = (-\ln \eta)^n (\mathrm{e}/2)^n$ in (3.48), it readily follows that the former (ITLR) is much faster than the latter (SLR), especially when γ is large.

Remark 3.27. Note that by minimizing (3.94) we can obtain the VM optimal solution, say ${}_*\nu$. It can be shown that similar to (3.93) we have, for small η,

$${}_*\nu \approx \frac{c}{\eta} , \tag{3.96}$$

where $c = 1.59362\ldots$ is the solution to the equation

$$2 - (2 - c)\mathrm{e}^c = 0 .$$

Hence, this is an example where the VM and CE optimal solutions do not coincide asymptotically. However, both solutions give bounded relative error.

The following proposition illustrates the usefulness of ITLR for estimating small probabilities, for *any* distribution. In the results below the univariate ITLR method is used with a $\mathsf{Beta}(\nu, 1)$ change of measure. It is important to realize that this CoM may not be appropriate for similar problems concerning multivariate random variables. Indeed the $\mathsf{Beta}(\nu, 1)$ CoM may give exponential complexity, whereas a $\mathsf{Beta}(1, \nu)$ CoM could give polynomial complexity.

Proposition 3.28. *Let $X = L(1 - Z)$, with $Z \sim \mathsf{U}(0,1)$, for some monotone increasing function L on $(0,1)$. Then, estimating $\ell = \mathbb{P}(X \geqslant \gamma)$ via ITLR using the $\{\mathsf{Beta}(\nu, 1), \nu > 0\}$ family of distributions gives an LR estimator with bounded relative error.*

Proof. The proof uses similar arguments to the ones used in Example 3.26. First, we write $\ell = \mathbb{P}(X \geqslant \gamma)$ as $\ell = \mathbb{P}(Z \geqslant 1 - \eta)$, with $\eta = L^{-1}(\gamma)$. Hence, if we estimate ℓ via the IS density

$$h(z;\nu) = \nu z^{\nu-1} , \tag{3.97}$$

then the optimal CE parameter is given, analogously to (3.93), by

$$\nu^* = \frac{\eta}{\eta + (1-\eta)\ln(1-\eta)} \approx \frac{2}{\eta} ,$$

as $\eta \to 0$. Moreover, the corresponding SCV satisfies

$$N \times \kappa^2 \approx \frac{e^2 - 1}{4} - 1 \approx 0.597264 . \tag{3.98}$$

Note that this is independent of η (and hence γ). Thus, the estimator has bounded relative error.

Example 3.29 (Example 3.14 continued). Let $X \sim \mathsf{Weib}(a, u^{-1})$. That is, X has cdf F given by

$$F(x) = 1 - \mathrm{e}^{-(u^{-1}x)^a}, \quad x \geqslant 0 .$$

We wish to estimate $\ell = \mathbb{P}(X \geqslant \gamma) = \mathrm{e}^{-(u^{-1}\gamma)^a}$ for large γ. Using

$$L(z) = u\,(-\ln z)^{\frac{1}{a}} , \quad z \in (0,1) ,$$

we can write $\ell = \mathbb{P}(Z \geqslant 1-\eta)$, with $Z \sim \mathsf{U}(0,1)$ and $\eta = \mathrm{e}^{-(u^{-1}\gamma)^a}$. Hence, by Proposition 3.28 we can efficiently estimate ℓ via ITLR using the $\mathsf{Beta}(\nu, 1)$ density, yielding an estimator with bounded relative error given in (3.98). Note that this is true for *any* shape parameter $a > 0$, including the heavy-tail case $0 < a < 1$.

3.9 Equivalence Between SLR and TLR

As we have seen the TLR method can be viewed as a universal device involving an additional transformation step as compared to SLR. Its main advantage is computational, since using TLR one can readily transform any underlying problem with any (arbitrary) input pdf, such as Weibull, to an equivalent one, such as the exponential, for which the updating of the parameter vector can be performed analytically. In this section we illustrate that seemingly different TLR and SLR methods can in fact be probabilistically equivalent. So, in the cases below we cannot get a more accurate estimate while switching from SLR to TLR.

Let $X_1, X_2, \ldots, X_n$ be i.i.d. $\mathsf{Weib}(a, u^{-1})$ distributed and consider the estimation of a general rare-event probability $\ell = \mathbb{P}(S(\mathbf{X}) \geqslant \gamma)$ for large γ using importance sampling. We consider three methods.

(1) SLR with $\mathsf{Weib}(a, v^{-1})$ twisting, fixed a

The first method is a straightforward change of the Weibull scale parameter, as in Example 3.14. In particular, we consider the change of measure

$$X_n \sim \mathsf{Weib}(a, u^{-1}) \longrightarrow \mathsf{Weib}(a, v^{-1}), \quad v \geqslant u .$$

Note that the problem is of the form discussed in Remark 3.15, but by symmetry we know that the components of the reference vector must be equal. This leads to slightly different updating formulas, namely:

$$\widehat{v}_t = \left(\frac{\sum_{k=1}^N I_{\{S(\mathbf{X}_k) \geqslant \widehat{\gamma}_t\}} W(\mathbf{X}_k; u, \widehat{v}_{t-1})\, n^{-1} \sum_{i=1}^n X_{ki}^a}{\sum_{k=1}^N I_{\{S(\mathbf{X}_k) \geqslant \widehat{\gamma}_t\}} W(\mathbf{X}_k; u, \widehat{v}_{t-1})} \right)^{1/a} . \tag{3.99}$$

(2) ITLR with $\mathsf{Beta}(1, \nu)$ twisting

In the second method we estimate ℓ via the ITLR method. First, write $X_i \sim \mathsf{Weib}(a, u^{-1})$ as

$$X_i = u\,(-\ln(1 - Z_i))^{1/a} ,$$

with the Z_i i.i.d. $\mathsf{U}(0,1) = \mathsf{Beta}(1,1)$. We now apply a change of measure on the distribution of Z_i:

$$Z_i \sim \mathsf{Beta}(1,1) \longrightarrow \mathsf{Beta}(1,\nu) \quad 0 < \nu \leqslant 1 .$$

Define $\widetilde{S}(\mathbf{Z}) = S(\mathbf{X})$. The CE updating formula is, similar to (3.90),

$$\widehat{\nu}_t = -\frac{\displaystyle\sum_{i=1}^N I_{\{\widetilde{S}(\mathbf{Z}_i) \geqslant \widehat{\gamma}_t\}} \widetilde{W}(\mathbf{Z}_i; 1, \widehat{\nu}_{t-1})}{\displaystyle\sum_{i=1}^N I_{\{\widetilde{S}(\mathbf{Z}_i) \geqslant \widehat{\gamma}_t\}} \widetilde{W}(\mathbf{Z}_i; 1, \widehat{\nu}_{t-1})\, n^{-1} \sum_{j=1}^n \ln(1 - Z_{ij})} , \tag{3.100}$$

where Z_{ij} is the j-th component of $\mathbf{Z}_i$.

It is interesting to compare the present ITLR method with the previous Weibull change of measure. Since, Z_i can be written as $Z_i = 1 - (1 - U_i)^{1/\nu}$, with $U_i \sim \mathsf{U}(0,1)$, we have

$$\begin{aligned} X_i &= u \left(-\ln \left(\{1 - U_i\}^{1/\nu} \right) \right)^{1/a} \\ &= \frac{u}{\nu^{1/a}} \left(-\ln (1 - U_i) \right)^{1/a} , \end{aligned}$$

so that under the change of measure $Z_i \sim \mathsf{Beta}(1,1) \longrightarrow \mathsf{Beta}(1,\nu)$ we have that $X_i \sim \mathsf{Weib}(a, u^{-1}\nu^{1/a})$. Let us compare the behavior of the SLR and ITLR estimators for $v = u\nu^{-1/a}$. First of all, observe that

$$W(\mathbf{X}; u, v) = \prod_{i=1}^{n} \frac{au^{-1}(u^{-1}X_i)^{a-1}\mathrm{e}^{-(u^{-1}X_i)^a}}{av^{-1}(v^{-1}X_i)^{a-1}\mathrm{e}^{-(v^{-1}X_i)^a}}$$
$$= \prod_{i=1}^{n} \frac{1}{\nu\,(1-Z_i)^{\nu-1}} = \widetilde{W}(\mathbf{Z}; 1, \nu)\,.$$

This shows that

$$\sum_{i=1}^{N} I_{\{S(\mathbf{X}_i)\geqslant\gamma\}} W(\mathbf{X}_i; u, v) = \sum_{i=1}^{N} I_{\{\widetilde{S}(\mathbf{Z}_i)\geqslant\gamma\}} \widetilde{W}(\mathbf{Z}_i; 1, \nu)\,.$$

In other words, the SLR estimator is *identical* to the ITLR estimator, provided we take $v = u\,\nu^{-1/a}$. Note also that, in the same way, the CE updating formulas and their deterministic counterparts are equivalent, in the sense that $\widehat{v}_t = u(\widehat{\nu}_t)^{-1/a}$ and $v_t = u(\nu_t)^{-1/a}$.

(3) TLR with $\mathsf{Exp}(\lambda)$ twisting

Let us finally apply the TLR method with an "exponential change of measure." We now write $X_i \sim \mathsf{Weib}(a, u^{-1})$ as

$$X_i = uZ_i^{1/a}\,,$$

with the Z_i i.i.d. $\mathsf{Exp}(1)$, and apply the change of measure

$$Z_i \sim \mathsf{Exp}(1) \longrightarrow \mathsf{Exp}(\lambda), \quad 0 < \lambda \leqslant 1\,.$$

With $\widetilde{S}(\mathbf{Z}) = S(\mathbf{X})$ the CE updating formula is given by

$$\widehat{\lambda}_t = \frac{\displaystyle\sum_{i=1}^{N} I_{\{\widetilde{S}(\mathbf{Z}_i)\geqslant\widehat{\gamma}_t\}} \widetilde{W}(\mathbf{Z}_i; 1, \widehat{\lambda}_{t-1})}{\displaystyle\sum_{i=1}^{N} I_{\{\widetilde{S}(\mathbf{Z}_i)\geqslant\widehat{\gamma}_t\}} \widetilde{W}(\mathbf{Z}_i; 1, \widehat{\lambda}_{t-1})\, n^{-1} \sum_{j=1}^{n} Z_{ij}}\,, \tag{3.101}$$

where Z_{ij} is the j-th component of $\mathbf{Z}_i$.

Since, Z_i can be written as $Z_i = \lambda^{-1}\ln(1-U_i)$, with $U_i \sim \mathsf{U}(0,1)$, we have

$$X_i = u\,\lambda^{-1/a}\,\ln(1-U_i)^{1/a}\,,$$

so that under this change of measure $X_i \sim \mathsf{Weib}(a, u^{-1}\lambda^{1/a})$. Repeating the arguments of the ITLR method above, we find that this approach is equivalent to the two methods above, provided that we take $\lambda = \nu = (u/v)^a$.

Remark 3.30. **(Sum of independent random variables)** The special case where $S(\mathbf{X}) = X_1 + \cdots + X_n$, and the X_i are i.i.d. with a subexponential distribution was studied in both [14] and [91] via various methods, as explained in the introduction. In particular for the heavy-tail Weibull case, Juneja and Shahabuddin [91] (see their Theorem 3.2) proved that the change of measure

$$X_i \sim \mathsf{Weib}(a,1) \longrightarrow \mathsf{Weib}(a,\eta^{1/a}) \tag{3.102}$$

provides a logarithmically efficient estimator, in the sense of (1.14), when we choose

$$\eta = c\,\gamma^{-a}\,, \tag{3.103}$$

no matter how c is chosen. On the other hand [14] proposed an importance sampling distribution independent of γ which is consistent with the fact that $\eta \to 0$. In Appendix 3.13 we prove for the case $n = 2$ the somewhat stronger result that the estimator is in fact polynomial and that the variance of the estimator is minimized for $c = 2$; we conjecture that for general n the variance minimal (VM) parameter is

$$_*\eta = n\,\gamma^{-a}\,.$$

In [15] it is shown that for large γ the optimal CE parameter, η^* say, is indeed given by $_*\eta$ above. More precisely, the following argument is used: Let $X_1, X_2, \ldots$ be i.i.d. random variables distributed according to a subexponential distribution with respect to the Lebesgue measure. For large γ the asymptotic distribution of $X_1, \ldots, X_n$ given $X_1 + \cdots + X_n \geqslant \gamma$ is such that with probability $1/n$ one of the random variables $X_1, \ldots, X_n$, say X_i, is distributed according to the conditional distribution of $(X_i \,|\, X_i \geqslant \gamma)$, and the other $\{X_j, j \neq i\}$ are distributed according to the original distribution. In particular, consider $X_1, X_2, \ldots \sim \mathsf{Weib}(a,1)$, with $0 < a < 1$. The optimal CE parameter η^* for the SLR change of measure (3.102) can be written as

$$\eta^* = 1/\mathbb{E}\left[n^{-1} \sum_{i=1}^{n} X_i^a \,\middle|\, X_1 + \cdots + X_n \geqslant \gamma \right]\,.$$

Using the fact that the Weibull distribution with $0 < a < 1$ is subexponential it follows that for large γ

$$\begin{aligned} &\mathbb{E}\left[n^{-1} \sum_{i=1}^{n} X_i^a \,\middle|\, X_1 + \cdots + X_n \geqslant \gamma \right] \\ &\qquad\qquad \approx n^{-1}(n-1)\mathbb{E}X^a + n^{-1}\mathbb{E}\left[X^a \middle| X \geqslant \gamma\right]\,. \end{aligned} \tag{3.104}$$

Since the conditional distribution of X given $X \geqslant \gamma$ has pdf $a x^{a-1} \mathrm{e}^{-(x^a - \gamma^a)}$, $x \geqslant \gamma$ the expectation in (3.104) is given by

$$\frac{n-1}{n} + \frac{1}{n}\,(1+\gamma^a)\,.$$

It follows that for large γ

$$\eta^* \approx \frac{n}{n+\gamma^a}\,. \tag{3.105}$$

Similar results are obtained for the Pareto distribution.

Remark 3.31 (Other changes of measure). Various other ideas for selecting a good change of measure for the problem $\ell = \mathbb{P}(X_1 + \cdots + X_n \geqslant \gamma)$ with i.i.d. heavy-tail X_i are considered in [15]. We mention the idea of estimating ℓ via

$$\mathbb{P}(X_1 + \cdots + X_n \geqslant \gamma, X_1 > X_2, \ldots, X_1 > X_n) = \ell/n\,, \tag{3.106}$$

where the left-hand side can be estimated via a change of measure in which only the distribution of X_1 is changed. This leads to a significant variance reduction.

3.10 Stationary Waiting Time of the GI/G/1 Queue

In this section we give an application of the CE method to a *dynamic* model, as opposed to the *static* models considered so far. We will see that the formalism carries through without much alteration.

Consider a stable GI/G/1 queue starting with customer $n = 1$ arriving at an empty system. Let the interarrival time between customer n and $n+1$ be denoted by $A_n \sim f^A$, $n = 1, 2, \ldots$ and let the service time of customer n be denoted by $B_n \sim f^B$. We assume that all the service and interarrival times are independent. Let S_n denote the *actual waiting time* of the n-th customer; hence, by definition $S_1 = 0$. The stochastic process $\{S_n, n \geqslant 1\}$ satisfies the celebrated *Lindley equation* (see for example [12])

$$S_{n+1} = (S_n + X_n)^+\,,$$

with $X_n = B_n - A_n$, $i = 1, 2 \ldots$. For a stable system the random variables $\{S_n\}$ converge in distribution to the *steady-state* waiting time, S say.

We are interested in estimating $\ell = \mathbb{P}(S \geqslant \gamma)$ via importance sampling. We consider two methods.

The regenerative method

Using the regenerative method, see for example [148], we can write

$$\ell = \frac{\mathbb{E} \sum_{n=1}^{\sigma} I_{\{S_n \geqslant \gamma\}}}{\mathbb{E}\,\sigma}\,, \tag{3.107}$$

where σ is the number of customers during the first busy period, that is

$$\sigma = \inf\{n > 1 \,:\, S_n = 0\} - 1\,.$$

Define τ as

$$\tau = \inf\{n > 1 \,:\, S_n \geqslant \gamma\}\,,$$

In other words, τ is the first time that the process $\{S_n\}$ exceeds level γ, if at all.

Consider now the following *switching* change of measure [148].

$$A_n \sim f^A \longrightarrow \widetilde{f}^A \quad \text{and} \quad B_n \sim f^B \longrightarrow \widetilde{f}^B \,, \quad \text{for } n = 1, \ldots, \min(\tau, \sigma) \,.$$

In other words, the IS distribution changes *dynamically* within the cycles. In particular, we initially use the IS densities $\widetilde{f}^A$ and $\widetilde{f}^B$ for the interarrival and service times until the process $\{S_n\}$ exceeds level γ, after which we switch back to the original densities; see Chapter 9 of [148] By doing so the process $\{S_n\}$ naturally returns to the regenerative state.

Under this change of measure the likelihood ratio of a sample $A_1, \ldots, A_n$, $B_1, \ldots, B_n$ satisfies

$$W_n = \begin{cases} W_{n-1} \dfrac{f^A(A_n) f^B(B_n)}{\widetilde{f}^A(A_n) \widetilde{f}^B(B_n)}, & n \leqslant \min(\tau, \sigma) \\ W_\tau, & n \geqslant \min(\tau, \sigma) \,. \end{cases} \tag{3.108}$$

From [148], we can write

$$\ell = \frac{\mathbb{E} W_\sigma \sum_{n=1}^{\sigma} I_{\{S_n \geqslant \gamma\}}}{\mathbb{E}\, \sigma} = \frac{\mathbb{E} \sum_{n=1}^{\sigma} I_{\{S_n \geqslant \gamma\}} W_n}{\mathbb{E}\, \sigma} \,. \tag{3.109}$$

Note that the denominator of (3.109) can be easily estimated via CMC (no change of measure here). The numerator of (3.109) (num) can be estimated as

$$\widehat{\text{num}} = \frac{1}{N} \sum_{i=1}^{N} \sum_{n=1}^{\sigma_i} I_{\{S_{in} \geqslant \gamma\}} W_{in} \,, \tag{3.110}$$

where S_{in} and W_{in} are the waiting time of the n-th customer and the corresponding likelihood ratio, for iteration i.

Now consider the special case $A_1, A_2, \ldots \sim \mathsf{Weib}(a_1, u_1^{-1})$ and $B_1, B_2, \ldots \sim \mathsf{Weib}(a_2, u_2^{-1})$. Using the TLR method, we may write

$$X_n = u_2 \left(Z_n^{(2)}\right)^{1/a_2} - u_1 \left(Z_n^{(1)}\right)^{1/a_1} ,$$

with $Z_n^{(k)} \sim \mathsf{Exp}(1)$, $k = 1, 2$, $n = 1, 2, \ldots$, so that

$$S_{n+1} = \left(S_n + u_2 \left(Z_n^{(2)}\right)^{1/a_2} - u_1 \left(Z_n^{(1)}\right)^{1/a_1}\right)^+ , \tag{3.111}$$

with $S_1 = 0$. Consider the following particular case of the switching change of measure described above:

$$Z_n^{(1)} \sim \mathsf{Exp}(1) \longrightarrow \mathsf{Exp}(v_1^{-1}) \text{ and } Z_n^{(2)} \sim \mathsf{Exp}(1) \longrightarrow \mathsf{Exp}(v_2^{-1}), n \leqslant \min(\tau, \sigma) \,.$$

Then (3.108) is given by

$$W_n = \begin{cases} W_{n-1} \prod_{k=1}^{2} v_k \, \mathrm{e}^{-(1-v_k^{-1}) Z_n^{(k)}}, & n \leqslant \min(\tau, \sigma) \\ W_\tau, & n \geqslant \min(\tau, \sigma) . \end{cases} \tag{3.112}$$

Since the $Z_n^{(k)}$ are independent and have an exponential distribution we can apply again the standard CE technique to determine/estimate the optimal reference parameters v_1^* and v_2^* for the estimator (3.110) and achieve variance reduction. In particular, if we define

$$H(\mathbf{Z}) = \sum_{n=1}^{\sigma} I_{\{S_n \geqslant \gamma\}} ,$$

with $\mathbf{Z} = (Z_1^{(1)}, Z_1^{(2)}, \ldots, Z_\sigma^{(1)}, Z_\sigma^{(2)})$, then, similar to Example 3.5, we have

$$v_k^* = \frac{\mathbb{E}_{\mathbf{v}} H(\mathbf{Z}) \, W_\tau \sum_{n=1}^{\tau} Z_n^{(k)}}{\mathbb{E}_{\mathbf{v}} H(\mathbf{Z}) \, W_\tau \, \tau} , \qquad k = 1, 2,$$

for any reference vector $\mathbf{v} = (v_1, v_2)$. Note that in a multilevel CE procedure the updating rule for the level γ_t is not the usual quantile rule. Instead γ_t should be chosen such that during each regeneration cycle at least a fraction ϱ of the customers has a waiting time $\geqslant \gamma$.

Random Walk

It is well known (see for example page 173 of [148]) that the steady-state waiting time for this queueing system has the same distribution as the supremum of the random walk $\{Y_n, \, n = 1, \ldots\}$, where $Y_1 = 0$ and

$$Y_{n+1} = Y_n + X_n , \quad n \geqslant 1 ,$$

with $X_i = B_i - A_i$, $i = 1, 2 \ldots$, and the A_i and B_i the same as before. Thus ℓ in (3.107) is the same as

$$\ell = \mathbb{P}(\sup_n Y_n \geqslant \gamma) . \tag{3.113}$$

Similar to (3.111) let us now (re-)define

$$S_{n+1} = S_n + u_2 \left(Z_i^{(2)} \right)^{1/a_2} - u_1 \left(Z_i^{(1)} \right)^{1/a_1} , \tag{3.114}$$

with $Z_i^{(k)} \sim \mathsf{Exp}(1)$, $k = 1, 2$. Then, with $S = \sup_n S_n$, the estimation of (3.113) (under the original pdfs $\{\mathsf{Weib}(a_1, u_1^{-1})\}$ and $\{\mathsf{Weib}(a_2, u_2^{-1})\}$) is equivalent to the estimation of

$$\ell = \mathbb{P}(S \geqslant \gamma) .$$

Thus, alternatively to $I_{\{\sup_n Y_n \geqslant \gamma\}}$, which employs Weibull random variables, we can simulate the random variable $I_{\{\sup_n S_n \geqslant \gamma\}}$ to estimate ℓ, which employs $\mathsf{Exp}(1)$ random variables $Z^{(1)}$ and $Z^{(2)}$. We can again apply the standard CE technique to find the optimal IS reference parameter.

To proceed, define τ as the first time $\{S_n\}$ exceeds level γ or falls below some low level $-L$, that is

$$\tau = \inf\{n > 0 \; : \; S_n \geqslant \gamma \text{ or } S_n < -L\} \; . \tag{3.115}$$

Consider the IS change of measure with $Z_i^{(k)} \sim \mathsf{Exp}(v_k^{-1})$. Typically, we look for an IS change of measure under which the queue has a positive drift. In that case $S_\tau \geqslant \gamma$ with high probability. For $-L$ small enough we may write to a very close approximation

$$\ell \approx \mathbb{P}(S_\tau \geqslant \gamma) \; .$$

It will be clear how we estimate the probability above: we run N samples of $S_1, \dots, S_\tau$ and evaluate the estimator

$$\widehat{\ell} = \frac{1}{N} \sum_{i=1}^{N} I_{\{S_{\tau_i} \geqslant \gamma\}} W_{\tau_i} \; ,$$

where

$$W_\tau = \prod_{k=1}^{2} \prod_{n=1}^{\tau} v_{t-1,k} \, \mathrm{e}^{-(1-v_{t-1,k}^{-1}) Z_n^{(k)}} \; .$$

Applying the CE Algorithm 3.4.1 it is readily seen that the deterministic updating rules for $\mathbf{v}_t = (v_{t,1}, v_{t,2})$ are

$$v_{t,k} = \frac{\mathbb{E}_{\mathbf{v}_{t-1}} I_{\{S_\tau \geqslant \gamma_t\}} W_\tau \sum_{n=1}^{\tau} Z_n^{(k)}}{\mathbb{E}_{\mathbf{v}_{t-1}} I_{\{S_\tau \geqslant \gamma_t\}} W_\tau \, \tau} \; ,$$

with $v_{0,k} = 1$, $k = 1, 2$. This leads to the simulated updating rules

$$\widehat{v}_{t,k} = \frac{\sum_{i=1}^{N} I_{\{S_{i\tau_i} \geqslant \widehat{\gamma}_t\}} W_{i\tau_i} \sum_{n=1}^{\tau_i} Z_{kn}^{(i)}}{\sum_{i=1}^{N} I_{\{S_{i\tau_i} \geqslant \widehat{\gamma}_t\}} W_{i\tau_i} \, \tau_i} \; ,$$

where the simulation is run under $\mathbf{v}_{t-1}$. Note that the updating rules for the regenerative method and the random walk method are very similar. Indeed, it is reasonable to expect that the optimal CE parameters for the two methods should coincide for large γ; numerical results indicate that this is indeed the case. Finally we remark that some care should be taken with the choice of the low level $-L$. Typically, under the CE optimal parameter the system becomes *unstable* and hence $-L$ can be safely set to $-\infty$, but for the first iteration the system is still stable and hence $-L$ has to be chosen not too small in order to save CPU time.

Remark 3.32. It is important to set L in any simulation involving (3.115) large enough to obtain a valid estimator for the steady-state waiting time probabilities. The choice of L is somewhat arbitrary. An alternative approach is to take $L = 0$ and let ℓ correspond to the probability that the waiting time process exceeds level γ during a busy period. This is called the *transient setting* in [148], section 9.3.2. In our numerical results we will consider examples of both cases.

Remark 3.33 (Exponential complexity). Although we show in [15] that both methods described above have in general exponential complexity, the numerical results in Section 3.12 show that in practice the CE method yields an excellent speed up (variance reduction), for probabilities down to, say, 10^{-10}.

3.11 Big-Step CE Algorithm

Consider Algorithm 3.4.1. The algorithm is designed to find a reference vector $\mathbf{v}_T$ such that

$$\mathbb{P}_{\mathbf{v}_T}(S(\mathbf{X}) \geqslant \gamma) \geqslant \varrho \,,$$

for a given (large) γ and ϱ not too small, for example, 0.01 or 0.1. However, the number of levels required (T) may be quite large. Here we propose the following simple alternative, called the *big-step CE algorithm* which uses only *two* levels.

Algorithm 3.11.1 (Big-step CE Algorithm)

1. *Generate a random sample* $\mathbf{X}_1, \ldots, \mathbf{X}_N$ *from the original distribution* $f(\cdot; \mathbf{u})$. *Let* $\widehat{\gamma}_1$ *be the* $(1-\varrho)$*-quantile of the performances, assuming that this is less than* γ. *Otherwise put* $\gamma_1 = \gamma$.
2. *Apply the following gradient procedure:*

$$\widehat{\mathbf{v}}_1 = \mathbf{u} + \beta \nabla \widehat{D}(\mathbf{u}; \mathbf{u}, \widehat{\gamma}_1) \,, \tag{3.116}$$

where $\nabla \widehat{D}(\mathbf{u}; \mathbf{u}, \widehat{\gamma}_1)$, *denotes the gradient at* $\mathbf{u}$ *of the function* $\mathbf{v} \mapsto \widehat{D}(\mathbf{v}; \mathbf{u}, \widehat{\gamma}_1)$ *defined by*

$$\widehat{D}(\mathbf{v}; \mathbf{u}, \widehat{\gamma}_1) = \frac{1}{N} \sum_{i=1}^{N} I_{\{S(\mathbf{X}_i) \geqslant \widehat{\gamma}_1\}} \ln f(\mathbf{X}_i; \mathbf{v}) \,, \tag{3.117}$$

using the sample $\mathbf{X}_1, \ldots, \mathbf{X}_N$ *of Step 2, and* β, *called the "big-step parameter" is chosen (say by trial and error or by the bisection method) such that*

$$\frac{1}{N} \sum_{i=1}^{N} I_{\{S(\mathbf{X}_i) \geqslant \gamma\}} \geqslant \varrho \,, \tag{3.118}$$

for some sample from $f(\cdot; \widehat{\mathbf{v}}_1)$.

3. *Generate a sample* $\mathbf{X}_1, \ldots, \mathbf{X}_{N_1}$ *from the density* $f(\cdot; \widehat{\mathbf{v}}_1)$ *and solve the stochastic program*

$$\max_{\mathbf{v} \in \mathcal{V}} \left\{ \frac{1}{N_1} \sum_{i=1}^{N_1} I_{\{S(\mathbf{X}_i) \geqslant \gamma\}} W(\mathbf{X}_i; \mathbf{u}, \widehat{\mathbf{v}}_1) \ln f(\mathbf{X}_i, \mathbf{v}) \right\} .$$

Repeat this step for several additional iterations, until the solution has converged. Denote the desired estimate by $\widehat{\mathbf{v}}_2$.

4. *Estimate the rare-event probability* ℓ *using the estimate* (3.36), *with* $\mathbf{v}$ *replaced by* $\widehat{\mathbf{v}}_2$.

The rationale of the big-step method is that to estimate the optimal reference parameter $\mathbf{v}^*$ using CE it is only necessary to obtain, one way or the other, a reference vector $\widehat{\mathbf{v}}_1$ such that

$$\mathbb{P}_{\widehat{\mathbf{v}}_1}(S(\mathbf{X}) \geqslant \gamma) \geqslant \varrho .$$

This reference parameter is then used to estimate the *optimal* reference parameter for the estimation of the rare event $\{S(\mathbf{X}) \geqslant \gamma\}$. The algorithm sets out to find such a $\widehat{\mathbf{v}}_1$ satisfying (3.118) by searching in the direction of the optimal reference parameter for estimating $\mathbb{P}_{\mathbf{u}}(S(\mathbf{X}) \geqslant \widehat{\gamma}_1)$. Numerical results with the big-step Algorithm 3.11.1 are given Tables 3.10 and 3.11 of Section 3.12.

3.12 Numerical Results

This section presents simulation studies for the rare-event probability $\ell = \mathbb{P}(S(\mathbf{X}) \geqslant \gamma)$ for several static and queueing models with both light- and heavy-tail distributions. The specific choices for γ can be inferred from the tables. We shall employ both the SLR (3.36) and TLR estimators.

Unless otherwise specified we set in all our experiments with Algorithm 3.4.1 the rarity parameter $\varrho = 0.01$, the sample size for Steps 2–4 of the algorithm $N = 10^4$ and the final sample size $N_1 = 5 \cdot 10^5$. Note that estimator $\widehat{\ell}$ is of the form $\widehat{\ell} = N^{-1} \sum_{i=1}^{n} Z_i$; see (2.26). The SCV κ^2 of each Z_i is estimated as

$$\widehat{\kappa^2} = \frac{N^{-1} \sum_{i=1}^{N} Z_i^2 - (\widehat{\ell})^2}{(\widehat{\ell})^2} .$$

The relative error of the estimator is thus estimated by $\mathrm{RE} = \sqrt{\widehat{\kappa^2}/N}$. Assuming asymptotic normality of the estimator — which we verified numerically in each case — the confidence intervals (CI) now follow in a standard way. For example, a 95% CI for ℓ is given by

$$\widehat{\ell} \pm 1.96\, \widehat{\ell}\, \mathrm{RE} .$$

For a quite moderate probability like $\ell = 10^{-3}$, we typically compare the CE results with the corresponding CMC results.

3.12.1 Sum of Weibull Random Variables

Our first model concerns five i.i.d. $\mathsf{Weib}(a, u^{-1})$ random variables with $a = 5$ and $a = 0.2$, respectively. For both cases we selected $u = 1$. We wish to estimate

$$\mathbb{P}(X_1 + \cdots + X_5 \geqslant \gamma) .$$

Tables 3.2 and Tables 3.3 present, for the cases $a = 5$ and $a = 0.2$, respectively, the performance of Algorithm 3.4.1 for the TLR method

$$X_i = uZ_i^{1/a}, \quad Z_i \sim \mathsf{Exp}(1) \longrightarrow \mathsf{Exp}(v_i^{-1}) \tag{3.119}$$

which is equivalent to the (one-parameter) SLR method

$$X_i \sim \mathsf{Weib}(a, u^{-1}) \longrightarrow \mathsf{Weib}(a, v_i^{-1/a}) .$$

Table 3.2. The evolution of $\widehat{\mathbf{v}}_t$ for estimating the optimal parameter $\mathbf{v}^*$ with the TLR method (3.119), with $a = 5$. The estimated probability is $\widehat{\ell} = 1.6694 \cdot 10^{-9}$, the relative error $RE = 0.011763$, and $\widehat{\kappa^2} = 62.2$.

t	$\widehat{\gamma}_t$	$\widehat{v}_{1t}$	$\widehat{v}_{2t}$	$\widehat{v}_{3t}$	$\widehat{v}_{4t}$	$\widehat{v}_{5t}$
0	-	1	1	1	1	1
1	5.7	2.37	2.42	2.54	2.49	2.46
2	6.7	5.52	4.91	4.84	4.97	5.20
3	7.0	6.06	6.04	6.03	5.93	5.89
4	7.0	5.99	5.96	6.02	6.00	5.99
5	7.0	5.95	5.90	6.03	6.04	5.98
6	7.0	5.95	5.98	6.04	5.93	5.98
7	7.0	6.03	5.93	6.01	6.01	5.95
8	7.0	6.00	6.08	6.02	5.90	5.95

Table 3.3. The evolution of $\widehat{\mathbf{v}}_t$ for estimating the optimal parameter $\mathbf{v}^*$ with the TLR method (3.119), with $a = 0.2$. The estimated probability is $\widehat{\ell} = 6.54 \cdot 10^{-7}$, relative error $RE = 0.0278$, and $\widehat{\kappa^2} = 386$.

t	$\widehat{\gamma}_t$	$\widehat{v}_{1t}$	$\widehat{v}_{2t}$	$\widehat{v}_{3t}$	$\widehat{v}_{4t}$	$\widehat{v}_{5t}$
0	-	1	1	1	1	1
1	9.7e+003	2.45	2.25	2.55	1.97	2.12
2	6.4e+005	3.06	3.70	4.28	3.54	4.62
3	1.0e+006	3.68	5.82	3.92	3.34	4.35
4	1.0e+006	4.37	3.88	4.13	4.62	3.67
5	1.0e+006	4.13	4.47	4.11	3.77	4.37
6	1.0e+006	4.15	4.53	3.98	3.94	3.99
7	1.0e+006	4.10	4.22	4.40	4.11	4.16
8	1.0e+006	4.18	4.39	4.35	4.53	4.11

Note that in both cases Algorithm 3.4.1 reaches the desired level γ after three iterations, but we have continued iterating Steps 2–4 of Algorithm 3.4.1 in view of Remark 3.9. We see that the parameter vector $\widehat{\mathbf{v}}_t$ stabilizes very quickly. Note also that we could have taken the average of the reference parameter at each iteration as a more accurate estimate for the optimal reference parameter.

The asymptotic value for optimal reference parameter v in the heavy-tail case is, see (3.105), given by

$$\frac{1}{\eta^*} = \frac{1+\gamma^a}{n} .$$

In particular for Table 3.3 we obtain a value of $(1 + 10^{1.2})/5 \approx 3.4$, which is not too far from the observed value of around 4.2. Note that for the light-tail case the above formula does not hold. Tables 3.4 and 3.5 present, for the same cases $a = 5$ and $a = 0.2$ as above, the performance of Algorithm 3.4.1 for the two-parameter SLR method

$$X_i \sim \mathsf{Weib}(a, u^{-1}) \longrightarrow \mathsf{Weib}(b_i, v_i^{-1/b_i}) \tag{3.120}$$

of Remark 3.16.

Table 3.4. The evolution of $\widehat{\mathbf{b}}_t$ and $\widehat{\mathbf{v}}_t$ for estimating the optimal parameters $\mathbf{b}^*$ and $\mathbf{v}^*$ with the two-parameter SLR method (3.120). The estimated probability is $\widehat{\ell} = 1.6570 \cdot 10^{-9}$, the relative error $RE = 0.0041$, and $\widehat{\kappa^2} = 8.4$.

t	$\widehat{\gamma}_t$	$\widehat{b}_{1t}$	$\widehat{v}_{1t}$	$\widehat{b}_{2t}$	$\widehat{v}_{2t}$	$\widehat{b}_{3t}$	$\widehat{v}_{3t}$	$\widehat{b}_{4t}$	$\widehat{v}_{4t}$	$\widehat{b}_{5t}$	$\widehat{v}_{5t}$
0	-	5	1	5	1	5	1	5	1	5	1
1	5.19	7.6	2.8	7.1	2.7	7.0	2.6	7.4	3.0	7.3	2.7
2	5.87	9.1	8.7	8.9	7.9	9.9	10.4	10.1	10.1	10.5	10.5
3	6.41	11.2	28.5	11.8	37.4	12.1	39.8	11.2	34.3	12.2	34.5
4	6.86	12.3	70.6	12.0	106.4	14.3	153.1	14.6	238.9	14.3	158.2
5	7.00	14.4	231.5	14.1	250.1	12.9	109.0	11.2	92.4	13.8	179.2
6	7.00	14.1	201.9	13.7	206.5	13.6	167.1	12.8	128.6	14.3	246.6
7	7.00	14.0	211.9	13.9	209.5	14.2	206.0	13.3	167.5	14.0	205.5
8	7.00	14.2	202.6	13.2	193.8	13.9	183.3	12.7	133.9	13.4	193.8
9	7.00	14.0	194.4	13.3	195.3	14.2	201.3	13.0	146.1	13.7	195.6
10	7.00	14.2	200.7	13.6	191.7	13.2	185.0	12.5	124.8	14.1	202.6

We see that both the one- and two-parameter methods give very accurate results for both heavy- and light-tail Weibull distributions, and that the TLR updating performs similar to its two-parameter counterpart, although repeated measurements indicate that for the cases above the RE is about two times smaller for the two-parameter TLR method.

Table 3.5. The evolution of $\widehat{\mathbf{b}}_t$ and $\widehat{\mathbf{v}}_t$ for estimating the optimal parameters $\mathbf{b}^*$ and $\mathbf{v}^*$ with the two-parameter SLR method (3.120). The estimated probability is $\widehat{\ell} = 6.5964 \cdot 10^{-7}$, the relative error $RE = 0.014723$, and $\widehat{\kappa^2} = 108.3$.

t	$\widehat{\gamma}_t$	$\widehat{b}_{1t}$	$\widehat{v}_{1t}$	$\widehat{b}_{2t}$	$\widehat{v}_{2t}$	$\widehat{b}_{3t}$	$\widehat{v}_{3t}$	$\widehat{b}_{4t}$	$\widehat{v}_{4t}$	$\widehat{b}_{5t}$	$\widehat{v}_{5t}$
0	-	0.2	1	0.2	1	0.2	1	0.2	1	0.2	1
1	971.28	0.17	1.55	0.18	1.69	0.18	1.63	0.17	1.48	0.18	1.52
2	28750	0.15	1.76	0.15	2.09	0.15	1.89	0.15	1.75	0.14	1.40
3	461370	0.12	1.86	0.13	1.84	0.12	1.43	0.12	1.53	0.13	2.38
4	1000000	0.12	1.50	0.13	2.17	0.12	1.83	0.13	1.62	0.11	1.93
5	1000000	0.12	1.59	0.12	1.66	0.12	1.92	0.11	1.66	0.12	1.99
6	1000000	0.13	1.68	0.12	2.02	0.12	1.96	0.12	1.83	0.13	1.91
7	1000000	0.12	1.72	0.13	1.97	0.12	1.87	0.12	1.77	0.12	1.87
8	1000000	0.12	1.81	0.12	2.05	0.12	1.90	0.13	1.94	0.12	1.67
9	1000000	0.12	1.95	0.12	1.70	0.13	1.88	0.12	1.66	0.12	1.88

3.12.2 Sum of Pareto Random Variables

Here we repeat the experiments of Tables 3.2 and 3.3 for the Pareto case. Specifically, we now let the X_i have a Pareto pdf $f(x) = a\,u^{-1}(1+xu^{-1})^{-(1+a)}$ and consider the TLR change of measure

$$X_i = u\left(\mathrm{e}^{Z_i/a} - 1\right), \quad Z_i \sim \mathsf{Exp}(1) \longrightarrow \mathsf{Exp}(v_i^{-1}) \,. \tag{3.121}$$

This is equivalent to the SLR change of measure

$$X_i \sim \mathsf{Pareto}(a, u^{-1}) \longrightarrow \mathsf{Pareto}(av_i^{-1}, u^{-1}) \,. \tag{3.122}$$

Tables 3.6 and 3.7 present the performance of the TLR method for $a = 5$ and $a = 0.2$, respectively. For both cases we selected $u = 1$ and took $N = 2 \cdot 10^5$ and $N_1 = 10^6$.

Table 3.6. The evolution of $\widehat{\mathbf{v}}_t$ for estimating the optimal parameter v^* with the TLR method for $a = 5$. The estimated probability is $\widehat{\ell} = 5.22 \cdot 10^{-7}$, the relative error $RE = 0.0238$, and $\widehat{\kappa^2} = 570.98$.

t	$\widehat{\gamma}_t$	$\widehat{v}_{1t}$	$\widehat{v}_{2t}$	$\widehat{v}_{3t}$	$\widehat{v}_{4t}$	$\widehat{v}_{5t}$
0	-	1	1	1	1	1
1	2.14	1.90	1.88	1.88	1.93	1.93
2	5.56	2.95	2.94	2.93	2.93	2.96
3	13.06	3.67	3.62	3.46	3.68	3.87
4	22.41	4.50	3.99	4.19	4.30	3.89
5	25.00	3.61	5.35	3.92	4.52	3.88
6	25.00	4.02	4.24	4.40	4.36	4.45
7	25.00	4.44	4.30	4.26	4.09	4.27
8	25.00	4.38	4.18	4.11	4.09	4.63
9	25.00	4.27	4.07	4.29	4.47	4.25
10	25.00	4.38	4.33	4.41	4.28	3.92

Table 3.7. The evolution of $\widehat{\mathbf{v}}_t$ for estimating the optimal parameter $\mathbf{v}^*$ with the TLR method for $a = 0.2$. The estimated probability is $\widehat{\ell} = 4.86 \cdot 10^{-7}$, the relative error $RE = 0.0267$, and $\widehat{\kappa^2} = 716.74$.

t	$\widehat{\gamma}_t$	$\widehat{v}_{1t}$	$\widehat{v}_{2t}$	$\widehat{v}_{3t}$	$\widehat{v}_{4t}$	$\widehat{v}_{5t}$
0	-	1	1	1	1	1
1	2.6e+008	1.74	1.75	1.75	1.74	1.74
2	4.6e+014	2.32	2.34	2.33	2.38	2.34
3	4.9e+019	2.78	2.86	2.72	2.89	2.82
4	4.9e+023	3.22	3.24	2.99	3.26	3.23
5	6.7e+026	3.56	3.47	3.26	3.48	3.58
6	1.3e+029	3.80	4.02	3.29	3.74	3.53
7	1.0e+031	3.60	4.09	3.74	4.11	3.78
8	4.5e+032	3.74	4.05	3.91	3.39	4.67
9	2.8e+033	4.00	4.72	3.78	3.81	4.48
10	1e+035	4.48	3.97	4.12	4.57	3.86
11	1e+035	4.16	4.35	4.57	3.99	4.11
12	1e+035	4.37	4.49	4.16	4.13	4.00
13	1e+035	4.14	4.00	4.25	4.11	4.54
14	1e+035	4.12	4.24	4.44	4.16	4.24
15	1e+035	4.30	4.16	4.53	4.18	4.30

Although in this case the TLR change of measure (3.121) does not seem as "natural" as the SLR one, where a or u is changed, we can see, however, that again a good variance reduction is obtained. An advantage of (3.121) is that only one line of the code for the Weibull case needed to be changed. Besides, as we saw, the SLR method where only a is allowed to change is identical to the TLR method described above.

3.12.3 Stochastic Shortest Path

Our second model concerns a *stochastic shortest path* problem of Section 2.2.1. That is, we have the weighted graph given in Figure 2.1, with random weights $X_1, \ldots, X_5$. Suppose the random variables $X_1, \ldots, X_5$ are independent and have a $\mathsf{Weib}(a_i, u_i^{-1})$ distribution, $i = 1, \ldots, 5$. Let $S(\mathbf{X})$ be the length of the shortest path from node A to node B. Note that there are four possible paths. We wish to estimate from simulation the probability $\ell = \mathbb{P}(S(\mathbf{X}) \geqslant \gamma)$ that the length of the shortest path $S(\mathbf{X})$ will exceed some fixed γ. We consider the light- and heavy-tail cases $a_i = 5$ and $a_i = 0.2, i = 1, \ldots, 5$. In both cases $\mathbf{u} = (0.25, 0.4, 0.1, 0.3, 0.2)$. Tables 3.8 and 3.9 present the performance of Algorithm 3.4.1 with the TLR method (3.119), for the cases $a = 5$ and $a = 0.2$ respectively. The results are self-explanatory.

Table 3.8. The evolution of $\widehat{\mathbf{v}}_t$ for estimating the optimal parameter $\mathbf{v}^*$ with the TLR method and $a = 5$. The estimated probability is $\widehat{\ell} = 1.20 \cdot 10^{-10}$, the relative error 0.044.

t	$\widehat{\gamma}_t$	$\widehat{v}_{1t}$	$\widehat{v}_{2t}$	$\widehat{v}_{3t}$	$\widehat{v}_{4t}$	$\widehat{v}_{5t}$
0	-	1	1	1	1	1
1	0.568	2.491	1.530	1.267	1.748	1.931
2	0.650	4.257	2.152	1.543	2.431	2.977
3	0.706	6.052	2.705	1.896	3.294	4.153
4	0.752	8.125	3.476	2.260	4.128	5.360
5	0.792	10.356	4.074	2.630	4.994	6.687
6	0.800	10.293	4.126	2.850	5.519	7.460
7	0.800	10.712	4.265	2.520	5.090	7.109
8	0.800	10.550	4.125	2.565	5.310	7.383
9	0.800	10.897	4.377	2.577	5.277	7.096

Table 3.9. The evolution of $\widehat{\mathbf{v}}_t$ for estimating the optimal parameter $\mathbf{v}^*$ with the TLR method and $a = 0.2$. The estimated probability is $\widehat{\ell} = 1.09 \cdot 10^{-11}$, the relative error 0.026.

t	$\widehat{\gamma}_t$	$\widehat{v}_{1t}$	$\widehat{v}_{2t}$	$\widehat{v}_{3t}$	$\widehat{v}_{4t}$	$\widehat{v}_{5t}$
0	-	1	1	1	1	1
1	6.760	2.005	1.906	1.166	1.857	1.912
2	159.419	3.067	2.911	1.038	2.499	2.619
3	1070.002	4.226	3.940	1.052	3.029	3.211
4	4173.601	5.320	4.930	0.854	3.598	3.901
5	11663.017	6.877	6.333	1.118	3.730	3.867
6	34307.081	9.237	8.434	1.078	3.461	3.548
7	100000.000	7.030	6.623	0.842	7.762	7.658
8	100000.000	11.309	10.660	1.043	3.227	3.474
9	100000.000	14.038	13.035	0.981	1.126	1.189
10	100000.000	14.261	13.008	0.979	1.066	1.035

Tables 3.10 and 3.11 present the performance of the big-step Algorithm 3.11.1 for the same data as in Tables 3.8 and 3.9, respectively. It is readily seen that the big-step algorithm speeds up substantially the process; it converges, in fact, after the first iteration.

Table 3.10. The evolution of $\widehat{\mathbf{v}}_t$ for estimating the optimal parameter $\mathbf{v}^*$ with the TLR method using the big-step Algorithm 3.11.1 and $a = 5$. Estimated probability $\ell = 1.24 \cdot 10^{-10}$, relative error $RE = 0.0438$, Matlab CPU 15.2 seconds.

t	$\widehat{\gamma}_t$	$\widehat{v}_{1t}$	$\widehat{v}_{2t}$	$\widehat{v}_{3t}$	$\widehat{v}_{4t}$	$\widehat{v}_{5t}$
0	-	1	1	1	1	1
1	0.8	7.12	4.99	4.36	5.43	5.79
2	0.8	10.8	4.24	2.70	5.20	7.05
3	0.8	11.1	4.35	2.57	5.13	7.12
4	0.8	11.1	4.37	2.54	5.18	6.86
5	0.8	11.0	4.42	2.86	5.18	6.71
6	0.8	10.7	4.24	2.89	5.30	7.00

Table 3.11. The evolution of $\widehat{\mathbf{v}}_t$ for estimating the optimal parameter $\mathbf{v}^*$ with the TLR method using the big-step Algorithm 3.11.1 and $a = 0.2$. Estimated probability $\ell = 1.082 \cdot 10^{-11}$, relative error $RE = 0.026$, Matlab CPU 14.8 seconds.

t	$\widehat{\gamma}_t$	$\widehat{v}_{1t}$	$\widehat{v}_{2t}$	$\widehat{v}_{3t}$	$\widehat{v}_{4t}$	$\widehat{v}_{5t}$
0	-	1	1	1	1	1
1	10000	7.02	7.83	5.31	7.76	7.72
2	10000	10.9	9.91	1.22	4.93	5.31
3	10000	12.8	12.1	1.24	2.67	2.91
4	10000	14.2	13.1	1.08	1.01	1.04
5	10000	14.1	13.0	0.975	0.995	1.10
6	10000	14.2	13.0	0.995	0.991	1.01

3.12.4 GI/G/1 Queue

Our third model is the GI/G/1 queue with interarrival time distribution $\mathsf{Weib}(a_1, u_1^{-1})$ and service time distribution $\mathsf{Weib}(a_2, u_2^{-1})$, discussed in Section 3.10. Note that the traffic intensity of the queue — the expected service time divided by the expected interarrival — is thus given by

$$\frac{u_2 \Gamma(1 + 1/a_2)}{u_1 \Gamma(1 + 1/a_1)} .$$

We first consider the estimation of the probability that the stationary waiting time in the queue exceeds some fixed level γ, using the *random walk* method described in Section 3.10. In particular, with A_i and B_i the interarrival and service times, we use the TLR change of measure

$$\begin{aligned} A_i &= u_1 \left(Z_i^{(1)}\right)^{1/a_1}, \qquad Z_i^{(1)} \sim \mathsf{Exp}(1) \longrightarrow \mathsf{Exp}(v_1^{-1}) \\ B_i &= u_2 \left(Z_i^{(2)}\right)^{1/a_2}, \qquad Z_i^{(2)} \sim \mathsf{Exp}(1) \longrightarrow \mathsf{Exp}(v_2^{-1}) . \end{aligned} \tag{3.123}$$

Table 3.12 illustrates the evolution of Algorithm 3.4.1 for determining the CE optimal parameters v_1 and v_2 to be used in the TLR estimator. In this particular case the parameters are $a_1 = 0.5$, $u_1 = 1$, $a_2 = 0.5$ and $u_2 = 0.5$, which gives a traffic intensity of 0.5. The level to be exceeded is $\gamma = 80$. The sample size used in Steps 1–4 was $N = 50,000$. The rarity parameter ϱ was set to 0.01.

We have repeated Steps 2–4 four more times after reaching γ to show the accuracy of the estimation of the true optimal CE parameter. (The corresponding estimate and RE for this case are given in Table 3.13.)

Table 3.12. The evolution of Algorithm 3.4.1 using the TLR method for the $GI/G/1$ with the following parameters: $a_1 = 0.5$, $u_1 = 1$, $a_2 = 0.5$, $u_2 = 0.5$.

t	$\widehat{\gamma}_t$	$\widehat{v}_{1t}$	$\widehat{v}_{2t}$
0	-	1	1
1	39.5	0.774073	1.39477
2	80	0.796896	1.44949
3	80	0.813729	1.42962
4	80	0.810056	1.40465
5	80	0.799487	1.43608
6	80	0.801236	1.44118

It is interesting to note that after one iteration the system becomes unstable, so that $\widehat{\gamma}_t$ in Step 2 of the CE algorithm reaches level γ in just *two* iterations. This is in accordance with the *instability property* of the CE algorithm described and analyzed in [46]. As a consequence, the choice of the rarity parameter does not matter very much.

Tables 3.13–3.16 summarize some performance characteristics of the TLR estimation procedure as a function of γ, for various light- and heavy-tail cases. In all cases we set $N = 10^4$ and $N_1 = 5 \cdot 10^5$. Also, the rarity parameter ϱ was set to 0.1 (in fact any parameter $\varrho < 1$ would be ok) and the level $-L$ was set low enough to -100.

In all tables we report the optimal CE parameters (the original ones are 1), the estimate of the probability, the RE and the CPU time in seconds.

Table 3.13. Simulation results for method 2 for the waiting time probabilities of a GI/G/1 queue with heavy-tail interarrival and service time distributions, as a function of γ. The traffic intensity is 0.5.

$a_1 = 0.5, u_1 = 1,\quad a_2 = 0.5, u_2 = 0.5$						
γ	20	40	60	80	100	120
v_1^*	0.78	0.79	0.80	0.80	0.80	0.81
v_2^*	1.36	1.38	1.40	1.41	1.43	1.45
$\widehat{\ell}$	$7.139 \cdot 10^{-2}$	$1.152 \cdot 10^{-2}$	$2.08 \cdot 10^{-3}$	$4.25 \cdot 10^{-4}$	$8.99 \cdot 10^{-5}$	$2.08 \cdot 10^{-5}$
RE	0.002	0.0036	0.0067	0.016	0.020	0.045
sec	149	264	396	467	587	696

Table 3.14. Simulation results for method 2 for the waiting time probabilities of a GI/G/1 queue with light-tail interarrival and service time distributions, as a function of γ. The traffic intensity is 0.75.

$a_1 = 2, u_1 = 1,\quad a_2 = 2, u_2 = 0.75$						
γ	3	6	9	12	15	18
v_1^*	0.56	0.56	0.56	0.56	0.56	0.56
v_2^*	1.57	1.58	1.58	1.58	1.58	1.59
$\widehat{\ell}$	$1.031 \cdot 10^{-2}$	$1.63 \cdot 10^{-4}$	$2.60 \cdot 10^{-6}$	$4.15 \cdot 10^{-8}$	$6.63 \cdot 10^{-10}$	$1.56 \cdot 10^{-11}$
RE	0.0017	0.0027	0.0040	0.0053	0.013	0.016
sec	101	154	210	274	338	398

Table 3.15. Simulation results for method 2 for the waiting time probabilities of an M/M/1 queue, as a function of γ. The traffic intensity is 0.75.

$a_1 = 1, u_1 = 2,\quad a_2 = 1, u_2 = 1.5$						
γ	20	40	60	80	100	120
v_1^*	0.75	0.75	0.75	0.75	0.75	0.75
v_2^*	1.33	1.33	1.33	1.33	1.33	1.33
$\widehat{\ell}$	$2.676 \cdot 10^{-2}$	$9.539 \cdot 10^{-4}$	$3.404 \cdot 10^{-5}$	$1.214 \cdot 10^{-6}$	$4.333 \cdot 10^{-8}$	$1.546 \cdot 10^{-9}$
RE	0.00036	0.00039	0.00040	0.00040	0.00038	0.00053
sec	160	429	509	558	691	828

Table 3.16. Simulation results for method 2 for the waiting time probabilities of an M/G/1 queue, with heavy-tail service distribution, as a function of γ. The traffic intensity is 0.5.

$a_1 = 1, u_1 = 1,\quad a_2 = 0.5, u_2 = 0.25$						
γ	10	20	30	40	50	60
v_1^*	0.81	0.83	0.84	0.85	0.85	0.89
v_2^*	1.64	1.68	1.71	1.65	1.73	1.77
$\widehat{\ell}$	$2.83 \cdot 10^{-2}$	$3.55 \cdot 10^{-3}$	$5.63 \cdot 10^{-4}$	$1.05 \cdot 10^{-4}$	$2.60 \cdot 10^{-5}$	$7.07 \cdot 10^{-6}$
RE	0.003	0.0067	0.012	0.017	0.047	0.093
sec	108	190	224	306	335	407

The results seem to indicate that the RE increases (sub)linearly, but there is not sufficient evidence to conclude that the estimators are polynomial, except in the M/M/1 case, where the RE remains constant. In the latter case we have the well-known optimal (exponential) change of measure where the service and interarrival rates are interchanged. What is clearer is that for the light-tail case we can estimate much smaller probabilities than for the heavy-tail case, for a given accuracy (RE) and simulation effort. It is interesting to note that for the second experiment (with $a_1 = a_2 = 2$) quite small probabilities can be efficiently estimated despite the fact that the TLR estimator is not logarithmically efficient. Specifically, the only logarithmically efficient estimator is obtained by an exponential change of measure [150, 18]. The TLR change of measure for this case is obviously not an exponential change of measure.

Note also that for both light-tail cases the reference parameters seem to have "converged," but not yet for the two heavy-tail cases. Also the estimates for the reference parameters seem more noisy in the heavy-tail case. In both the light- and heavy-tail case we observed that the estimates for the probabilities stabilized quite quickly (for moderate sample sizes). However, we also observed that accurate estimates for the variance of the estimator were much more difficult to obtain in the heavy-tail case than in the light-tail case.

We have repeated the experiments in Tables 3.13–3.16 for method 1, the *switching regenerative method*, using $N_1 = 5 \cdot 10^5$ regeneration cycles and using exactly the same CE parameters v_1^* and v_2^* as reported for method 2. The results were very similar to those of method 2. Tables 3.17 and 3.18 give the results for two of these experiments. We also ran the model with crude Monte Carlo, that is method 1 with $v_1 = v_2 = 1$, increasing the number of cycles to $5 \cdot 10^6$ in order to obtain execution times of the same order as the other methods. The CMC estimates were in close agreement with the IS estimates, and the IS estimates consistently gave a significant variance reduction, although less pronounced in the heavy-tail case.

Table 3.17. Simulation results for method 1 for the waiting time probabilities of a GI/G/1 queue with heavy-tail interarrival and service time distributions, as a function of γ. The traffic intensity is 0.5.

$a_1 = 0.5, u_1 = 1,\quad a_2 = 0.5, u_2 = 0.5$						
γ	20	40	60	80	100	120
$\widehat{\ell}$	$7.19 \cdot 10^{-2}$	$1.161 \cdot 10^{-2}$	$2.08 \cdot 10^{-3}$	$4.38 \cdot 10^{-4}$	$8.63 \cdot 10^{-5}$	$2.01 \cdot 10^{-5}$
RE	0.0087	0.011	0.017	0.029	0.034	0.071
sec	59	80	109	135	170	200

Table 3.18. Simulation results for method 1 for the waiting time probabilities of a GI/G/1 queue with light-tail interarrival and service time distributions, as a function of γ. The traffic intensity is 0.75.

$a_1 = 2, u_1 = 1, \quad a_2 = 2, u_2 = 0.75$						
γ	3	6	9	12	15	18
$\widehat{\ell}$	$1.028 \cdot 10^{-2}$	$1.63 \cdot 10^{-4}$	$2.58 \cdot 10^{-6}$	$4.12 \cdot 10^{-8}$	$6.76 \cdot 10^{-10}$	$1.05 \cdot 10^{-11}$
RE	0.0064	0.0084	0.011	0.019	0.020	0.021
sec	51	91	167	173	212	398

We also conducted various experiments in the *transient setting* (that is taking $L = 0$, see Remark 3.32, and using Pareto arrival and service times). Tables 3.19 and 3.20 present two examples. Table 3.21 presents an example using Pareto arrival and Weibull service time. For the Pareto case a similar TLR change of measure as in (3.121) was used. In all tables τ is as in (3.115) with $L = 0$ and SCV stands for the squared coefficient of variation for the random variable IW in the TLR estimator.

Table 3.19. Transient simulation results as function of γ for the GI/G/1 queue with the interarrival time distribution $\mathsf{Pareto}(0.5, 0.4)$ and service distribution $\mathsf{Pareto}(0.5, 0.36)$. The traffic intensity is 0.9. For $\gamma = 40$ the probability was checked by CMC estimation: $\widehat{\ell} = 1.78 \cdot 10^{-2}$

γ	40	120	160	240	300	360
N	$5 \cdot 10^4$	$5 \cdot 10^4$	$5 \cdot 10^4$	$5 \cdot 10^4$	$5 \cdot 10^4$	$5 \cdot 10^4$
N_1	$5 \cdot 10^5$	$5 \cdot 10^5$	$5 \cdot 10^5$	$5 \cdot 10^5$	10^6	10^6
$\widehat{\ell}$	1.76e-002	1.49e-003	4.78e-004	5.32e-005	1.00e-005	1.81e-006
RE	0.0068	0.013	0.016	0.023	0.021	0.026
$\widehat{\tau}$	94.26	655.96	1137.86	2075.22	3305.66	3703.51
SCV	23.46	87.73	129.57	283.18	430.81	664.02

Table 3.20. Transient simulation results as function of γ for the GI/G/1 queue with the interarrival time distribution $\mathsf{Pareto}(3, 0.75)$ and service distribution $\mathsf{Pareto}(3, 1)$. The traffic intensity is 0.75.

γ	25	50	80	120	250	350
N	10^5	10^5	10^5	10^5	10^5	10^5
N_1	$5 \cdot 10^5$	$5 \cdot 10^5$	$5 \cdot 10^5$	$5 \cdot 10^5$	$5 \cdot 10^5$	$5 \cdot 10^5$
$\widehat{\ell}$	1.25e-003	8.72e-005	9.66e-006	2.51e-006	2.59e-007	7.54e-008
RE	0.011	0.029	0.041	0.047	0.053	0.051
$\widehat{\tau}$	75.60	88.26	69.75	79.20	92.18	93.76
SCV	61.81	427.33	841.23	1082.71	1387.39	1301.01

Table 3.21. Transient simulation results as function of γ for GI/G/1 queue with the interarrival distribution $\mathsf{Weib}(2,1)$ and service distribution $\mathsf{Pareto}(2.5,1)$. The traffic intensity is 0.75225.

γ	20	50	130	160	300	400
N	$5 \cdot 10^4$	$5 \cdot 10^4$	$5 \cdot 10^4$	$5 \cdot 10^4$	$5 \cdot 10^4$	$5 \cdot 10^4$
N_1	$2 \cdot 10^5$	$3 \cdot 10^5$	$3 \cdot 10^5$	$3 \cdot 10^5$	$3 \cdot 10^5$	$3 \cdot 10^5$
$\widehat{\ell}$	2.37e-003	2.20e-004	1.05e-005	8.25e-006	1.33e-006	5.75e-007
RE	0.016	0.030	0.033	0.029	0.028	0.027
$\widehat{\tau}$	18.38	17.21	16.03	14.44	16.08	14.92
SCV	51.94	275.31	326.70	258.69	239.94	220.05

3.12.5 Two Non-Markovian Queues with Feedback

As a final example, we consider the network depicted in Figure 3.3. It consists of two queues in tandem, where customers departing from the second queue either leave the network (with probability p), or go back to the first queue (with probability $1 - p$). We are interested in estimating the transient probability that the total number of customers in the network exceeds some high level, 50 in this example, during one busy cycle. This model was also considered in [45], using only light-tail distributions and applying IS with exponential twisting.

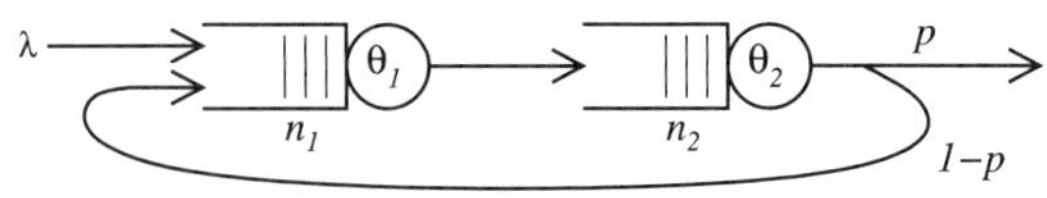

Fig. 3.3. Two queues in tandem with feedback

In the experiments reported below the interarrival time distribution is a two-stage Erlang distribution, with exponential parameter $\lambda = 0.2$. The service time distribution of the first queue is uniform on $[0, 3.333]$. In the second queue the service time distribution is $\mathsf{Weib}(a,c)$. In Table 3.22 we consider the light-tail case with $a = 2$ and $c = 0.354491$, which gives a mean service time of 2.5, while in Table 3.23 we consider the heavy-tail case with $a = 0.8$ and $c = 0.453201$, which gives again mean service time of 2.5. We note that this is the same mean service time as in [45]. In the tables, θ is the exponential twisting parameter for the uniform distribution. The λ column gives the evolution of reference parameter for the Erlang interarrivals, and similarly for c and p.

Table 3.22. Simulation results for the non-Markovian network for the light-tail case with $a = 2$, $c = 0.35449$, $\gamma = 50$, and sample sizes $N = N_1 = 10^4$. The estimated probability is $\widehat{\ell} = 1.62\,10^{-25}$, the relative error $RE = 0.018$.

t	$\widehat{\gamma}_t$	$\widehat{\lambda}_t$	$\widehat{\theta}_t$	$\widehat{c}_t$	$\widehat{p}_t$
0	3.0	0.200000	0.000000	0.354491	0.5
1	50	0.342317	-0.023671	0.294095	0.177778
2	50	0.363233	0.000000	0.315648	0.225282
3	50	0.360159	0.000000	0.320599	0.234336
4	50	0.360873	-0.003051	0.320986	0.234113
5	50	0.358857	-0.003623	0.320894	0.235779
6	50	0.360186	-0.000707	0.320591	0.234769
7	50	0.359469	-0.003483	0.320718	0.234796

We see that the optimal CE parameters are estimated quite accurately for a relatively small N. Since the second queue is the bottleneck *state independent tilting*, changing the parameters irrespective of the state of the queue, seems to work nicely, and the TLR method seems to deliver an accurate estimate of a very small probability. No numerical results are available for validation; therefore, we repeated the experiment a number of times. The fact that we obtained similar estimates gives confidence.

Table 3.23. Simulation results for the non-Markovian network for the heavy-tail case with $a = 0.8$, $c = 0.453201$, $\gamma = 50$, and sample sizes $N = N_1 = 10^5$. The estimated probability is $\widehat{\ell} = 4.323\,10^{-18}$, the relative error $RE = 0.0079$.

t	$\widehat{\gamma}_t$	$\widehat{\lambda}_t$	$\widehat{\theta}_t$	$\widehat{c}_t$	$\widehat{p}_t$
0	3.0	0.200000	0.000000	0.453201	0.5
1	50	0.300620	0.000000	0.263503	0.3019
2	50	0.301135	0.000000	0.263982	0.3031
3	50	0.301291	-0.000000	0.264346	0.3026
4	50	0.300832	0.000000	0.263580	0.3031
5	50	0.301350	-0.000000	0.263770	0.3029
6	50	0.300620	0.000000	0.263503	0.3019
7	50	0.301135	0.000000	0.263982	0.3031

For this heavy-tail case a similar picture emerges: the estimates for the reference parameters are quite stable, and a small probability can be estimated with reasonable accuracy. However, when we repeat this for a smaller a ($a = 0.5$) the results were not so satisfactory, indicating that a (much) larger sample size is required.

3.13 Appendix: The Sum of Two Weibulls

As noted in Remark (3.30) for the sum of n heavy-tail Weibulls, the change of measure given by (3.102) for any constant c in (3.103) gives an SLR estimator which is logarithmically efficient. A proof of this is given in Theorem 3.2 of [91]. In this appendix we prove that for the case $n = 2$ and for large γ the best — that is, minimum variance — choice for c is $c = n = 2$ and that the estimator is not only logarithmically efficient, but in fact polynomial. We conjecture that in general $c = n$. We show explicitly that the relative error grows (for $n = 2$) as γ^{2a}, and we conjecture that in general it grows as γ^{na}. The proof below uses the representation of the change of measure, but it could as easily have been given via an SLR approach. Most of the results hold for the light- ($a \geqslant 1$) and heavy-tail $a < 1$ case, except when the subexponentiality property is used for the heavy-tail case. Without loss of generality we take $u = 1$.

Thus the problem is as follows: Let X_1, X_2 be i.i.d. $\mathsf{Weib}(a,1)$ distributed; estimate

$$\ell = \mathbb{P}(X_1 + X_2 \geqslant \gamma) = \mathbb{P}(Z_1^{1/a} + Z_2^{1/a} \geqslant \gamma)\,,$$

with $Z_i \sim \mathsf{Exp}(1)$, independent. Consider the exponential change of measure $Z_i \sim \mathsf{Exp}(1) \longrightarrow \mathsf{Exp}(1-\theta)$, where $0 \leqslant \theta < 1$ is the exponential twisting parameter. Let $\mathbb{E}_\theta$ denote the corresponding expectation operator. Thus $\mathbb{E}_0$ corresponds to the original $\mathsf{Exp}(1)$ distribution. We have

$$\ell = \mathbb{E}_\theta I_{\{Z_1^{1/a}+Z_2^{1/a}\geqslant\gamma\}} W\,.$$

Here $W = W(\theta)$ is shorthand notation for the likelihood ratio

$$W(\theta) = \mathrm{e}^{-\theta(Z_1+Z_2)+2\zeta(\theta)} = \frac{\mathrm{e}^{-\theta(Z_1+Z_2)}}{(1-\theta)^2}\,,$$

where we have used the fact that the cumulant function for this exponential family is given by $\zeta(\theta) = \ln(1/(1-\theta)) = -\ln(1-\theta)$.

There is no simple formula for ℓ as a function of a and γ, but it is not difficult to verify that

$$\begin{aligned}\ell(\gamma) &= \left(2\iint_{A_1} + \iint_{A_2} + 2\iint_{A_3}\right) \mathrm{e}^{-(z_1+z_2)}\,dz_1 dz_2 \\ &= \exp\left(-\gamma^a\,2^{1-a}\right) + 2\int_0^{(\gamma/2)^a} \exp\left(-\left\{\gamma - x^{1/a}\right\}^a - x\right)\,dx\,,\end{aligned}$$

where the regions A_1, A_2 and A_3 are given in Figure 3.4.

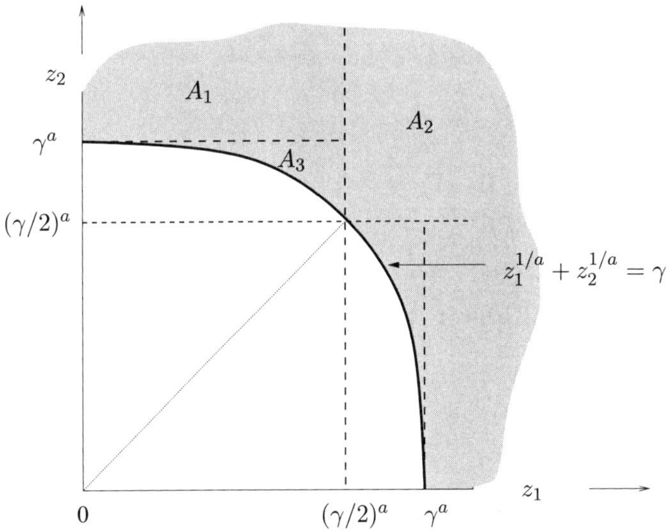

Fig. 3.4. ℓ is equal to the integral of $\mathrm{e}^{-(z_1+z_2)}$ over the shaded region.

Let us mention some known facts about ℓ. First, for the heavy-tail case $a < 1$ it is well known that the Weibull distribution is *subexponential*, which means that the sum of n i.i.d. Weibull random variables satisfies

$$\lim_{\gamma\to\infty} \frac{\mathbb{P}(X_1 + \cdots + X_n \geqslant \gamma)}{\mathbb{P}(X_1 \geqslant \gamma)} = n \ .$$

In particular, for our $n = 2$ case we have that

$$\lim_{\gamma\to\infty} \frac{\ell(\gamma)}{2\mathrm{e}^{-\gamma^a}} = 1 \ .$$

For $a = 1$ it is not difficult to see that

$$\ell = \mathrm{e}^{-\gamma}(\gamma + 1) \ .$$

For $a > 1$ one can show that

$$\lim_{\gamma\to\infty} \frac{\ell(\gamma)}{\mathrm{e}^{-2(\gamma/2)^a}\, \gamma^{a/2}} = c(a) \ ,$$

for some constant $c(a)$, decreasing as a increases. For example, for $a = 2$, $c(a) = \sqrt{\pi/2}$ and for $a = 3$, $c(a) = \sqrt{3\pi}/4$.

Let us now turn to the complexity properties of the TLR estimator, as a function of γ. These are, as always, determined by the second moment (under

θ) of the random variable $IW = I_{\{Z_1^{1/a}+Z_2^{1/a} \geqslant \gamma\}} W(\theta)$. Using a simplified notation we have

$$\begin{aligned}
\mathbb{E}_\theta (IW)^2 &= \mathbb{E}_\theta \, IW^2 \\
&= \mathbb{E}_0 \, IW \\
&= \mathbb{E}_0 \, I \frac{\mathrm{e}^{-\theta(Z_1+Z_2)}}{(1-\theta)^2} \qquad (3.124) \\
&= \left(2 \iint_{A_1} + \iint_{A_2} + 2 \iint_{A_3} \right) \frac{\mathrm{e}^{-(1+\theta)(z_1+z_2)}}{(1-\theta)^2} \, d\mathbf{z} \, .
\end{aligned}$$

We wish to show that the SCV increases at most polynomially in γ, for a certain choice of θ. This is equivalent to showing that $\mathbb{E}_\theta (IW)^2/\ell^2$ increases at most polynomially in γ. We do this by considering the contributions of the three integrals in (3.124) individually.

Define $D_i = \iint_{A_i} \frac{\mathrm{e}^{-(1+\theta)(z_1+z_2)}}{(1-\theta)^2} \, d\mathbf{z}, i = 1, 2, 3$. The easiest of these is D_2; namely

$$D_2 = \left(\frac{\mathrm{e}^{-(1+\theta)(\gamma/2)^a}}{1-\theta^2} \right)^2 .$$

It follows that, for fixed θ,

$$\lim_{\gamma \to \infty} \frac{D_2}{\ell^2} = \frac{4}{(1-\theta^2)^2} \lim_{\gamma \to \infty} \mathrm{e}^{-\gamma^a \{(1+\theta) 2^{1-a} - 2\}} = 0 \, ,$$

provided that $1 + \theta > 2^a$, or equivalently $1 - \theta < 2 - 2^a$.

Second, we have

$$D_1 \leqslant \widetilde{D}_1 = \int_0^\infty \int_{\gamma^a}^\infty \frac{\mathrm{e}^{-(1+\theta)(z_1+z_2)}}{(1-\theta)^2} \, d\mathbf{z} = \frac{1}{(1-\theta^2)^2} \mathrm{e}^{-\gamma^a(1+\theta)} \, .$$

The contribution of D_1 to the SCV is therefore bounded by

$$\frac{\widetilde{D}_1}{\ell^2} \approx \frac{1}{2\,(1-\theta^2)^2} \mathrm{e}^{-\gamma^a(1+\theta-2)} \, .$$

As a consequence, this contribution remains polynomial in γ if we choose $\theta = 1 - c\gamma^{-a}$, for any c. In that case

$$\frac{\widetilde{D}_1}{\ell^2} \approx \frac{\mathrm{e}^c \gamma^{4a}}{4c^2(c - 2\gamma^a)^2} \, .$$

If we minimize this with respect to c, we obtain for fixed γ the minimal argument

$$c^* = \gamma^a - \sqrt{\gamma^{2a} + 4} + 2 \, .$$

For large γ we have thus $c \approx 2$. This suggests we take

$$\theta = 1 - 2\gamma^{-a} .$$

Obviously, with this choice of θ the contribution of D_2 to the SCV tends 0, as γ increases. It follows that

$$\frac{2D_1 + D_2}{\ell^2} = \gamma^{2a}\frac{e^2}{64} + o\left(\gamma^{2a}\right) .$$

It remains to show that the contribution of D_3 remains polynomial. We have

$$\begin{aligned}\frac{D_3}{\ell^2} &\approx \frac{e^{2\gamma^a}}{2}\int_0^{(\gamma/2)^a}\int_{\left(\gamma - z_1^{1/a}\right)^a}^{\gamma^a} \frac{e^{-(1+\theta)(z_1+z_2)}}{(1-\theta)^2}\, d\mathbf{z} \\ &= \frac{d_3}{2(1-\theta)^2(1+\theta)} ,\end{aligned}$$

where

$$d_3 = \int_0^{(\gamma/2)^a} e^{2\gamma^a}\, e^{-(1+\theta)z}\left\{e^{-(1+\theta)(\gamma - z^{1/a})^a} - e^{-(1+\theta)\gamma^a}\right\} dz > 0 .$$

For fixed z and $\theta = 1 - 2\gamma^{-a}$ write the integrand of d_3 as $e^{-(1+\theta)z}g(z,\gamma)$, where

$$\begin{aligned} g(z,\gamma) &= e^{2\gamma^a}\left\{e^{-(2-2\gamma^{-a})\gamma^a(1-z^{1/a}/\gamma)^a} - e^{-(2-2\gamma^{-a})\gamma^a}\right\} \\ &= \exp\left\{\gamma^a\left[2 - 2\left(1 - \frac{z^{1/a}}{\gamma}\right)^a\right] + 2\left(1 - \frac{z^{1/a}}{\gamma}\right)^a\right\} - e^2 \end{aligned}$$

decreases monotonically to 0 as $\gamma \to \infty$. By the monotone convergence theorem, it follows that $d_3 \to 0$ as well, as $\gamma \to \infty$. Hence, we have $D_3/\ell^2 = o\left(\gamma^{2a}\right)$.

Concluding, for $a < 1$ we have proved that with the exponential twist $\theta = 1 - 2\gamma^{-a}$ the SCV of the TLR estimator increases proportionally to γ^{2a}, as $\gamma \to \infty$, that is

$$\kappa^2(\gamma) = \mathcal{O}(\gamma^{2a}) \text{ as } \gamma \to \infty . \tag{3.125}$$

It is interesting to note that κ^2 decreases with a, that is as the tail of Weibull pdf becomes heavier.

We conjecture that for arbitrary n the optimal twisting parameter is asymptotically $\theta^* \approx 1 - n\gamma^{-a}$ and that the SCV increases proportionally to γ^{na}, as $\gamma \to \infty$.

3.14 Exercises

1. Consider the estimation of the probability $\ell = \mathbb{P}(X \geqslant \gamma)$, $0 \leq \gamma \leq 1$, where $X \sim \mathsf{Beta}(\nu_0, 1)$. That is, the pdf of X is given by

$$f(x;\nu_0) = \nu_0\, x^{\nu_0 - 1}, \quad 0 \leqslant x \leqslant 1\,.$$

Suppose we wish to estimate ℓ via the SLR method. That is, through the change of measure

$$X \sim \mathsf{Beta}(\nu_0, 1) \longrightarrow \mathsf{Beta}(\nu, 1),$$

for some ν, using the SLR estimator

$$\widehat{\ell} = \frac{1}{N} \sum_{i=1}^{N} I_{\{X_i \geq \gamma\}} \frac{\nu_0\, X_i^{\nu_0 - 1}}{\nu X_i^{\nu - 1}}\,. \tag{3.126}$$

a) Using the cross-entropy program

$$\max_{\nu} D(\nu) = \max_{\nu} \mathbb{E}_{\nu_0} I_{\{X \geqslant \gamma\}} \ln f(X;\nu) \tag{3.127}$$

find the optimal reference parameter ν^* analytically. Show that

$$\nu^* \approx \frac{2}{1-\gamma}, \qquad \text{as } \gamma \to 1\,. \tag{3.128}$$

We could also estimate ℓ with TLR method, for example via the (TLR) change of measure

$$X = Z^{1/\nu_0}, \quad Z \sim \mathsf{U}(0,1) \longrightarrow \mathsf{Beta}(\xi, 1)\,,$$

for some ξ.

b) Verify from Example 3.26 that

$$\xi \approx \frac{2}{1 - \gamma^{\nu_0}}\,. \tag{3.129}$$

c) Show that this TLR change of measure is equivalent to the SLR change of measure above — in the sense of Section 3.9 — if we take $\nu = \nu_0 \xi$.
d) Show that (3.129) implies (3.128), as $\gamma \to 1$.
e) Use Proposition 3.28 to show that for $\gamma \approx 1$ the square coefficient of variation of $\widehat{\ell}$ in (3.126)

$$\kappa^2 = \frac{\mathrm{Var}(\widehat{\ell})}{\ell^2} \approx \frac{1}{N} \frac{\mathrm{e}^2 - 1}{4}\,,$$

using the reference parameter $\nu^* = 2/(1-\gamma)$. Find also the κ^2 for the CMC estimator, (without change of measure) and compare the two.
f) Write a computer program to estimate (for the SLR case) both ν^* and ℓ using Algorithm 3.4.1. Set $\varrho = 10^{-2}$, $\nu_0 = 2$, $N = 10^3$ (for estimating ν^*) and $N_1 = 10^5$ (for estimating ℓ), respectively. Select γ such that $\ell = 10^{-20}$. Present a table similar to Table 2.1 of the tutorial chapter.

2. Implement the CE method for the three different cases (SLR, ITLR and TLR) in Section 3.9, taking $n = 1$. Set $\varrho = 10^{-2}$, $u = 1$ and $N = 10^3$. Select N_1 large enough such that the estimated relative error is less than 0.05. Select γ (by trial and error) such that $\ell \leq 10^{-10}$.

3. Reproduce the results of Tables 3.8 and 3.9, by modifying the code of the first toy example in Appendix A.1.

4. Reproduce the results of Tables 3.2 and 3.3, by changing only the performance function in the code of the above example.

5. Reproduce the results of Tables 3.6 and 3.7, by altering in the code for Exercise 4 only the change of measure, from (3.119) to (3.121).

6. Repeat Exercise 4 with the big-step method.

7. **Max-cut: Rare-event probability estimation**. Consider the synthetic max-cut problem of Section 2.5.1, with cost matrix (c_{ij}) of the form (2.38). Let $\mathbf{p}_0 = (1/2, \ldots, 1/2)$ and $\mathbf{X} \sim \mathsf{Ber}(\mathbf{p}_0)$. We wish to estimate the probability $\ell = \mathbb{P}_{\mathbf{p}_0}(S(\mathbf{X}) \geqslant \gamma)$ for some large γ. This can be done efficiently via (3.36) using $\mathbf{X}_i \sim \mathsf{Ber}(\mathbf{p})$ instead of $\mathbf{X}_i \sim \mathsf{Ber}(\mathbf{p}_0)$. To find the optimal $\mathbf{p}$ run the CE Algorithm 3.4.1 for 20 iterations, reporting the estimates of γ_t and $\mathbf{p}_t$ and the rare-event probability $\ell_t = \mathbb{P}_{\mathbf{p}_0}(S(\mathbf{X}) \geqslant \gamma_t)$. Take $n = 50$, $\varrho = 0.01$, $N = 10\,n$. Eliminate the stopping rule, since γ is not specified. Discuss your results.

8. Consider the graph of Figure 2.1, with random weights $X_1, \ldots, X_5$. Suppose the weights are independent random variables, with $X_i \sim f(x; u_i)$. Set $u_1 = u_2 = 0.2$ and generate the rest of the parameters u_3, u_4, u_5 randomly from the uniform distribution $\mathsf{U}(0.5, 1.0)$. Let $S(\mathbf{X})$ be the total length of the shortest path from node A to node B. Set $\varrho = 0.01$, $N = 1000$ and run Algorithm 3.4.1 for the following cases
   a) $f(x; v_i) = v_i^{-1} \mathrm{e}^{-x/v_i}, \quad x \geqslant 0.$ (exponential)
   b) $f(x; v_i) = \mathrm{e}^{-v_i} v_i^x / x!, \quad x = 0, 1, \ldots.$ (Poisson)
   c) $f(x; v_i) = c\,\mathrm{e}^{-x/v_i}, \quad 0 \leqslant x \leqslant 1.$ (truncated exponential)
   d) $f(x; v_i) = v_i^x\,(1 - v_i)^{1-x}, \quad x = 0, 1.$ (Bernoulli)

   When estimating ℓ using the LR estimator $\widehat{\ell}$ choose the final sample size N_1 large enough to ensure that the relative error $\kappa < 0.05$. Also, make sure that γ is chosen such that the rare-event probability ℓ satisfies $10^{-5} < \ell < 10^{-6}$. For the truncated exponential and the Bernoulli case run Algorithm 3.4.1 without fixing γ in advance. By doing so, $\widehat{\gamma}_t$ in Algorithm 3.4.1 will attempt to continually increase γ. As a result, the parameters v_1 and v_2 associated with the random variables X_1 and X_2, called the bottleneck parameters, will produce degenerated marginal distributions. In particular, for $i = 1, 2$ we obtain $v_i \to \infty$ and $v_i = 1$ for the truncated exponential and the Bernoulli case, respectively.

9. Verify (3.104) and (3.105).

10.* Consider the *double exponential* density

$$g(x;\boldsymbol{\theta}) = \frac{b}{2}\,\mathrm{e}^{-b|x-a|}, \quad x \in \mathbb{R}, \quad \boldsymbol{\theta} = (a,b)\ .$$

The optimal solution of (3.65) follows from maximization of

$$\sum_i I_i W_i \ln(b) - b \sum_i I_i W_i\,|X_i - a|\ ,$$

using similar abbreviations as in (3.60). For each fixed b, the optimal a follows from minimization of

$$\sum_i I_i W_i\,|X_i - a|\ ,$$

which is very akin to determining the median of a discrete distribution. In particular, let $X_{(1)}, \ldots, X_{(N)}$ denote the order statistics of $X_1, \ldots, X_n$, and for each $X_{(i)}$ let $I_{(i)}W_{(i)}$ be the corresponding product in $\{I_1W_1, \ldots, I_NW_N\}$. Now define K as the first index such that

$$\frac{\sum_{i=K}^{N} I_{(i)}W_{(i)}}{\sum_{i=1}^{N} I_{(i)}W_{(i)}} \leqslant \frac{1}{2}\ .$$

Verify that the CE optimal parameters $\widehat{a}$ and $\widehat{b}$ are given by

$$\widehat{a} = X_{(K)}$$

and

$$\widehat{b} = \frac{\sum_i I_i W_i}{\sum_i I_i W_i\,|X_i - \widehat{a}|}\ .$$

11.* A GI/G/1 queue with exponential interarrival times is called an M/G/1 queue. Consider an M/G/1 queue with $\mathsf{Exp}(\lambda)$-distributed interarrival times and $\mathsf{Weib}(a, u^{-1})$-distributed service times. Let ϱ denote the traffic intensity:

$$\varrho = \frac{\mathbb{E}B}{1/\lambda} = \lambda\, u\, \Gamma(1 + 1/a)\ .$$

The famous *Pollaczek-Khinchine* formula [12] states that the steady-state waiting time Z in an M/G/1 queue satisfies

$$\mathbb{P}(Z \geqslant \gamma) = (1-\varrho) \sum_{m=0}^{\infty} \varrho^m (1 - \widetilde{F}_B^{*m}(\gamma))\ , \qquad (3.130)$$

where $\widetilde{F}_B^{*m}$ is the cdf of the sum of m stationary *excess* service times, that is, $\widetilde{F}_B^{*m}$ is the m-fold convolution of the cdf $\widetilde{F}_B$ with density

$$\widetilde{f}_B(x) = \frac{1 - F_B(x)}{\mathbb{E}B}, \quad x \geqslant 0,$$

where F_B is the cdf of the service time. Let M be a "geometric" random variable with the following distribution

$$\mathbb{P}(M = m) = (1 - \varrho)\varrho^n, \quad n = 0, 1, 2, \ldots, \tag{3.131}$$

and let $X_1, X_2, \ldots$ be i.i.d. with cdf $\widetilde{F}_B$ and independent of M. Then we may write (3.130) as the *random sum*

$$\mathbb{P}(Z \geqslant \gamma) = \mathbb{P}(S_M \geqslant \gamma),$$

with $S_M = X_1 + \cdots + X_M$.

a) Describe how we can estimate $\mathbb{P}(Z \geqslant \gamma)$ via CMC.
b) In order to apply CMC we need to draw from the cdf $\widetilde{F}_B$. Verify that if $Y \sim \mathsf{Gam}(1/a, 1)$, then $uY^{1/a} \sim \widetilde{F}_B$.
c) Describe how we can estimate $\mathbb{P}(Z \geqslant \gamma)$ via IS, by sampling from the excess lifetime distribution of $\mathsf{Weib}(a, v^{-1})$, for some v.
d) Derive the likelihood ratio of $(X_1, \ldots, X_m)$ for this change of measure.
e) Verify the CE updating rule for this change of measure:

$$\widehat{v}_t = \left(\frac{\sum_{i=1}^N I_{\{S_{iM_i} \geqslant \widehat{\gamma}_t\}} W_{iM_i} \sum_{n=1}^{M_i} X_{in}^a}{\sum_{i=1}^N I_{\{S_{iM_i} \geqslant \widehat{\gamma}_t\}} W_{iM_i} \, M_i} \right)^{1/a}.$$

12. Implement the root finding Algorithm 3.7.1 and verify that for the toy example in Section 2.2.1 the root of the equation

$$10^{-5} = \mathbb{P}(S(\mathbf{X}) \geqslant \gamma)$$

is given by $\gamma \approx 2.40$.

13.* Consider the random walk $\{S_n, n = 1, 2, \ldots\}$ with $S_n = X_1 + \cdots + X_n$, such that each X_k takes values 1 and -1 with probabilities p and $q = 1-p$, respectively. Define the NEF $\{f(\cdot; \theta), \theta \geqslant 0\}$ by

$$f(x; \theta) = \mathrm{e}^{\theta x - \zeta(\theta)} h(x), \quad x \in \{-1, 1\}, \quad \theta \geqslant 0,$$

with $h(1) = p$, $h(-1) = q$ and

$$\zeta(s) = \ln \mathbb{E} \mathrm{e}^{s X_1} = \ln(p\mathrm{e}^s + q\mathrm{e}^{-s}). \tag{3.132}$$

We introduce a change of measure for each X_k via the NEF above. Note that for $\theta = 0$ we get our original distribution.

a) Verify that the change of measure parameterized by θ is equivalent to X_k taking values 1 and -1 with probabilities

$$\tilde{p} = \frac{p\mathrm{e}^\theta}{p\mathrm{e}^\theta + q\mathrm{e}^{-\theta}} \quad \text{and} \quad \tilde{q} = \frac{q\mathrm{e}^{-\theta}}{p\mathrm{e}^\theta + q\mathrm{e}^{-\theta}}. \tag{3.133}$$

Define the probability measure $\tilde{\mathbb{P}}_i$ such that under this measure the random walk $\{S_n, n = 0, 1, \ldots\}$ starts in i, and the parameters p and q are changed to $\tilde{p}$ and $\tilde{q}$ in (3.133). The corresponding expectation operator is denoted by $\tilde{\mathbb{E}}_i$. For the original measure we use similarly $\mathbb{P}_i$ and $\mathbb{E}_i$. Let A be the event that the random walk reaches $\gamma - i$ before $-i$ and let T be the first time that it reaches $\gamma - i$ *or* $-i$. Let $\ell_i = \mathbb{P}_i(A) = \mathbb{P}_i(S_T = \gamma - i)$.

b) Prove that

$$\ell_i = \frac{1 - (q/p)^i}{1 - (q/p)^\gamma}, \quad i = 1, \ldots, \gamma . \tag{3.134}$$

This is known as the *gambler's ruin* [73].

c) Verify that the likelihood ratio of $(X_1, \ldots, X_n)$ with respect to $\mathbb{P}_i$ and $\tilde{\mathbb{P}}_i$ is

$$\mathrm{e}^{\theta S_n + n\kappa(\theta)} .$$

We may estimate ℓ_i by simulating independent copies of the random variable $Z = I_A W_T$ under $\tilde{\mathbb{P}}_i$, and then taking the average. To find the optimal CE parameter $\theta = \theta^*$, reparameterize the NEF via the mean $v = \zeta'(\theta)$.

d) Show that the optimal CE parameter v^* satisfies

$$v^* = \frac{\mathbb{E}_i I_A S_T}{\mathbb{E}_i I_A T} .$$

e) Demonstrate that a zero-variance way to simulate ℓ_i is to use IS with a state-dependent change of measure in which the p and q are replaced with

$$p_k \propto p\,\ell_{k+1} \quad \text{and} \quad q_k \propto q\,\ell_{k-1}, \quad k = 1, \ldots, \gamma - 1 , \tag{3.135}$$

where $\propto$ is the symbol for proportionality.

More information on CE method for the random walks and queueing networks can be found in [46].

14* **Network reliability.** Consider a network of unreliable links, labeled $\{1, \ldots, m\}$, modeling for example a communication network. With each edge i we associate its "state" X_i, such that $X_i = 1$ if link i works and $X_i = 0$ otherwise. We assume that $X_i \sim \mathsf{Ber}(p_i)$ and that all X_i are independent. Let $\mathbf{X} = (X_1, \ldots, X_m)$. We are interested in estimating the *reliability* r of the network, expressed as the probability that certain nodes, say in some set $\mathcal{K}$, are connected. Thus,

$$r = \mathbb{E}\,\varphi(\mathbf{X}) = \sum_{\mathbf{x} \in \mathcal{S}} \varphi(\mathbf{x})\,\mathbb{P}(\mathbf{X} = \mathbf{x}) , \tag{3.136}$$

where

$$\varphi(\mathbf{x}) = \begin{cases} 1 & \text{if } \mathcal{K} \text{ is connected,} \\ 0 & \text{otherwise.} \end{cases}$$

This is the standard formulation of the reliability of unreliable systems; see for example [62]. For highly reliable networks, that is, $p_i \approx 1$, or equivalently, $q_i = 1 - p_i \approx 0$, it is more useful to analyze or estimate the system *unreliability*

$$\overline{r} = 1 - r \; .$$

a) Consider the bridge network of Figure 2.1. Assume that the system works if A and B are connected. Express the unreliability of the system in terms of the link unreliabilities $\{q_i\}$.
b) Explain how we can simulate $\overline{r}$ via CMC, and show that this is very inefficient when $\overline{r}$ is small.

A simple CE modification [87] can substantially speed up the estimation of $\bar{r}$, when this quantity is small. The idea is to translate the original problem (estimating $\overline{r}$), which involves independent Bernoulli random variables $X_1, \ldots, X_m$, into an estimation problem involving independent exponential random variables $Y_1, \ldots, Y_m$. Specifically, imagine that we have a time-dependent system in which at time 0 all edges have failed and are under repair, and let $Y_1, \ldots, Y_m$, with $Y_i \sim \mathsf{Exp}(u_i^{-1})$ and $u_i = 1/\lambda(i) = -1/\ln q_i$ be the independent *repair times* of the edges. Note that, by definition

$$\mathbb{P}(Y_i \geqslant 1) = \mathrm{e}^{-1/u_i} = q_i \quad i = 1, \ldots, m \; .$$

Now, for each $\mathbf{Y} = (Y_1, \ldots, Y_m)$ let $S(\mathbf{Y})$ be the random time at which the system "comes up" (the nodes in $\mathcal{K}$ become connected). Then, we can write

$$\overline{r} = \mathbb{P}(S(\mathbf{Y}) \geqslant 1) \; .$$

Hence, we have written the estimation of $\overline{r}$ in the standard rare-event formulation of (3.1) and thus can directly apply the main CE Algorithm 3.4.1.

c) Explain how the CE method can be employed using the above representation.

Finally, we remark that a more sophisticated and significantly faster CE method, based on *graph-evolution* models [53], is given in [87].

15. Let Y_1, Y_2 be i.i.d. $\mathsf{U}(0,1)$-distributed random variables. We wish to estimate

$$\ell = \mathbb{P}\left((-\ln(Y_1))^{1/a} + (-\ln(Y_2))^{1/a} \geqslant \gamma\right) = \int_0^1 \int_0^1 I_{\{(y_1,y_2) \in A\}} \, dy_1 \, dy_2 \; ,$$

for some fixed $a > 0$ and $\gamma > 0$, where $A = \{(y_1, y_2) \in [0,1]^2 : (-\ln(y_1))^{1/a} + (-\ln(y_2))^{1/a} \geqslant \gamma\}$. Let $b = 1 - \mathrm{e}^{-(\gamma/2)^2}$ and define B to be the region $[0,1]^2 \setminus [0,b]^2$.

a) Show that $A \subset B$.

b) By (a) we may estimate ℓ via IS using the change of measure whereby $\mathbf{Y} = (Y_1, Y_2)$ is uniform on B, rather than $[0,1]^2$. Verify that the conditions for (3.5) are satisfied.
c) Show that the likelihood ratio for this change of measure is

$$W(\mathbf{y}) = 1 - b^2 \,, \quad \mathbf{y} \in B \,.$$

d) Show that the SCV of $Z = I_{\{\mathbf{Y} \in A\}} W(\mathbf{Y})$ is given by

$$(1 - b^2)/\ell - 1 \,.$$

In other words, this IS estimator has exponential complexity.

4

Combinatorial Optimization via Cross-Entropy

4.1 Introduction

In this chapter we show how the CE method can be easily transformed into an efficient and versatile randomized algorithm for solving optimization problems, in particular *combinatorial* optimization problems.

Most combinatorial optimization problems, such as deterministic and stochastic (noisy) scheduling, the travelling salesman problem (TSP), the maximal cut (max-cut) problem, the longest path problem (LPP), optimal buffer allocation in a production line, and optimization of topologies and configurations of computer communication and traffic systems are NP-hard problems. Well-known stochastic methods for combinatorial optimization problems are simulated annealing [1, 2, 38, 138], initiated by Metropolis [118] and later generalized in [80] and [100], tabu search [66], and genetic algorithms [69]. For very interesting landmark papers on the simulated annealing method see [138]. For some additional references on both deterministic and stochastic (noisy) combinatorial optimization see [2, 4, 6, 8, 38, 86, 88, 89, 93, 94, 95, 100, 101, 102, 108, 109, 111, 123, 124, 127, 129, 130, 134, 135].

Recent randomized algorithms for combinatorial optimization include the *nested partitioning method* (NP) of Shi and Olafsson [158, 159], the *stochastic comparison method* of Gong [70], the method of Andradóttir [10, 11], and the *ant colony optimization* (ACO) meta-heuristic of Dorigo and colleagues [49, 77].

The basic idea behind the NP method is based on systematic partitioning of the feasible region into smaller subregions until some of the subregions contain only one point. The method then moves from one region to another based on information obtained by random sampling. It is shown that the NP algorithm converges to an optimal solution with probability one.

The stochastic comparison method is similar to that of simulated annealing, but does not require a neighborhood structure.

The method of Andradóttir can be viewed as a discrete stochastic approximation method. The method compares two neighboring points in each iteration and moves to the point that is found to be better. This method is shown to converge almost surely to a local optimum. Andradóttir [11] also developed a similar method for finite noisy optimization which converges almost surely to a global optimum.

The ant algorithms are based on ant colony behavior. It is known that ant colonies are able to solve shortest-path problems in their natural environment by relying on a rather simple biological mechanism: while walking, ants deposit on the ground a chemical substance, called *pheromone*. Ants have a tendency to follow these pheromone trails. Within a fixed period, shorter paths between nest and food can be traversed more often than longer paths, and so they obtain a higher amount of pheromone, which, in turn, tempts a larger number of ants to choose them and thereby to reinforce them again. The above behavior of real ants has inspired many researchers to use the ant system models and algorithms in which a set of artificial ants cooperate for food by exchanging information via pheromone depositing either on the edges or on the vertices of the graph. Gutjahr [76, 75, 77] was the first to prove the convergence of the ant colonies algorithm.

This chapter deals mainly with applications of CE to deterministic COPs — as opposed noisy COPs, which will be discussed in Chapter 6 — with a particular emphasis on the max-cut problem and the TSP. The CE method for combinatorial optimization was motivated by the work [144], where an adaptive (variance minimization) algorithm for estimating probabilities of rare events for stochastic networks was presented. It was modified in [145] to solve continuous multi-extremal and combinatorial optimization problems. Continuous multi-extremal optimization is discussed in Chapter 5.

The main idea behind using CE for COPs is to first associate with each COP a rare-event estimation problem — the so-called *associated stochastic problem* (ASP) — and then to tackle this ASP efficiently using an adaptive algorithm similar to Algorithm 3.4.1. The principle outcome of this approach is the construction of a random sequence of solutions which converges probabilistically to the optimal or near-optimal solution of the COP.

As soon as the ASP is defined, the CE method involves the following two iterative phases:

1. Generation of a sample of random data (trajectories, vectors, etc.) according to a specified random mechanism.
2. Updating the parameters of the random mechanism, typically parameters of pdfs, on the basis of the data, to produce a "better" sample in the next iteration.

The ASP for COPs can usually be formulated in terms of graphs. Depending on the particular problem, we introduce the randomness in the ASP either in (a) the nodes or (b) the edges of the graph. In the former case, we speak of *stochastic node networks* (SNNs), in the latter case of *stochastic edge networks* (SENs). Notice that a similar terminology is used in Wagner, Lindenbaum, and Bruckstein [169] for the *graph covering* problem, called *vertex ant walk* (VAW) and *edge ant walk* (EAW), respectively.

Examples of SNN problems are the max-cut problem, the buffer allocation problem, and clustering problems. Examples of SEN problems are the travelling salesman problem, the quadratic assignment problem, different deterministic and stochastic scheduling problems and the clique problem.

In this and the subsequent chapters we demonstrate numerically the high accuracy of the CE method by applying it to different case studies from the World Wide Web. In particular, we show that the relative error of CE is typically within the limits of 1–2% of the best known solution. For some instances CE even produced better solutions than the best ones known. For this reason we purposely avoided comparing CE with other alternative heuristics such as simulated annealing and genetic algorithms. Our goal here is to demonstrate its reliability, simplicity, and high speed of convergence. Moreover, we wish to establish some mathematical foundations for further research and promote it for new applications. We note that various issues dealing with convergence have been discussed in Section 3.6. For other convergence results see [85, 113, 145].

Remark 4.1 (Comparison). To compare CE with some other optimization method, one could use the following criterion: Let

$$\delta = w_1 \frac{\widehat{\gamma}_T^{(1)}}{\widehat{\gamma}_T^{(2)}} + w_2 \frac{\tau^{(1)}}{\tau^{(2)}} , \tag{4.1}$$

where $\widehat{\gamma}_T^{(i)}$ is the optimal value of the objective function obtained by the method i, $i = 1, 2$, $\tau^{(i)}$ denotes the CPU time by the method i, $i = 1, 2$ and w_1 and w_2 are weight factors, with $w_1 + w_2 = 1$ If $\delta < 1$ then the first method is considered more efficient, and visa versa if $\delta > 1$. We leave it up to the decision maker to chose the parameter vector (w_1, w_2).

The rest of this chapter is organized as follows: In Section 4.2 we formulate the optimization framework for the CE method and present its main algorithm. In Sections 4.3 and 4.4 we discuss separately the SNN- and SEN-type combinatorial optimization problems. In Sections 4.5 through 4.8 we apply the CE method to the max-cut and partition problem, the TSP, and the quadratic assignment problem (QAP), respectively. Numerical results for SNN and SEN are presented in Sections 4.9 and 4.10, respectively. Finally, in Section 4.11 we give various auxiliary results, including faster tour generating algorithms for the TSP and a detailed discussion of sufficient conditions for convergence of a CE-like algorithm in which the updating is based on a single best sample, rather than the usual ϱN best samples.

4.2 The Main CE Algorithm for Optimization

The main idea of the CE method for optimization can be stated as follows. Suppose we wish to maximize some performance function $S(\mathbf{x})$ over all states $\mathbf{x}$ in some set $\mathcal{X}$. Let us denote the maximum by γ^*, thus

$$\gamma^* = \max_{\mathbf{x}\in\mathcal{X}} S(\mathbf{x}) . \tag{4.2}$$

First, we randomize our deterministic problem by defining a family of pdfs $\{f(\cdot;\mathbf{v}), \mathbf{v}\in\mathcal{V}\}$ on the set $\mathcal{X}$. Next, we associate with (4.2) the following estimation problem

$$\ell(\gamma) = \mathbb{P}_{\mathbf{u}}(S(\mathbf{X}) \geqslant \gamma) = \mathbb{E}_{\mathbf{u}} I_{\{S(\mathbf{X})\geqslant\gamma\}} , \tag{4.3}$$

the so-called *associated stochastic problem* (ASP). Here, $\mathbf{X}$ is a random vector with pdf $f(\cdot;\mathbf{u})$, for some $\mathbf{u}\in\mathcal{V}$ (for example, $\mathbf{X}$ could be a Bernoulli random vector) and γ is a known or unknown parameter. Note that there are in fact two possible estimation problems associated with (4.3). For a given γ we can estimate ℓ, or alternatively, for a given ℓ we can estimate γ, the root of (4.3). Let us consider the problem of estimating ℓ for a certain γ close to γ^*. Then, typically $\{S(\mathbf{X}) \geqslant \gamma\}$ is a rare event, and estimation of ℓ is a nontrivial problem. Fortunately, we can use the CE formulas (3.37)–(3.41) to solve this efficiently by making adaptive changes to the probability density function according to the Kullback-Leibler cross-entropy and thus, creating a sequence $f(\cdot;\mathbf{u}), f(\cdot;\mathbf{v}_1), f(\cdot;\mathbf{v}_2), \ldots$ of pdfs that are "steered" in the direction of the theoretically optimal density. The following proposition shows that the CE optimal density is often the atomic density at $\mathbf{x}^*$.

Proposition 4.2. Let γ^* be the maximum value of a real-valued function S on a finite set $\mathcal{X}$. Suppose that the corresponding maximizer is *unique*, say $\mathbf{x}^*$ and that the class of densities $\{f(\cdot;\mathbf{v})\}$ to be used in the CE program (3.23) contains the *atomic* or *degenerate* density with mass at $\mathbf{x}^*$:

$$\delta_{\mathbf{x}^*}(\mathbf{x}) = \begin{cases} 1, & \text{if } \mathbf{x} = \mathbf{x}^* , \\ 0, & \text{otherwise.} \end{cases} \tag{4.4}$$

Then, the solutions of both VM and CE programs (3.15) and (3.23) for estimating $\mathbb{P}(S(\mathbf{X}) \geqslant \gamma^*)$ coincide and correspond with $\delta_{\mathbf{x}^*}$.

Proof. Let ${}_*\mathbf{v}$ be such that

$$f(\cdot;{}_*\mathbf{v}) = \delta_{\mathbf{x}^*}(\cdot) . \tag{4.5}$$

Since the variance of the LR estimator $\widehat{\ell}$ given in (3.6), under ${}_*\mathbf{v}$, is *zero*, it is obvious that ${}_*\mathbf{v}$ is the optimal reference parameter for (3.15), with $H(\mathbf{x}) = I_{\{S(\mathbf{x})\geqslant\gamma^*\}}$. Since, by assumption, $\{f(\cdot;\mathbf{v})\}$ contains $f(\cdot;{}_*\mathbf{v})$, it is also immediate that $\mathcal{D}(\delta_{\mathbf{x}^*}, f(\cdot; \mathbf{v}^*)) = 0$ for $\mathbf{v}^* = {}_*\mathbf{v}$; in other words the CE and VM solutions coincide. □

Proposition 4.2 demonstrates the importance of finite support distributions when γ^* is the maximum value of S: The solution of both VM and CE programs to estimate $\mathbb{P}(S(\mathbf{X}) \geqslant \gamma^*)$ are often the same, *regardless of the distribution of* $\mathbf{X}$. In particular this holds when the degenerate density $\delta_{\mathbf{x}^*}$ given in (4.4) belongs to the family $\{f(\cdot; \mathbf{v})\}$ of the ASP, that is, when there exists some $\mathbf{v}^* \in \mathcal{V}$, such that (4.5) holds. This is typically the case for finite support distributions. This has important implications for combinatorial optimization. In particular, by associating the underlying combinatorial optimization problem (optimizing S) with a rare-event probability of the type $\mathbb{P}(S(X) \geqslant \gamma^*)$ we can obtain an estimate of the "degenerated vector" $\mathbf{v}^*$ via a CE algorithm similar to Algorithm 3.4.1. This in turn gives an estimate of the true optimal solution of the COP.

It is worth mentioning that the assumption of uniqueness of the maximizer of S in Proposition 4.2 can be artificially enforced by imposing some ordering on the finite set $\mathcal{X}$, say the lexicographical order. Then, we can define a function Z on $\mathcal{Y}$ as $Z(\mathbf{x}) = S(\mathbf{x}) - \varepsilon(\mathbf{x})$, where $\varepsilon(\mathbf{x})$ is a small perturbation, proportional to the rank of $\mathbf{x}$ and small enough to ensure that $Z(\mathbf{x}^1) > Z(\mathbf{x}^2)$ if $S(\mathbf{x}^1) > S(\mathbf{x}^2)$. In that case, the degenerate measure has mass at the element with highest rank within the set of maximizers of S.

Having defined the ASP, we wish to generate, analogously to Chapter 3, a sequence of tuples $\{(\widehat{\gamma}_t, \widehat{\mathbf{v}}_t)\}$, which converges quickly to a small neighborhood of the optimal tuple $(\gamma^*, \mathbf{v}^*)$. More specifically, we initialize by setting $\mathbf{v}_0 = \mathbf{u}$, choosing a not very small ϱ, say $\varrho = 10^{-2}$, and based on the fundamental CE formulas (3.37)–(3.41) we proceed as follows:

1. **Adaptive updating of** γ_t. For a fixed $\mathbf{v}_{t-1}$, let γ_t be the $(1-\varrho)$-quantile of $S(\mathbf{X})$ under $\mathbf{v}_{t-1}$. That is, γ_t satisfies (3.37) and (3.38).
   A simple estimator $\widehat{\gamma}_t$ of γ_t can be obtained by drawing a random sample $\mathbf{X}_1, \ldots, \mathbf{X}_N$ from $f(\cdot; \mathbf{v}_{t-1})$ and evaluating the sample $(1-\varrho)$-quantile of the performances
   $$\widehat{\gamma}_t = S_{(\lceil (1-\varrho)N \rceil)}\,, \tag{4.6}$$
   as in (3.39).
2. **Adaptive updating of** $\mathbf{v}_t$. For fixed γ_t and $\mathbf{v}_{t-1}$, derive $\mathbf{v}_t$ from the solution of the program
   $$\max_{\mathbf{v}} D(\mathbf{v}) = \max_{\mathbf{v}} \mathbb{E}_{\mathbf{v}_{t-1}} I_{\{S(\mathbf{X}) \geqslant \gamma_t\}} \ln f(\mathbf{X}; \mathbf{v})\,. \tag{4.7}$$
   The stochastic counterpart of (4.7) is as follows: for fixed $\widehat{\gamma}_t$ and $\widehat{\mathbf{v}}_{t-1}$, derive $\widehat{\mathbf{v}}_t$ from the following program
   $$\max_{\mathbf{v}} \widehat{D}(\mathbf{v}) = \max_{\mathbf{v}} \frac{1}{N} \sum_{i=1}^{N} I_{\{S(\mathbf{X}_i) \geqslant \widehat{\gamma}_t\}} \ln f(\mathbf{X}_i; \mathbf{v})\,. \tag{4.8}$$

It is important to observe that in contrast to the formulas (3.40) and (3.41) (for the rare-event setting), formulas (4.7) and (4.8) do not contain the likelihood ratio terms W. The reason is that in the rare-event setting the initial (nominal) parameter $\mathbf{u}$ is specified in advance and is an essential part of the estimation problem. In contrast, the initial reference vector $\mathbf{u}$ in the ASP is quite arbitrary. In fact, it is beneficial to change the ASP as we go along. In effect, by dropping the W term, we efficiently estimate at each iteration t the CE optimal reference parameter vector $\mathbf{v}_t$ for the rare-event probability $\mathbb{P}_{\mathbf{v}_t}(S(\mathbf{X}) \geqslant \gamma_t) \geq \mathbb{P}_{\mathbf{v}_{t-1}}(S(\mathbf{X}) \geqslant \gamma_t)$. Of course, we could also include here the W terms, but numerical experiments suggest that this will often lead to less reliable (noisy) estimates of $\mathbf{v}_t$ and $\mathbf{v}^*$.

Remark 4.3 (Smoothed Updating). Instead of updating the parameter vector $\mathbf{v}$ directly via the solution of (4.8) we use the following *smoothed* version

$$\widehat{\mathbf{v}}_t = \alpha \tilde{\mathbf{v}}_t + (1-\alpha)\widehat{\mathbf{v}}_{t-1} , \tag{4.9}$$

where $\tilde{\mathbf{v}}_t$ is the parameter vector obtained from the solution of (4.8), and α is called the *smoothing parameter*, with $0.7 < \alpha \leq 1$. Clearly, for $\alpha = 1$ we have our original updating rule. The reason for using the smoothed (4.9) instead of the original updating rule is twofold: (a) to smooth out the values of $\widehat{\mathbf{v}}_t$, (b) to reduce the probability that some component $\widehat{v}_{t,i}$ of $\widehat{\mathbf{v}}_t$ will be zero or one at the first few iterations. This is particularly important when $\widehat{\mathbf{v}}_t$ is a vector or matrix of *probabilities*. Note that for $0 < \alpha < 1$ we always have that $\widehat{v}_{t,i} > 0$, while for $\alpha = 1$ one might have (even at the first iterations) that either $\widehat{v}_{t,i} = 0$ or $\widehat{v}_{t,i} = 1$ for some indices i. As result, the algorithm will converge to a wrong solution.

Thus, the main CE optimization algorithm, which includes smoothed updating of parameter vector $\mathbf{v}$, can be summarized as follows.

Algorithm 4.2.1 (Main CE Algorithm for Optimization)

1. *Choose some* $\widehat{\mathbf{v}}_0$. *Set* $t = 1$ *(level counter).*
2. *Generate a sample* $\mathbf{X}_1, \ldots, \mathbf{X}_N$ *from the density* $f(\cdot; \mathbf{v}_{t-1})$ *and compute the sample* $(1-\varrho)$*-quantile* $\widehat{\gamma}_t$ *of the performances according to* (4.6).
3. *Use the* **same** *sample* $\mathbf{X}_1, \ldots, \mathbf{X}_N$ *and solve the stochastic program* (4.8). *Denote the solution by* $\widetilde{\mathbf{v}}_t$.
4. *Apply* (4.9) *to smooth out the vector* $\widetilde{\mathbf{v}}_t$.
5. *If for some* $t \geq d$, *say* $d = 5$,

$$\widehat{\gamma}_t = \widehat{\gamma}_{t-1} = \cdots = \widehat{\gamma}_{t-d} , \tag{4.10}$$

then **stop** *(let* T *denote the final iteration); otherwise set* $t = t + 1$ *and reiterate from Step 2.*

Remark 4.4 (Parameters setting). In our numerical studies with stochastic optimization problems we typically take $\varrho = 10^{-2}$, and the smoothing parameter $\alpha = 0.7$. We choose $N = C\,R$, where C is a constant, say $C = 5$, and R is the number of auxiliary distributional parameters introduced into the ASP and which need to be eventually estimated in the sense of degenerated values as per (4.4). For example, in the max-cut problem we take $R = n$ since, as we shall see below, we associate with each vertex an independent $\mathsf{Ber}(p_k)$, $k = 1, \ldots, n$ random variable. Similarly, in the TSP we take $R = n^2$ since here we associate an $n \times n$ probability matrix with the given $n \times n$ cost or distance matrix between the cities. Note that we assume that $n \geq 100$. Depending on whether we are dealing with a maximization or minimization problem the parameter $\eta = \varrho N$ corresponds to the number of sample performances S_j, $j = 1, \ldots, N$, that lie either in the upper or lower 100ϱ percentage. We will refer to the latter as *elite* samples. The choice of $\varrho = 10^{-2}$ allows Algorithm 4.5.2 to avoid local extrema and to settle down in the global one with high probability.

The suggestion to take $\alpha = 0.7, C = 5$ and $\varrho = 10^{-2}$ should be viewed merely as a rule of thumb. For some problems this combination of parameters will lead to the global optimum, but for other problems (α, C, ϱ) might need to be set for example to $(0.7, 5, 10^{-1})$ or $(0.3, 8, 10^{-2})$. A useful technique to find the "right" parameters is to run a number of small test problems that are derived from the original problem. For more details see Remark 4.13.

In analogy to Algorithm 3.4.2 we also present the deterministic version of Algorithm 4.2.1, which will be used below.

Algorithm 4.2.2 (Deterministic Version)

1. *Choose some* $\mathbf{v}_0$. *Set* $t = 1$.
2. *Calculate* γ_t *as*

$$\gamma_t = \max\left\{s \,:\, \mathbb{P}_{\mathbf{v}_{t-1}}(S(\mathbf{X}) \geqslant s) \geqslant \varrho\right\}. \tag{4.11}$$

3. *Calculate* $\mathbf{v}_t$ *as*

$$\mathbf{v}_t = \operatorname*{argmax}_{\mathbf{v}} \mathbb{E}_{\mathbf{v}_{t-1}} I_{\{S(\mathbf{X}) \geqslant \gamma_t\}} \ln f(\mathbf{X}; \mathbf{v}). \tag{4.12}$$

4. *If for some* $t \geq d$, *say* $d = 5$,

$$\gamma_t = \gamma_{t-1} = \cdots = \gamma_{t-d}, \tag{4.13}$$

then **stop** *(let T denote the final iteration); otherwise set $t = t + 1$ and reiterate from Step 2.*

The main advantage of the CE method versus variance minimization is that in most combinatorial optimization problems the updating of $\mathbf{v}_t$ (Step 3) can often be done *analytically*, that is, there is no need for numerical optimization. Note also that in general there are many ways to randomize the objective function $S(\mathbf{x})$ and thus to generate samples from $\mathscr{X}$, while estimating the rare-event probability $\ell(\gamma)$ of the ASP. Note finally that it is not always immediately clear which way of generating the sample will yield better results or easier updating formulas; this issue is more of an art than a science and will be discussed in some detail for various COPs in this and following chapters. We proceed next with adaptive updating of $\mathbf{v}_t$ for SNN-type and SEN-type optimization problems.

4.3 Adaptive Parameter Updating in Stochastic Node Networks

SNN-type problems can be formulated as follows. Consider a complete graph with n nodes, labeled $1, 2, \ldots, n$. Let $\mathbf{x} = (x_1, \ldots, x_n)$ be a vector which assigns some characteristic x_i to each node i. The vector $\mathbf{x}$ can be thought of as a "coloring" of the nodes. A typical example is the max-cut problem; see Section 2.5.1 and Section 4.5, where we have only two colors, $x_i \in \{0, 1\}$, which determine how the nodes are partitioned into two sets. In Figure 4.1 a node coloring is illustrated for the case $n = 5$ nodes and $m = 3$ colors, thus $x_i \in \{1, 2, 3\}$.

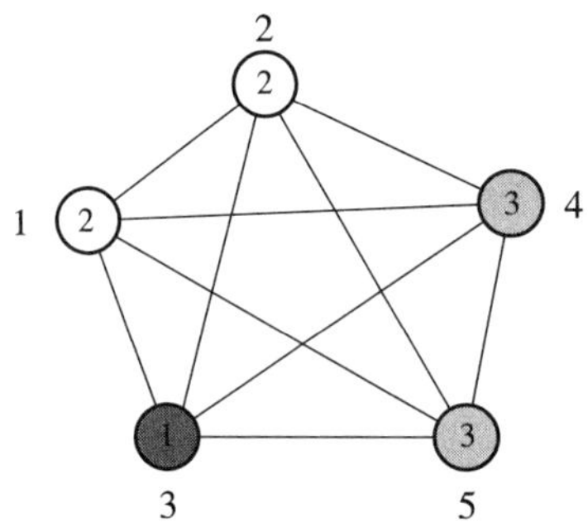

Fig. 4.1. Stochastic Node Network.

We assume that to each vector $\mathbf{x}$ is associated a cost $S(\mathbf{x})$, which we wish to optimize — say maximize — over all $\mathbf{x} \in \mathscr{X}$ by using Algorithm 4.2.1. Let us assume for simplicity that

$$\mathscr{X} = \{1, \ldots, m\}^n$$

The simplest mechanism for "randomizing" our deterministic problem into an ASP (4.3) is to draw the components of $\mathbf{X} = (X_1, \ldots, X_n)$ independently according to a discrete distribution $\mathbf{p} = \{p_{ij}\}$, where $p_{ij} = \mathbb{P}(X_i = j)$, $i = 1, \ldots, n$, $j = 1, \ldots, m$.

It follows directly from Example 3.6 that the analytical solution of the program (4.8) for the parameters p_{ij} is given by

$$\widehat{p}_{t,ij} = \frac{\sum_{k=1}^{N} I_{\{S(\mathbf{X}_k) \geqslant \widehat{\gamma}_t\}} I_{\{X_{ki}=j\}}}{\sum_{k=1}^{N} I_{\{S(\mathbf{X}_k) \geqslant \widehat{\gamma}_t\}}} . \tag{4.14}$$

Similarly, the deterministic updating formulas are

$$p_{t,ij} = \frac{\mathbb{E}_{\mathbf{p}_{t-1}} I_{\{S(\mathbf{X}) \geqslant \gamma_t\}} I_{\{X_i=j\}}}{\mathbb{E}_{\mathbf{p}_{t-1}} I_{\{S(\mathbf{X}) \geqslant \gamma_t\}}} , \tag{4.15}$$

where $\mathbf{p}_{t-1} = \{p_{t-1,ij}\}$ denotes the reference parameter (probability distribution) at the $(t-1)$-st iteration.

4.3.1 Conditional Sampling

In many SNN-type problems we wish to optimize some function S not over the whole set $\mathcal{X} = \{1, \ldots, m\}^n$ but rather over some *subset* $\mathcal{Y}$ of $\mathcal{X}$. For example, we may wish to consider only colorings of the graph in Figure 4.1 that use all three colors. Or, for the max-cut problem we may wish to consider only equal-sized node partitions. To accommodate these types of problems, we now formulate an important *generalization* of the sample generation and updating formulas above. Suppose we wish to maximize some function S, over a set

$$\mathcal{Y} \subset \{1, \ldots, m\}^n .$$

Define a new function $\tilde{S}(\mathbf{x}) = S(\mathbf{x})$, for $\mathbf{x} \in \mathcal{Y}$ and $\tilde{S}(\mathbf{x}) = -\infty$, else. Clearly, maximization of S over $\mathcal{Y}$ is equivalent to maximization of $\tilde{S}$ over $\mathcal{X}$. For the latter problem we can use the CE approach discussed above. That is, we draw the components of $\mathbf{X}$ independently according to a discrete distribution $\mathbf{p}$ and update the parameters p_{ij} according to (4.15), with S replaced by $\tilde{S}$. By conditioning on the event $\{\mathbf{X} \in \mathcal{Y}\}$, these updating rules can be written as

$$p_{t,ij} = \frac{\mathbb{E}_{\mathbf{p}_{t-1}} \left[I_{\{S(\mathbf{X}) \geqslant \gamma_t\}} I_{\{X_i=j\}} \,\middle|\, \mathbf{X} \in \mathcal{Y} \right]}{\mathbb{E}_{\mathbf{p}_{t-1}} \left[I_{\{S(\mathbf{X}) \geqslant \gamma_t\}} \,\middle|\, \mathbf{X} \in \mathcal{Y} \right]} , \tag{4.16}$$

where we have used the fact that $\tilde{S}(\mathbf{x}) = S(\mathbf{x})$ on $\mathcal{Y}$. In other words, the updating rule for this more general SNN problem is very similar to the original updating rule (4.15). The only difference is that we need to draw a sample from the conditional distribution of $\mathbf{X} \sim \mathbf{p}$ given $\mathbf{X} \in \mathcal{Y}$, instead of from the distribution $\mathbf{p}$. This can be done, for example, by using the acceptance–rejection method, or by more direct methods, see for example Sections 4.5 and 4.6.

4.3.2 Degenerate Distribution

Assume that the optimal solution γ^* to (4.2) for a certain SNN problem is unique, with corresponding argument $\mathbf{x}^*$. Then, using reasoning similar to Proposition 4.2 and (4.4), we can see that the CE and VM optimal distribution $\mathbf{p}$ for the ASP (4.3) is the *optimal degenerate* distribution $\mathbf{p}^*$ with

$$p_{ij}^* = \begin{cases} 1 \text{ if } & x_i^* = j, \\ 0 \text{ otherwise.} \end{cases} \tag{4.17}$$

As an example, consider the SNN in Figure 4.1, and suppose that the optimal vector $\mathbf{x}^*$ is given as shown, that is $\mathbf{x}^* = (2, 2, 1, 3, 3)$. Then we have $p_{12}^* = p_{22}^* = p_{31}^* = p_{43}^* = p_{53}^* = 1$ and all the other entries are 0.

As we already mentioned, we shall approximate the unknown true solution $(\gamma^*, \mathbf{p}^*)$ by the sequence of tuples $\{(\widehat{\gamma}_t, \widehat{\mathbf{p}}_t)\}$ generated by Algorithm 4.2.1.

4.4 Adaptive Parameter Updating in Stochastic Edge Networks

SEN-type problems can be formulated in a similar fashion to SNN-type problems. The main difference is that instead of assigning characteristics (colors, values) to nodes, we assign them to *edges* instead. For example, consider the first graph of Figure 4.2. Here we assign to each edge (i, j) a value x_{ij} which is either 1 (bold edge) or 0. Let $\mathbf{x} = \{x_{ij}\}$. Suppose each edge (i, j) has a weight or cost c_{ij}. A possible SEN optimization problem could be to find the maximum spanning tree, that is to maximize $S(\mathbf{x}) = \sum_{i,j} x_{ij} c_{ij}$ over the set $\mathscr{X}$ of all spanning trees. We can generate random vectors $\mathbf{X} \in \mathscr{X}$ via (conditional) independent sampling from a Bernoulli distribution in exactly the same way as described for SNN networks.

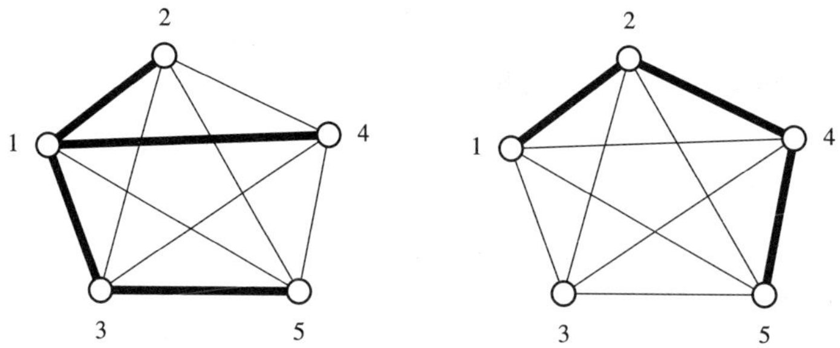

Fig. 4.2. Stochastic Edge Networks.

4.4.1 Parameter Updating for Markov Chains

A different SEN-type problem is illustrated in the second graph of Figure 4.2. Here we wish to construct a random *path* $\mathbf{X}$ through a graph with n nodes, labeled $1, 2, \ldots, n$. This random path can be represented in several ways. For example, the path through the sequence of nodes $x_1 \to x_2 \to \ldots \to x_m$ can be represented as the vector $\mathbf{x} = (x_1, \ldots, x_m)$. The second graph of Figure 4.2 depicts the $m = 4$-dimensional vector $\mathbf{x} = (1, 2, 4, 5)$ that represents the path $1 \to 2 \to 4 \to 5$ through a graph with $n = 5$ nodes. From now on we identify the path with its corresponding path vector.

A natural way to generate the path $\mathbf{X}$ is to draw $X_1, \ldots, X_m$ according to a *Markov chain.* Thus, in this case our random mechanism of generating samples from $\mathscr{X} = \{1, \ldots, n\}^m$ is determined by a *one-step transition matrix* $P = (p_{ij})$, such that

$$\mathbb{P}(X_{k+1} = j \,|\, X_k = i) = p_{ij} \ , \quad k = 1, \ldots, m-1 \ .$$

Suppose for simplicity that the initial distribution of the Markov chain is fixed. For example, the Markov chain always starts at 1. The logarithm of the density of $\mathbf{X}$ is given by

$$\ln f(\mathbf{x}; P) = \sum_{r=1}^{m} \sum_{i,j} I_{\{\mathbf{x} \in \mathscr{X}_{ij}(r)\}} \ln p_{ij} \ ,$$

where $\mathscr{X}_{ij}(r)$ is the set of all paths in $\mathscr{X}$ for which the r-th transition is from node i to j. The deterministic CE updating rules for this mechanism follow from (4.7), with P taking the role of the reference vector $\mathbf{v}$, with the extra condition that the rows of P need to sum up to 1. Using Lagrange multipliers we obtain analogously to (2.44)–(2.46) the updating formula

$$p_{t,ij} = \frac{\mathbb{E}_{P_{t-1}} I_{\{S(\mathbf{X}) \geqslant \gamma_t\}} \sum_{r=1}^{m} I_{\{\mathbf{X} \in \mathscr{X}_{ij}(r)\}}}{\mathbb{E}_{P_{t-1}} I_{\{S(\mathbf{X}) \geqslant \gamma_t\}} \sum_{r=1}^{m} I_{\{\mathbf{X} \in \mathscr{X}_{i}(r)\}}} \ , \tag{4.18}$$

where $\tilde{\mathscr{X}}_i(r)$ is the set of paths for which the r-th transition starts from node i. The corresponding estimator is:

$$\widehat{p}_{t,ij} = \frac{\sum_{k=1}^{N} I_{\{S(\mathbf{X}_k) \geqslant \widehat{\gamma}_t\}} \sum_{r=1}^{m} I_{\{\mathbf{X}_k \in \mathscr{X}_{ij}(r)\}}}{\sum_{k=1}^{N} I_{\{S(\mathbf{X}_k) \geqslant \widehat{\gamma}_t\}} \sum_{r=1}^{n} I_{\{\mathbf{X}_k \in \mathscr{X}_{i}(r)\}}} \ . \tag{4.19}$$

This has a very simple interpretation. To update p_{ij} we simply take the fraction of times that the transitions from i to j occurred, taking only those paths into account that have a performance greater than or equal to γ_t.

4.4.2 Conditional Sampling

It is important to realize that the derivation of the updating rules above assumes that the paths are generated via a Markov process and that S is optimized over the whole set $\mathscr{X}$. However, similar to the discussion of (4.16) for SNNs, many SEN-type problems involve optimization of S over a *subset* $\mathscr{Y}$ of $\mathscr{X}$. For example, in the second graph of Figure 4.2 one may wish to consider only nonintersecting paths; in particular, in the TSP only nonintersecting paths that visit all nodes and start and finish at the same node are allowed.

Fortunately, to tackle these kind of problems we may reason in exactly the same way as for (4.16). That is, we obtain for these problems the *same* updating formula (4.19), provided we sample from the conditional distribution of the original Markov chain given the event $\{\mathbf{X} \in \mathscr{Y}\}$. Note that for the TSP the sum $\sum_{r=1}^{m} I_{\{\mathbf{X}_k \in \mathscr{X}_i(r)\}}$ in (4.19) is equal to 1. As remarked earlier, sampling from the conditional distribution can be done via the acceptance–rejection method or via more direct techniques; examples are given in Section 4.7.

4.4.3 Optimal Degenerate Transition Matrix

Suppose that the optimal solution γ^* to (4.2) for a SEN problem is unique, with a corresponding path $\mathbf{x}^*$. Similar to (4.17) we can identify $\mathbf{x}^*$ with a "degenerate" transition matrix $P^* = (p^*_{ij})$. In particular, if the path $\mathbf{x}^*$ does not traverse the same link more than once, then for all i we have $p^*_{ij} = 1$ if $\mathbf{x}^*$ contains the link $i \to j$; and in that case $p^*_{ik} = 0$ for all $k \neq j$. For certain SENs, such as the TSP, the matrix P^* is truly "degenerate" (contains only zeros and ones), but for other problems such as the longest path problem some rows may remain unspecified.

4.5 The Max-Cut Problem

The maximal cut (max-cut) problem in a graph can be formulated as follows. Given a graph $G(V, E)$ with set of nodes $V = \{1, \dots, n\}$ and set of edges E between the nodes, partition the nodes of the graph into two arbitrary subsets V_1 and V_2 such that the sum of the weights (costs) c_{ij} of the edges going from one subset to the other is maximized. Note that some of the c_{ij} may be 0, indicating that there is, in fact, no edge from i to j. For simplicity we assume the graph is not directed, see also Remark 4.5. In that case C is a *symmetric* matrix.

As an example, consider the graph in Figure 4.3.

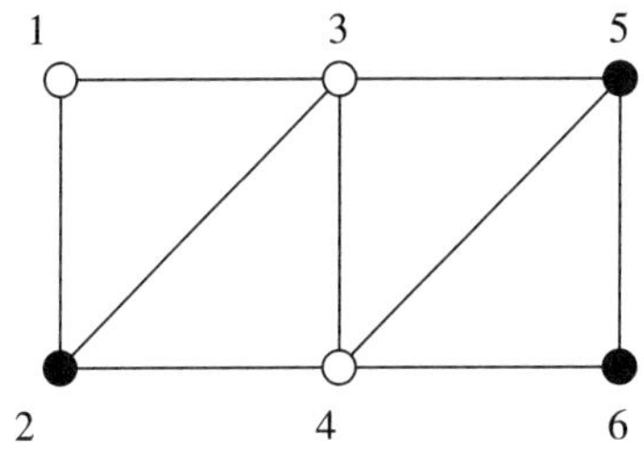

Fig. 4.3. A 6-node network with the cut $\{\{1,3,4\},\{2,5,6\}\}$.

with the corresponding cost matrix

$$C = \begin{pmatrix} 0 & c_{12} & c_{13} & 0 & 0 & 0 \\ c_{21} & 0 & c_{23} & c_{24} & 0 & 0 \\ c_{31} & c_{32} & 0 & c_{34} & c_{35} & 0 \\ 0 & c_{42} & c_{43} & 0 & c_{45} & c_{46} \\ 0 & 0 & c_{53} & c_{54} & 0 & c_{56} \\ 0 & 0 & 0 & c_{64} & c_{65} & 0 \end{pmatrix} \tag{4.20}$$

where $c_{ij} = c_{ji}$. Here the cut $\{\{1,3,4\},\{2,5,6\}\}$ has cost

$$c_{12} + c_{32} + c_{42} + c_{45} + c_{46} + c_{53} \; .$$

As explained in Section 2.5.1, a cut can be conveniently represented by its corresponding *cut vector* $\mathbf{x} = (x_1, \ldots, x_n)$, where $x_i = 1$ if node i belongs to same partition as 1, and 0 else. For example the cut in Figure 4.3 can be represented via the cut vector $(1,0,1,1,0,0)$. For each cut vector $\mathbf{x}$, let $\{V_1(\mathbf{x}), V_2(\mathbf{x})\}$ be the partition of V induced by $\mathbf{x}$, such that $V_1(\mathbf{x})$ contains the set of indices $\{i : x_i = 1\}$. If not stated otherwise we set $x_1 = 1 \in V_1$.

Let $\mathscr{X}$ be the set of all cut vectors $\mathbf{x} = (1, x_2, \ldots, x_n)$ and let $S(\mathbf{x})$ be the corresponding cost of the cut. Then,

$$S(\mathbf{x}) = \sum_{i \in V_1(\mathbf{x}),\, j \in V_2(\mathbf{x})} c_{ij} \; . \tag{4.21}$$

It is readily seen that the total number of cuts is

$$|\mathscr{X}| = 2^{n-1} - 1 \; . \tag{4.22}$$

Remark 4.5 (Directed graphs). It is important to note that we always assume the graph to be undirected. For completeness we mention that for a *directed* graph the cost of a cut $\{V_1, V_2\}$ includes both the cost of the edges from V_1 to V_2 and from V_2 to V_1. In this case the cost corresponding to a cut vector $\mathbf{x}$ is therefore

$$S(\mathbf{x}) = \sum_{i \in V_1(\mathbf{x}),\, j \in V_2(\mathbf{x})} c_{ij} + c_{ji} \; . \tag{4.23}$$

We proceed next with random cut generation and update the corresponding parameter vector $\{(\gamma_t, \mathbf{p}_t)\}$ using the CE Algorithm 4.2.1.

Since the max-cut problem is a typical SNN-type optimization problem the most natural and easiest way to generate the cut vectors is to let $X_2, \ldots, X_n$ be independent Bernoulli random variables with success probabilities $p_2, \ldots, p_n$.

Algorithm 4.5.1 (Random Cut Generation)

1. *Generate an n-dimensional random vector $\mathbf{X} = (X_1, \ldots, X_n)$ from $\mathsf{Ber}(\mathbf{p})$ with independent components, where $\mathbf{p} = (1, p_2, \ldots, p_n)$.*
2. *Construct the partition $\{V_1(\mathbf{X}), V_2(\mathbf{X})\}$ of V and calculate the performance $S(\mathbf{X})$ as in* (4.21).

It immediately follows from (4.14) that the updating formula in Algorithm 4.2.1 at the t-th iteration is given by

$$\widehat{p}_{t,i} = \frac{\sum_{k=1}^{N} I_{\{S(\mathbf{X}_k) \geqslant \widehat{\gamma}_t\}} \, I_{\{X_{ki}=1\}}}{\sum_{k=1}^{N} I_{\{S(\mathbf{X}_k) \geqslant \widehat{\gamma}_t\}}} \,, \tag{4.24}$$

$i = 2, \ldots, n$.

The resulting algorithm for estimating the pair $(\gamma^*, \mathbf{p}^*)$ can be presented as follows.

Algorithm 4.5.2 (Main Algorithm for Max-Cut)

1. *Choose an initial reference vector $\widehat{\mathbf{p}}_0$ with components $\widehat{p}_{0,1} = 1$, $\widehat{p}_{0,i} = \frac{1}{2}$, $i = 2, \ldots, n$. Set $t = 1$.*
2. *Generate a sample $\mathbf{X}_1, \ldots, \mathbf{X}_N$ of cut vectors via Algorithm 4.5.1, with $\mathbf{p} = \widehat{\mathbf{p}}_{t-1}$, and compute the sample $(1-\varrho)$-quantile $\widehat{\gamma}_t$ of the performances according to* (4.6).
3. *Use the* **same** *sample $\mathbf{X}_1, \ldots, \mathbf{X}_N$ to update $\widehat{\mathbf{p}}_t$ via* (4.24).
4. *Apply* (4.9) *to smooth out the vector $\widehat{\mathbf{p}}_t$.*
5. *If for some $t \geq d$, say $d = 5$,*

$$\widehat{\gamma}_t = \widehat{\gamma}_{t-1} = \cdots = \widehat{\gamma}_{t-d} \,, \tag{4.25}$$

then **stop** *(let T denote the final iteration); otherwise set $t = t + 1$ and reiterate from Step 2.*

As an alternative to the estimate $\widehat{\gamma}_T$ of γ^* and to the stopping rule in (4.25) one can consider the following:

4. (*) If for some $t = T \geq d$ and some d, say $d = 5$, (4.25) holds, stop and deliver

$$\widehat{\gamma}_T = \max_{0 \leq s \leq T} \widehat{\gamma}_s \tag{4.26}$$

as an estimate of γ^*. Otherwise, set $t = t + 1$ and go to Step 2.

Remark 4.6 (Parallel Computing). The CE method is ideally suited to parallel computation. Assume that we have r parallel processors and consider, for example, the max-cut problem. We can speed up Algorithm 4.5.1 for random cut generation and the associated calculation of the performance function S by a factor of r by generating on each processor N/r cuts instead of generating N cuts on a single processor. Since the time required to calculate $(\widehat{\gamma}_t, \widehat{\mathbf{p}}_t)$ is negligible compared to the time needed to generate the random cuts and evaluate the performances, parallel computation will speed up the CE Algorithm 4.5.2 by a factor of r, approximately. Similar speed-ups can be achieved for other combinatorial optimization problems.

The following toy example illustrates the performance of the deterministic version of Algorithm 4.5.2 step by step for a simple 5-node max-cut problem, where the total number of cuts is $|\mathcal{X}| = 2^{n-1} - 1 = 15$; see (4.22). The small size of the problem allows us to make all calculations analytically, that is using directly the deterministic Algorithm 4.2.2 (see (4.11), (4.12)) rather than their stochastic counterparts.

Example 4.7 (Illustration of Algorithm 4.2.2). Consider the 5-node graph presented in Figure 4.4. The 15 possible cuts and the corresponding cut values are given in Table 4.1.

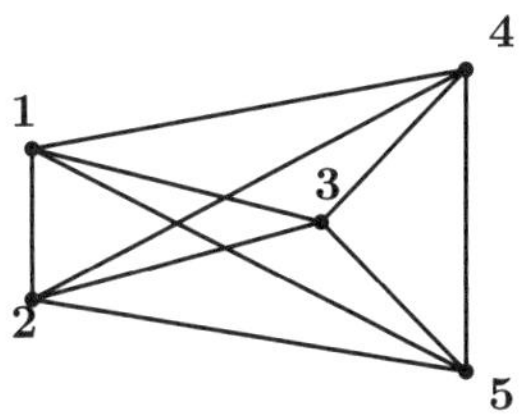

$$C = \begin{pmatrix} 0 & 1 & 3 & 5 & 6 \\ 1 & 0 & 3 & 6 & 5 \\ 3 & 3 & 0 & 2 & 2 \\ 5 & 6 & 2 & 0 & 2 \\ 6 & 5 & 2 & 2 & 0 \end{pmatrix} \quad (4.27)$$

Fig. 4.4. A 5-node network with cost matrix C.

Table 4.1. The 15 possible cuts of Example 4.7.

$\mathbf{X}$	V_1	V_2	$S(\mathbf{X})$
(1,0,0,0,0)	$\{1\}$	$\{2,3,4,5\}$	15
(1,1,0,0,0)	$\{1,2\}$	$\{3,4,5\}$	28
(1,0,1,0,0)	$\{1,3\}$	$\{2,4,5\}$	19
(1,0,0,1,0)	$\{1,4\}$	$\{2,3,5\}$	20
(1,0,0,0,1)	$\{1,5\}$	$\{2,3,4\}$	18
(1,1,1,0,0)	$\{1,2,3\}$	$\{4,5\}$	26
(1,1,0,1,0)	$\{1,2,4\}$	$\{3,5\}$	21
(1,1,0,0,1)	$\{1,2,5\}$	$\{3,4\}$	21
(1,0,1,1,0)	$\{1,3,4\}$	$\{2,5\}$	20
(1,0,1,0,1)	$\{1,3,5\}$	$\{2,4\}$	18
(1,0,0,1,1)	$\{1,4,5\}$	$\{2,3\}$	19
(1,1,1,1,0)	$\{1,2,3,4\}$	$\{5\}$	15
(1,1,1,0,1)	$\{1,2,3,5\}$	$\{4\}$	15
(1,1,0,1,1)	$\{1,2,4,5\}$	$\{3\}$	10
(1,0,1,1,1)	$\{1,3,4,5\}$	$\{2\}$	15
(1,1,1,1,1)	$\{1,2,3,4,5\}$	$\emptyset$	0

It follows that in this case the optimal cut vector is $\mathbf{x}^* = (1,1,0,0,0)$ with $S(\mathbf{x}^*) = \gamma^* = 28$.

We shall show next that in the deterministic Algorithm 4.2.2 adapted to the max-cut problem the parameter vectors $\mathbf{p}_0, \mathbf{p}_1, \ldots$ converge to the optimal $\mathbf{p}^* = (1,1,0,0,0)$ after two iterations, when $\varrho = 10^{-1}$ and $\mathbf{p}_0 = (1, 1/2, 1/2, 1/2, 1/2)$.

Iteration 1

In the first step of the first iteration, we have to determine γ_1 from

$$\gamma_t = \max \left\{\gamma \,:\, \mathbb{E}_{\mathbf{p}_{t-1}} I_{\{S(\mathbf{X}) \geqslant \gamma\}} \geqslant 0.1\right\} . \tag{4.28}$$

It is readily seen that under the parameter vector $\mathbf{p}_0$, $S(\mathbf{X})$ takes values, $\{0, 10, 15, 18, 19, 20, 21, 26, 28\}$ with probabilities $\{1/16, 1/16, 4/16, 2/16, 2/16, 2/16, 2/16, 1/16, 1/16\}$. Hence, we find $\gamma_1 = 26$. In the second step, we need to solve

$$\mathbf{p}_t = \operatorname*{argmax}_{\mathbf{p}} \mathbb{E}_{\mathbf{p}_{t-1}} I_{\{S(\mathbf{X}) \geqslant \gamma_t\}} \ln f(\mathbf{X}; \mathbf{p}) , \tag{4.29}$$

which has the solution

$$p_{t,i} = \frac{\mathbb{E}_{\mathbf{p}_{t-1}} I_{\{S(\mathbf{X}) \geqslant \gamma_t\}} X_i}{\mathbb{E}_{\mathbf{p}_{t-1}} I_{\{S(\mathbf{X}) \geqslant \gamma_t\}}} .$$

There are only two vectors $\mathbf{x}$ for which $S(\mathbf{x}) \geqslant 26$, namely $(1,1,1,0,0)$ and $(1,1,0,0,0)$, and both have probability $1/16$ under $\mathbf{p}_0$. Thus,

$$p_{1,i} = \begin{cases} \dfrac{2/16}{2/16} = 1 & \text{for } i = 1, 2 \,, \\ \dfrac{1/16}{2/16} = \dfrac{1}{2} & \text{for } i = 3 \,, \\ \dfrac{0}{2/16} = 0 & \text{for } i = 4, 5 \,. \end{cases}$$

Iteration 2

In the second iteration, $S(\mathbf{X})$ is 26 or 28 with probability $1/2$. Applying (4.28) and (4.29) again yields the optimal $\gamma_2 = 28$ and the optimal $\mathbf{p}_2 = (1,1,0,0,0)$, respectively.

Remark 4.8 (Alternative Stopping Rule). Note that the stopping rule (4.25), which is based on convergence of the sequence $\{\widehat{\gamma}_t\}$ to γ^*, stops Algorithm 4.5.2 when the sequence $\{\widehat{\gamma}_t\}$ does not change. As an alternative stopping rule we consider the one which is based on convergence of sequence of $\{\widehat{\mathbf{p}}_t\}$, rather than on convergence of $\{\widehat{\gamma}_t\}$. According to that stopping rule, Algorithm 4.5.2 stops when the sequence $\{\widehat{\mathbf{p}}_t\}$ is very close to the optimal "degenerated," (i.e, zero–one) reference vector $\mathbf{p}^*$, provided $\mathbf{p}^*$ is a unique solution.

Formally, let $\widehat{p}_{t,(1)} \leqslant \cdots \leqslant \widehat{p}_{t,(N)}$ denote the ordered components of the vector
$\widehat{\mathbf{p}}_t = (\widehat{p}_{t,1}, \ldots, \widehat{p}_{t,n})$. Let $\widehat{A}_t = (i_1, \ldots, i_n)$ be the corresponding sequence of indices. Thus,

$$\widehat{p}_{t,i_k} = \widehat{p}_{t,(k)} \;, \quad k = 1, \ldots, n \;.$$

Let τ denote the number of unities in the optimal degenerate (binary) reference vector. Denote by $\widehat{\tau}_t$ the minimal index i in $\widehat{A}_t$ for which $\widehat{p}_{t,i} > 1/2$. Thus, by definition, $\widehat{p}_{t,(i)} > 1/2$ for $i \geqslant \widehat{\tau}_t$ and $\widehat{p}_{t,(i)} \leqslant 1/2$ for $i < \widehat{\tau}_t$. It follows from the above that for each t, the index $\widehat{\tau}_t$ can be considered as an estimate of the true τ. For a given sequence $\widehat{A}_t$, let $\widehat{A}_t(k, \ldots, N)$ denote the sequence consisting of the last k elements of $\widehat{A}_t$. As our alternative stopping rule we use the following: Stop if for some $t = T \geq d$ and some d, say $d = 5$,

$$\widehat{A}_t(\widehat{\tau}_t, \ldots, N) = \widehat{A}_{t-1}(\widehat{\tau}_{t-1}, \ldots, N) = \cdots = \widehat{A}_{t-d}(\widehat{\tau}_{t-d}, \ldots, N), \;\; t > d \;. \tag{4.30}$$

In other words, we stop if the indices of the components of the vector $(\widehat{p}_{t,(1)}, \ldots, \widehat{p}_{t,(n)})$ that are greater than $1/2$ remain the same for d consecutive iterations.

The Max-Cut Problem with r Partitions

We can readily extend the max-cut procedure to the case where the node set V is partitioned into $r > 2$ subsets $\{V_1, \ldots, V_r\}$ such that the sum of edge weights between any pair of subsets V_a and V_b, $a, b = 1, \ldots, r$, $(a < b)$ is maximized. Thus for each partition $\{V_1, \ldots, V_r\}$ the value of the objective function is

$$\sum_{a=1}^{r} \sum_{b=a+1}^{r} \sum_{i \in V_a,\, j \in V_b} c_{ij} \;.$$

In this case one can follow the basic steps of Algorithm 4.2.1 using independent r-point distributions instead of independent Bernoulli distributions and update the probabilities exactly as in (4.14).

4.6 The Partition Problem

The partition problem is similar to the max-cut problem. The only difference is that the *size* of each class is *fixed* in advance. This has implications for the trajectory generation. Consider, for example, a partition problem in which V has to be partitioned into two *equal* sets, assuming n is even. We cannot simply use Algorithm 4.5.1 for the random cut generation, since most partitions generated in that way will have unequal size.

We describe next a simple algorithm for the generation of a random partition $\{V_1, V_2\}$ with exactly m elements in V_1 and $n - m$ elements in V_2. This is called a *bipartition* generation problem. Extension of the algorithm to r-partition generation is simple.

For a given vector $\mathbf{p} = (p_1, \ldots, p_n)$, with p_1 not fixed to 1 this time, we draw the partition vector $\mathbf{X} = (X_1, \ldots, X_n)$ according to $\mathbf{X} \sim \mathsf{Ber}(\mathbf{p})$ with independent components *conditional* upon the event that $\mathbf{X}$ has exactly m ones and $n-m$ zeros. To achieve this we introduce first an auxiliary mechanism which draws a random permutation $\Pi = (\Pi_1, \ldots, \Pi_n)$ of $(1, \ldots, n)$ uniformly over the space of all permutations. Drawing such a random permutation is very easy: simply let $U_1, \ldots, U_n$ be i.i.d. $\mathsf{U}(0,1)$−distributed. Then, arrange the U_i's in nondecreasing order, that is, as $U_{\Pi_1} \leqslant U_{\Pi_2} \leqslant \cdots \leqslant U_{\Pi_n}$, and take finally the indices $\Pi = (\Pi_1, \ldots, \Pi_n)$ as the required random permutation.

We demonstrate our algorithm first for a 5-node network assuming $m = 2$ and $m - n = 3$. Similar to the max-cut problem we assume that the vector $\mathbf{p} = (p_1, \ldots, p_5)$ in $\mathsf{Ber}(\mathbf{p})$ is given and then proceed as follows.

1. Generate a random permutation $\Pi = (\Pi_1, \ldots, \Pi_5)$ of $(1, \ldots, 5)$ uniformly over the space of all such permutations. Let $\Pi = (\pi_1, \ldots, \pi_5)$ be a particular outcome. Suppose, for example, that the permutation is $(\pi_1, \ldots, \pi_5) = (3, 5, 1, 2, 4)$.
2. Given $\Pi = (\pi_1, \ldots, \pi_5)$ generate independent Bernoulli random variables $X_{\pi_1}, X_{\pi_2}, \ldots$ from $\mathsf{Ber}(p_{\pi_1})$, $\mathsf{Ber}(p_{\pi_2}), \ldots$, respectively, *until either exactly* $m = 2$ unities or $n - m = 3$ zeros are generated. Note that the number of samples is a random variable with the range from $\min\{m, n - m\}$ to n. Suppose, for example, that the first four independent Bernoulli samples from the above $\mathsf{Ber}(p_3), \mathsf{Ber}(p_5), \mathsf{Ber}(p_1), \mathsf{Ber}(p_2)$ result in the following outcome $(0, 0, 1, 0)$. Since we already generated three zeros we can set $X_4 \equiv 1$ and deliver $\{V_1(\mathbf{X}), V_2(\mathbf{X})\} = \{(1, 4), (2, 3, 5)\}$ as the desired partition.
3. If in the previous step $m = 2$ unities are generated, set the remaining three elements to zero; if on the other hand three zeros are generated, set the remaining two elements to unities and deliver $\mathbf{X} = (X_1, \ldots, X_n)$ as the final partition vector. Construct the partition $\{V_1(\mathbf{X}), V_2(\mathbf{X})\}$ of V.

With this example at hand, the random partition generation algorithm can be written as follows:

Algorithm 4.6.1 (Random Partition Generation Algorithm)

1. *Generate a random permutation* $\Pi = (\Pi_1, \ldots, \Pi_n)$ *of* $(1, \ldots, n)$ *uniformly over the space of all such permutations.*
2. *Given* $\Pi = (\pi_1, \ldots, \pi_n)$*, independently generate Bernoulli random variables* X_{π_1}, $X_{\pi_2}, \ldots$ *from* $\mathsf{Ber}(p_{\pi_1}), \mathsf{Ber}(p_{\pi_2}), \ldots$*, respectively,* until m *unities or* $n - m$ *zeros are generated.*
3. *If in the previous step* m *unities are generated, set the remaining elements to zero; if on the other hand* $n-m$ *unities are generated, set the remaining elements to one. Deliver* $\mathbf{X} = (X_1, \ldots, X_n)$ *as the final partition vector.*
4. *Construct the partition* $\{V_1(\mathbf{X}), V_2(\mathbf{X})\}$ *of* V *and calculate the performance* $S(\mathbf{X})$ *according to* (4.21).

In effect, what Algorithm 4.6.1 does is sample from the conditional distribution of $\mathbf{X} \sim \mathsf{Ber}(\mathbf{p})$ conditional upon the fact that $\sum X_i = m$ or $n - m$. It follows, as in (4.16), that the updating formula for the reference vector $\mathbf{p}$ remains *exactly the same* as in (4.24).

Remark 4.9 (Alternative Stopping Rule). Similar to the max-cut case, we can formulate an alternative stopping rule, which is based on convergence of sequence of $\{\widehat{\mathbf{p}}_t\}$, rather than on convergence of $\{\widehat{\gamma}_t\}$. Since in the partition problem the number of nodes in the partition is fixed, the stopping criterion is simpler. Indeed, consider the bipartition problem where the sizes of the partitions are m and $n - m$, and the node 1 belongs to the partition of size m, $(X_1 = 1)$. Then in analogy with (4.30) we can use the following stopping criterion: Stop if for some $t = T \geq d$ and some d, say $d = 5$,

$$\begin{aligned} \widehat{A}_t(N - m + 1, \ldots, N) &= \widehat{A}_{t-1}(N - m + 1, \ldots, N) \\ &= \cdots = \widehat{A}_{t-d}(N - m + 1, \ldots, N), \ t > d\,. \end{aligned} \tag{4.31}$$

In other words, we stop when the indices of the first m largest components of the ordered vector $\widehat{\mathbf{p}}_t$ coincide d times in a row.

4.7 The Travelling Salesman Problem

The objective of the travelling salesman problem (TSP), see Section 2.5.2, is to find the shortest tour in a graph containing n nodes. We will assume that the graph is *complete*, that is each node is connected to each other node. The TSP can be formulated as follows. Find:

$$\min_{\mathbf{x}} S(\mathbf{x}) = \min_{\mathbf{x}} \left\{ \sum_{i=1}^{n-1} c_{x_i, x_{i+1}} + c_{x_n, x_1} \right\}, \tag{4.32}$$

where $\mathbf{x} = (x_1, x_2, \ldots, x_n)$ denotes a permutation of $(1, \ldots, n)$, x_i, $i = 1, 2, \ldots, n$ is the i-th city to be visited in the tour represented by $\mathbf{x}$, and c_{ij} is the distance (travelling cost) from city i to city j.

It will be convenient to assume that the graph is *complete*, that is, that each node is connected to each other node. Note that this does not lead to loss of generality since for a noncomplete graph we can always add additional edges with infinite cost without affecting the minimal tour. This is illustrated in Figure 4.5, with the corresponding cost matrix

$$C = \begin{pmatrix} \infty & c_{12} & c_{13} & \infty & \infty & c_{16} \\ c_{21} & \infty & c_{23} & c_{24} & \infty & \infty \\ c_{31} & c_{32} & \infty & \infty & c_{35} & \infty \\ \infty & c_{42} & \infty & \infty & c_{45} & c_{46} \\ \infty & \infty & c_{53} & c_{54} & \infty & c_{56} \\ c_{61} & \infty & \infty & c_{64} & c_{65} & \infty \end{pmatrix} . \tag{4.33}$$

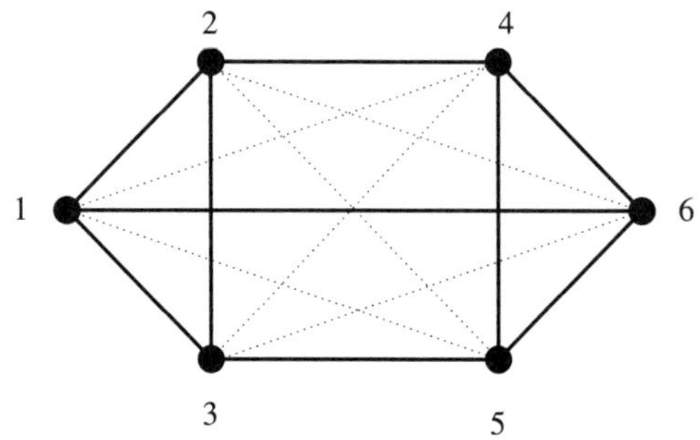

Fig. 4.5. A TSP with 6 cities. Dotted edges have ∞ cost.

Note that each permutation $\mathbf{x} = (x_1, \ldots, x_n)$ corresponds to a unique tour $x_1 \to x_2 \to \ldots \to x_n \to x_1$, so that there will be no confusion when we call $\mathbf{x}$ a "tour." For simplicity we always assume that $x_1 = 1$. For example, the TSP represented by (4.33) has only the following 6 finite-cost tours/permutations:

$$(1,2,3,5,4,6), \quad (1,2,4,6,5,3), \quad (1,3,2,4,5,6), \tag{4.34}$$

and

$$(1,6,4,5,3,2), \quad (1,3,5,6,4,2), \quad (1,6,5,4,2,3), \tag{4.35}$$

out of a total of $5! = 120$ possible tours.

Remark 4.10 (Forward and Backward Loop). For A TSP with a *symmetric* cost matrix C there are always at least *two* optimal tours. For example, if in Figure 4.5 the cost matrix is symmetric, and if $(1,2,3,5,4,6)$ is an optimal tour, then $(1,6,4,5,3,2)$ is also an optimal tour. We can call one tour the *forward* tour and the other the *backward* tour. To distinguish between the forward and the backward tour one can use the following simple rule: for a forward tour $x_2 < x_n$ and for a backward tour $x_2 > x_n$.

Let $\mathcal{X}$ be the set of all possible tours and let $S(\mathbf{x})$ in (4.32) be the length of a tour $\mathbf{x} \in \mathcal{X}$. For example, the length of the tour (1,2,3,5,4,6) is

$$S((1,2,3,5,4,6)) = c_{12} + c_{23} + c_{35} + c_{54} + c_{46} + c_{61} \, .$$

Our goal is to minimize S in (4.32) over the set $\mathcal{X}$ using the CE method. To minimize S we need to specify a mechanism for the generation of random tours $\mathbf{X} \in \mathcal{X}$, and, as usual, to adopt the basic cross-entropy Algorithm 4.2.1 to update the parameters of the distributions associated with this mechanism.

Taking into the account that the travelling salesman problem presents a SEN-type optimization problem we present now our first random trajectory generation algorithm, which is based on the transitions through the nodes of the network. More specifically, to generate a random tour $\mathbf{X} = (X_1, \ldots, X_n)$ we use an $n \times n$ transition probability matrix P and then proceed according to Algorithm 2.5.1, which is repeated below as Algorithm 4.7.1 for ease of reference. Recall that as usual we include in the trajectory generation algorithm an extra step for the objective function calculation.

Algorithm 4.7.1 (Trajectory Generation Using Node Transitions)

1. *Define* $P^{(1)} = P$ *and* $X_1 = 1$. *Let* $k = 1$.
2. *Obtain the matrix* $P^{(k+1)}$ *from* $P^{(k)}$ *by first setting the* X_k*-th column of* $P^{(k)}$ *to 0 and then normalizing the rows to sum up to 1. Generate* X_{k+1} *from the distribution formed by the* X_k*-th row of* $P^{(k+1)}$.
3. *If* $k = n - 1$ *then* **stop**; *otherwise set* $k = k + 1$ *and reiterate from Step 2.*
4. *Evaluate the length of the tour via* (4.32).

A more detailed description of this algorithm is given in Algorithm 4.11.1 in the appendix to this chapter.

Remark 4.11 (Starting City). Algorithm 4.7.1 can be easily modified by starting it from different cities, instead of always from city 1. For example, we could "cycle" through the starting cities for each consecutive tour generation. Alternatively, we could generate the initial city according to some random mechanism, for example, choose the starting city by drawing from a discrete uniform distribution on $\{1, \ldots, n\}$. A possible reason for not always wanting to start at 1 is that such a choice could introduce some bias toward the way in which the trajectories are generated. However, we have never observed this in practice.

It is crucial to understand that Algorithm 4.7.1 above is an example of the generating rule introduced in Section 4.4. Specifically, in Algorithm 4.7.1 we generate a stochastic process $\{X_1, \ldots, X_n\}$ according to the conditional distribution of the original Markov chain, *given* that the path $(X_1, \ldots, X_n)$ constitutes a genuine tour, that is, each node except for 1 is visited only once. Note that the stochastic process $\{X_1, \ldots, X_n\}$ *is no longer a Markov process.*

Note also that a Markov chain with transition matrix P would in general not yield tours because nodes could be repeated. The situation with $(X_1, \ldots, X_n)$ is somewhat akin to the classical urn problem in which a number m of balls is chosen from an urn with n balls numbered $1, \ldots n$. The original Markov chain would correspond to sampling with replacement, whereas the "conditional" process would correspond to sampling *without* replacement. For this reason we call the second process a *Markov chain with replacement* (MCWR), although formally it is not a Markov chain. As explained in Section 4.4.2 (and, in a more informal way, in Section 2.5.2), using the MCWR instead of the original Markov chain will not change the updating rules for the transition matrix P, because we are sampling from the conditional distribution of the Markov chain, given that the paths form a genuine tour.

Therefore, in analogy to (4.18) and (4.19) we update the components of P for the deterministic and stochastic case as

$$p_{t,ij} = \frac{\mathbb{E}_{P_{t-1}} I_{\{S(\mathbf{X}) \leqslant \gamma_t\}} I_{\{\mathbf{X} \in \mathcal{X}_{ij}\}}}{\mathbb{E}_{P_{t-1}} I_{\{S(\mathbf{X}) \leqslant \gamma_t\}}}, \tag{4.36}$$

and

$$\widehat{p}_{t,ij} = \frac{\sum_{k=1}^{N} I_{\{S(\mathbf{X}_k) \leqslant \widehat{\gamma}_t\}} \, I_{\{\mathbf{X}_k \in \mathcal{X}_{ij}\}}}{\sum_{k=1}^{N} I_{\{S(\mathbf{X}_k) \leqslant \widehat{\gamma}_t\}}}, \tag{4.37}$$

respectively. Here $\mathcal{X}_{ij}$ denotes the set of tours for which the transition from node i to node j is made.

Let us return to Algorithm (4.7.1) and consider the corresponding optimal degenerate transition matrix P^* (see Section 4.4.3). Suppose γ^* is the optimal length of the smallest tour, and there is only one such tour $\mathbf{x}^*$ with length γ^*. It is not difficult to see that for any P_{t-1} the solution of the CE program (4.8) is given by the optimal degenerate transition matrix (ODTM) $P^* = (p^*_{ij})$, given by

$$p^*_{ij} = \begin{cases} 1 \text{ if } & \mathbf{x}^* \in \mathcal{X}_{ij}, \\ 0 \text{ otherwise.} \end{cases}$$

As an example, consider the TSP in Figure 4.5. If, for instance, the trajectory

$$1 \to 2 \to 3 \to 5 \to 4 \to 6 \to 1$$

corresponds to the shortest tour, then we have

$$P^* = \begin{pmatrix} 0\,1\,0\,0\,0\,0 \\ 0\,0\,1\,0\,0\,0 \\ 0\,0\,0\,0\,1\,0 \\ 0\,0\,0\,0\,0\,1 \\ 0\,0\,0\,1\,0\,0 \\ 1\,0\,0\,0\,0\,0 \end{pmatrix}. \tag{4.38}$$

Remark 4.12 (Alias Method). The $(k+1)$-st node in Algorithm 4.7.1 is generated from a discrete n-point pdf (the X_k-th row of $P^{(k+1)}$). A straightforward way to do this is to use the inverse-transform technique; see Algorithm 1.7.2. However, this is quite time consuming, especially when n is large. Instead of calculating at each step the entries of $P^{(k+1)}$ from those of P, one can use as an alternative the acceptance–rejection method; see Section 1.7.4. Here, one selects the states in a tour such that for a given "previous" state i the "next" state j is generated according to the p_{ij} from the *original* P. And then, depending on whether state j was previously visited or not, one either rejects or accepts the associated transition $i \to j$. The efficiency of such an acceptance–rejection typically decreases dramatically towards the end of the tour. Indeed, let $n = 100$ and assume that in a specific tour we have already visited 90 different cities. In such case, before moving from city 90 to city 91 the acceptance–rejection method will reject on average 10 trials before making a successful one.

The trajectory generation via the acceptance–rejection method can be substantially increased by combining it with the *alias* method (see Section 1.7.2). Generation of random variables from an arbitrary discrete distribution by the alias method is much faster than by the inverse-transform method, especially for large n. For more details on the implementation of the alias method for SEN problems we refer to [113] and the appendix of this chapter. It was found empirically in [113] that the following combination is the fastest: For the first 85% of visited cities use a combination of the alias with the acceptance–rejection and for the remaining 15% of cities in the tour use the inverse-transform method.

We present now an alternative algorithm for trajectory generation in TSP, where alias method has been used as well. This algorithm is due to Margolin [113] and is called the *placement algorithm.* Recall that Algorithm 4.7.1 generates *transitions* from node to node, based on the transition matrix $P = (p_{ij})$. In contrast, in Algorithm 4.7.2 below a similar matrix

$$P = \begin{pmatrix} p_{(1,1)} & p_{(1,2)} & \cdots & p_{(1,n)} \\ p_{(2,1)} & p_{(2,2)} & \cdots & p_{(2,n)} \\ \vdots & \vdots & \vdots & \vdots \\ p_{(n,1)} & p_{(n,2)} & \cdots & p_{(n,n)} \end{pmatrix} . \tag{4.39}$$

generates *node placements*. Specifically, $p_{(i,j)}$ corresponds to the probability of node i being visited at the j-th place in a tour of n cities. In other words, $p_{(i,j)}$ can be viewed as probability that city (node) i is "arranged" to be visited at the j-th place in a tour of n cities.

More formally, a *node placement vector* is a vector $\mathbf{y} = (y_1, \ldots, y_n)$ such that y_i denotes the "place" of node i in the tour $\mathbf{x} = (x_1, \ldots, x_n)$. The precise meaning is given by the correspondence.

$$y_i = j \quad \Longleftrightarrow \quad x_j = i \,, \tag{4.40}$$

for all $i, j \in \{1, \ldots, n\}$. For example, the node placement vector $\mathbf{y} = (3,4,2,6,5,1)$ in a 6-node network, such as in Figure 4.5, defines uniquely the tour $\mathbf{x} = (6,3,1,2,5,4)$. The performance of each node placement $\mathbf{y}$ can be defined as $\widetilde{S}(\mathbf{y}) = S(\mathbf{x})$, where $\mathbf{x}$ is the unique path corresponding to $\mathbf{y}$.

Algorithm 4.7.2 (Trajectory Generation Using Node Placements)

1. *Define* $P^{(1)} = P$. *Let* $k = 1$.
2. *Generate* Y_k *from the distribution formed by the* k*-th row of* $P^{(k)}$. *Obtain the matrix* $P^{(k+1)}$ *from* $P^{(k)}$ *by first setting the* Y_k*-th column of* $P^{(k)}$ *to 0 and then normalizing the rows to sum up to 1.*
3. *If* $k = n$ *then* **stop**; *otherwise set* $k = k + 1$ *and reiterate from Step 2.*
4. *Determine the tour by* (4.40) *and evaluate the length of the tour* S *via* (4.32).

A more detailed description of this algorithm is given in Algorithm 4.11.2 in the appendix of this chapter.

Note that in contrast to Algorithm 4.7.1, where the random tours $\mathbf{X}$ are generated via the MCWR, we now generate the node placements $\mathbf{Y}$ using a conditional distribution formed by the corresponding row of the matrix P (see Step 2 of Algorithm 4.7.2). It follows from Section 4.3.1 that the updating formula for $\widehat{p}_{t,(i,j)}$ is given by

$$\widehat{p}_{t,(i,j)} = \frac{\displaystyle\sum_{k=1}^{N} I_{\{\tilde{S}(\mathbf{Y}_k) \leqslant \widehat{\gamma}_t\}} \, I_{\{Y_{ki}=j\}}}{\displaystyle\sum_{k=1}^{N} I_{\{\tilde{S}(\mathbf{Y}_k) \leqslant \widehat{\gamma}_t\}}} \,. \tag{4.41}$$

Our simulation results with the TSP and with other SEN models do not indicate any superiority among the two (Algorithm 4.7.2 and Algorithm 4.7.1) in terms of the efficiency (speed and the accuracy) of the main CE Algorithm 4.2.1.

For easy reference we present now the main CE algorithm for TSP.

Algorithm 4.7.3 (Main CE Algorithm for TSP)

1. *Choose an initial reference transition matrix* $\widehat{P}_0$*, say with all off-diagonal elements equal to* $1/(n-1)$*. Set* $t = 1$.
2. *Generate a sample* $\mathbf{X}_1, \ldots, \mathbf{X}_N$ *of tours via Algorithm 4.7.1, with* $P = \widehat{P}_{t-1}$*, and compute the sample* $(1-\varrho)$*-quantile* $\widehat{\gamma}_t$ *of the performances according to* (4.6).
3. *Use the* **same** *sample* $\mathbf{X}_1, \ldots, \mathbf{X}_N$ *to update* $\widehat{P}_t$ *via* (4.37).
4. *Apply* (4.9) *to smooth out the matrix* $\widehat{P}_t$.
5. *If for some* $t \geq d$*, say* $d = 5$,

$$\widehat{\gamma}_t = \widehat{\gamma}_{t-1} = \cdots = \widehat{\gamma}_{t-d} \, , \tag{4.42}$$

then **stop** *(let* T *denote the final iteration); otherwise set* $t = t + 1$ *and reiterate from Step 2.*

Recall that the goal of Algorithm 4.7.3 is to generate a sequence $\{(\widehat{\gamma}_t, \widehat{P}_t)\}$ which converges to a fixed point (γ^*, P^*), where γ^* is the optimal value of the objective function in the network and P^* is the ODTM, which uniquely defines the optimal tour $\mathbf{x}^*$. Also, it is essential that at each step of Algorithm 4.7.3 all components of P corresponding to the nodes that we have not yet visited are nonzero. To insure this, we apply the smoothed version (4.9) to $\widehat{p}_{t,ij}$.

Remark 4.13 (Parameter Setting). Similarly to Remark 4.4 we typically set for the TSP and for general SEN-type problems $\varrho = 10^{-2}$ and $\alpha = 0.7$. We take for SEN the sample size $N = \text{const} \cdot n^2$, like $N = 5n^2$, rather than $N = \text{const} \cdot n$ (for SNN), since the number of parameters to estimate in the probability matrix P is n^2, rather than n as in the max-cut problem, say.

A more ingenuous approach to selecting the parameters, which appears to work well, is to run the CE algorithm on a "miniature" of the original TSP, selecting at random a small percentage, say 10–20%, of the original nodes, while preserving the cost structure. For this miniature TSP good choices for α, C, and ϱ can be obtained quickly (e.g., by trial and error). Now, use *these* parameters for the larger TSP.

Remark 4.14 (Alternative stopping criterion). An alternative stopping criterion is based on the convergence of the sequence of matrices $\widehat{P}_0, \widehat{P}_1, \widehat{P}_2 \ldots$. Specifically, the algorithm will terminate at iteration T if for some integer d, for example, $d = 5$,

$$\xi_T(i) = \xi_{T-1}(i) = \cdots = \xi_{T-d}(i), \quad \text{for all } i = 1, \ldots, n \, , \tag{4.43}$$

where $\xi_T(i)$ denotes the index of the maximal element of the i-th row of $\widehat{P}_T$.

Remark 4.15. We show now how to calculate the minimal number of trajectories (sample size) $N_*(n)$ required in order to pass with high probability ζ *at least once* through all transitions $i \to j$. More specifically, the sample size $N_*(n)$ will guarantee that all transition probabilities in (4.37) will be positive with high probability, provided all off-diagonal elements of $\widehat{P}_0$ are $1/(n-1)$.

The quantity $N_*(n)$ can be calculated analytically by using the theory of *urn models.* More precisely, $N_*(n)$ can be calculated using the *inclusion–exclusion principle for occupancy problems* ([90], page 108). This is the same as finding the probability that all urns are occupied (no single transition probabilities in (4.37) will be zero), while distributing N balls into n cells (generating N trajectories through the network containing n nodes).

Instead of performing tedious calculations with such urn models (see [90]) we present below some simulation studies according to which we found that for $50 \le n \le 1000$, the quantity $N_*(n)$ can be approximated as $N_*(n) = C_n n$, where $C_n \approx e \cdot \ln n$. Table 4.2 presents the results of such simulation study with fixed $\zeta = 0.999$.

Table 4.2. $N_*(n)$ as a function of n for $\zeta = 0.999$.

n	$N_*(n)$	C_n	$\ln n$
50	400	8	3.9
100	900	9	4.6
250	2,500	10	5.5
500	6,000	12	6.2
750	9,750	13	6.6
1000	14,000	14	6.9

Note that the results of Table 4.2 are valid only under the assumption that the off-diagonal elements $P_{0,ij}$ of $\widehat{P}_0$ are equal. This typically holds only for the first iteration of Algorithm 4.7.3. It is still interesting to note that for a TSP with $(n-1)!$ trajectories, one needs only $N_*(n) = \mathcal{O}(n \ln n)$ trajectories to pass with high probability ζ *at least once* through all possible transitions. Note finally that in all our simulation results with TSP we took at each iteration of Algorithm 4.7.3 a sample of $N = \mathcal{O}(n^2)$ trajectories, which is larger than $N_*(n) = \mathcal{O}(n \ln n)$.

4.8 The Quadratic Assignment Problem

The quadratic assignment problem (QAP) has remained one of the challenges in combinatorial optimization. From a computational complexity point of view the QAP is one of the most difficult problems to solve, and it is still considered a computationally nontrivial task to solve modest size problems, say with $n = 20$.

The applications of the QAP include optimal allocation, scheduling, manufacturing, parallel and distributed computing and statistical data analysis. We consider the QAP in the context of *location theory* , where the objective is to find an assignment of a set of facilities to a set of locations such that the total cost of the assignment is minimized.

Mathematically, the QAP can be formulated as follows: given a set $V = \{1, 2, \ldots, n\}$ and three $n \times n$ input matrices $F = (f_{ij})$, $D = (d_{kl})$ and $E = (e_{il})$, minimize the function S given by

$$S(\mathbf{x}) = \sum_{i=1}^{n} \sum_{j=1}^{n} f_{ij}\, d_{x_i, x_j} + \sum_{i=1}^{n} e_{i, x_i} \tag{4.44}$$

over all *permutations* $\mathbf{x} = (x_1, \ldots, x_n)$ of V. The matrix $F = (f_{ij})$ is called the *flow* matrix, that is, f_{ij} is the flow of materials from facility i to facility j, $D = (d_{kl})$ is the *distance* matrix, that is, d_{kl} represents the distance from location k to location l, and $E = (e_{il})$ is the *direct cost* matrix, that is, e_{il} represents the cost of placing the facility i to location l. The matrix E is often set to zero and it does not appear in most case studies. However, it can be easily added to the objective function and incorporated in the CE algorithm described below.

Consider, for example, a three dimensional QAP with the following matrices D and F

$$D = \begin{pmatrix} 0 & 7 & 3 \\ 7 & 0 & 2 \\ 3 & 2 & 0 \end{pmatrix}, \qquad F = \begin{pmatrix} 0 & 3 & 9 \\ 5 & 0 & 6 \\ 4 & 3 & 0 \end{pmatrix}.$$

In this case, the total number of possible allocations is $n! = 3! = 6$, and the corresponding allocations are $(1, 2, 3)$; $(1, 3, 2)$; $(2, 1, 3)$; $(2, 3, 1)$; $(3, 1, 2)$; $(3, 2, 1)$.

For example, the first allocation implies that the first facility is allocated to place 1, the second one is allocated to the second place, and the third facility is allocated to the third place, and similarly for the remaining 5 allocations. The assignment costs (the objective function) of all 6 allocations are

$$\begin{aligned}
S((1,2,3)) &= d_{12}f_{12} + d_{13}f_{13} + d_{21}f_{21} + d_{23}f_{23} + d_{31}f_{31} + d_{32}f_{32} = 111 \\
S((1,3,2)) &= d_{12}f_{13} + d_{13}f_{12} + d_{21}f_{31} + d_{23}f_{32} + d_{31}f_{21} + d_{32}f_{23} = 133 \\
S((2,1,3)) &= d_{12}f_{21} + d_{13}f_{23} + d_{21}f_{12} + d_{23}f_{13} + d_{31}f_{32} + d_{32}f_{31} = 115 \\
S((2,3,1)) &= d_{12}f_{23} + d_{13}f_{21} + d_{21}f_{32} + d_{23}f_{31} + d_{31}f_{12} + d_{32}f_{13} = 110 \\
S((3,1,2)) &= d_{12}f_{31} + d_{13}f_{32} + d_{21}f_{13} + d_{23}f_{12} + d_{31}f_{23} + d_{32}f_{21} = 134 \\
S((3,2,1)) &= d_{12}f_{32} + d_{13}f_{31} + d_{31}f_{23} + d_{23}f_{21} + d_{31}f_{13} + d_{32}f_{12} = 118
\end{aligned}$$

It is well known that, with an appropriate choice of the coefficients of the matrices F and D, the TSP, the packing problem, and the maximum clique problem can be viewed as special cases of the QAP.

Note that both the trajectory generation algorithms and the main Algorithm 4.7.3 for TSP remain the same for QAP, provided the components c_{ij} of the cost matrix $C = (c_{ij})$ in TSP are replaced with

$$c_{ij} = \sum_{k=1}^{n} \sum_{l=1}^{n} f_{ij}\, d_{kl}\ .$$

Remark 4.16. The term "quadratic" in the QAP comes from the reformulation of the problem as an optimization problem with a quadratic objective function. The reasoning is as follows. First, observe that there is a one-to-one correspondence between the set of all permutations $\mathcal{X}$ and the set of $n \times n$ permutation matrices (y_{ik}), which must satisfy:

$$\sum_{i=1}^{n} y_{ik} = 1,\ k = 1, \ldots, n\ , \tag{4.45}$$

$$\sum_{k=1}^{n} y_{ik} = 1,\ i = 1, \ldots, n\ , \tag{4.46}$$

and

$$y_{ik} \in \{0,1\};\ i = 1, \ldots, n;\ k = 1, \ldots, n\ . \tag{4.47}$$

We can interpret y_{ik} as

$$y_{ik} = \begin{cases} 1 \text{ if facility } i \text{ is assigned to location } k, \\ 0 \text{ otherwise.} \end{cases} \tag{4.48}$$

Note that the cost of simultaneously assigning facility i to location k and facility j to location l is $f_{ij} d_{kl}$. As a result, taking (4.45)–(4.48) into account, we can rewrite the right-hand side of (4.44) in the following equivalent form

$$\sum_{i=1}^{n} \sum_{j=1}^{n} \sum_{k=1}^{n} \sum_{l=1}^{n} f_{ij}\, d_{kl}\, y_{ik}\, y_{jl} + \sum_{i=1}^{n} \sum_{k=1}^{n} e_{ik}\, y_{ik}\ . \tag{4.49}$$

4.9 Numerical Results for SNNs

Below we present numerical results for Algorithm 4.5.2 for the max-cut and partition problems. The emphasis will be on the *numerical convergence* of Algorithm 4.5.2, that is on the convergence of estimator $\widehat{\gamma}_t$ (see (4.25)) to the true unknown optimal value γ^*.

For the bipartition and the max-cut problem (with number of partitions $r = 2$) we always choose with the initial vector $\mathbf{p}_0 = (1, \frac{1}{2}, \ldots, \frac{1}{2})$. If not stated otherwise we set the parameters as follows (see also Remark 4.4): $\varrho = 10^{-2}$, $\alpha = 0.7$, $N = 5 \cdot n \cdot \frac{r}{2}$ and use the stopping rule (4.25) with the parameter $d = 5$.

Table 4.3 lists the main notation used in the tables for SNN problems.

Table 4.3. Notation used in tables for SNN-type experiments.

$\widehat{\gamma}_t$	the worst performance of the elite samples obtained at iteration t,
$S_{t,(N)}$	the best elite performance at iteration t,
$p\uparrow_{\min}$	$= \min\{\widehat{p}_{t,i} \,:\, \widehat{p}_{t,i} \geq 0.5,\ i = 1, \ldots, n\}$,
$p\downarrow_{\max}$	$= \max\{\widehat{p}_{t,i} \,:\, \widehat{p}_{t,i} < 0.5,\ i = 1, \ldots, n\}$,
ε	the *relative experimental error* of the final solution, that is,

$$\varepsilon = \frac{\gamma^* - \widehat{\gamma}_T}{\gamma^*}. \tag{4.50}$$

4.9.1 Synthetic Max-Cut Problem

We return to the synthetic max-cut problem of Section 2.5.1. Thus, our symmetric cost matrix is of the form

$$C = \begin{pmatrix} Z_{11} & C_{12} \\ C_{21} & Z_{22} \end{pmatrix}, \tag{4.51}$$

where Z_{11} and Z_{22} are square matrices of dimensions m and $n - m$, respectively, whose elements are strictly less than b, and C_{12} and C_{21} are matrices with identical elements $c > n(n - m)/m$. The optimal cut V^* is given by $V^* = \{\{1, \ldots, m\}, \{m + 1, \ldots, n\}\}$, and the optimal performance is

$$\gamma^* = c\, m\, (n - m) .$$

We generate the elements of the matrices Z_{ii} according to some fixed probability distribution with support in $[0, b)$, for example a uniform or Beta distribution. We write symbolically $Z \sim \mathsf{U}(a, b)$, if the Z_{ii} are generated via the $\mathsf{U}(a, b)$ distribution. A similar notation is used for other distributions. Note that only the elements above the diagonal of each Z_{ii} need to be generated, since the cost matrix is symmetric, with 0 diagonal elements.

Table 4.4 presents the relative experimental errors ε_i, $i = 1, \cdots, 6$ and the final iteration number T_i, $i = 1, \cdots, 6$ as a function of the sample size N for the *max-cut problem* with $n = 200$, $m = 100$, $c = 5$ and for the following 6 cases: $Z_1 = 4.0$, $Z_2 = 4.9$, $Z_3 \sim \mathsf{U}(1, 5)$, $Z_4 \sim \mathsf{U}(4.5, 5.0)$, $Z_5 \sim \mathsf{Beta}(100, 1, 1, 5)$, $Z_6 \sim \mathsf{Beta}(100, 1, 4.5, 5)$. Here $\mathsf{Beta}(\alpha, \beta, a, b)$ denotes the (generalized) Beta distribution on the interval (a, b), with pdf

$$\frac{\Gamma(\alpha+\beta)}{\Gamma(\alpha)\,\Gamma(\beta)}\left(\frac{1}{b-a}\right)^{\alpha+\beta-1}(b-x)^{\beta-1}(x-a)^{\alpha-1},\ a \leqslant x \leqslant b\,. \tag{4.52}$$

We found that for $N \geq 10\,n$ the relative experimental error of the final solution, denoted by ε in (4.50), equals zero, that is, Algorithm 4.5.2 is exact (always finds the optimal solution).

Table 4.4. The relative experimental errors $\varepsilon_i,\ i=1,\cdots,6$ (and the associated stopping times $T_i,\ i=1,\cdots,6$) as functions of the sample size N for the max-cut problems with $n=200$, $m=100$ and $c=5$.

N	$\varepsilon_1(T_1)$	$\varepsilon_2(T_2)$	$\varepsilon_3(T_3)$	$\varepsilon_4(T_4)$	$\varepsilon_5(T_5)$	$\varepsilon_6(T_6)$
200	0.078(11)	0.009(12)	0.179(11)	0.023(11)	0.004(11)	4.6e-4(12)
400	0.070(12)	0.005(14)	0.091(14)	0.014(14)	0.003(14)	4.2e-4(16)
800	0.013(15)	0.003(15)	0.072(17)	0.005(16)	0.002(17)	2.0e-4(19)
1200	0.004(15)	0.001(16)	0.027(16)	0.002(16)	5.0e-4(18)	1.5e-4(20)
1600	0.002(17)	0.000(16)	0.000(15)	0.001(16)	3.0e-4(19)	6.0e-5(20)
2000	0.000(16)	0.000(17)	0.000(15)	0.000(16)	0.000(18)	0.000(20)

Table 4.5 presents data similar to Table 4.4 for the associated CPU times in seconds on a Sun Enterprise 4000 workstation (12 CPU, 248 MHz).

Table 4.5. The CPU times for the same cases as in Table 4.4.

N	CPU_1	CPU_2	CPU_3	CPU_4	CPU_5	CPU_6
200	3.6	3.1	2.9	3.4	2.9	3.4
600	7.4	8.0	6.9	8.1	7.2	9.1
800	15.3	14.5	16.9	15.3	16.6	19.8
1200	25.4	23.0	26.4	23.1	26.1	32.5
1600	34.1	30.0	27.8	29.7	39.7	42.2
2000	35.2	37.5	35.1	37.7	48.5	51.3

Similar results were obtained with Algorithm 4.5.2 for *partition problems*, that is where instead of the random cut generation Algorithm 4.5.2 we use the random bipartition generation Algorithm 4.6.1.

Table 4.6 represents the evolution of Algorithm 4.5.2 for problem (4.51) with $n=200,\ \ m=100$, $c=1$ and $Z=0$. The relative experimental error is $\varepsilon=0$. It took 30 seconds of CPU time to find the optimal solution. Table 4.7 presents another evolution of Algorithm 4.5.2 for the max-cut problem as functions of t for the following parameters: $n=600$ with $m=n/10=60$, $c=$45, $Z\sim\mathsf{U}(4.5,5.0)$, $N=10\,n$ and $\varrho=0.01$.

Table 4.6. Evolution of Algorithm 4.5.2 for the artificial max-cut problem (4.51) with $n = 200,\ m = 100,\ c = 1,\ Z = 0,\ N = 1000,\ \alpha = 0.7,\ \varrho = 0.05,\ d = 5$.

t	$\widehat{\gamma}_t$	$S_{t,(N)}$	$p\uparrow_{\min}$	$p\downarrow_{\max}$
1	5084	5256	0.5140	0.4860
2	5072	5280	0.5014	0.4986
3	5108	5228	0.5004	0.4957
4	5252	5476	0.5073	0.4948
5	5512	6012	0.5009	0.4956
6	5874	6250	0.5031	0.4860
7	6276	6682	0.5040	0.4943
8	6736	7310	0.5250	0.4742
9	7232	8010	0.5215	0.4596
10	7730	8330	0.5579	0.4365
11	8276	8772	0.6228	0.3689
12	8696	9232	0.7270	0.3347
13	9140	9512	0.7641	0.2964
14	9512	9900	0.8172	0.2009
15	9800	10000	0.8472	0.0883
16	10000	10000	0.9542	0.0265

Table 4.7. Evolution of Algorithm 4.5.2 for the bipartition problem with $n = 600$, $m = 60$, $c = 45$, $Z \sim \mathsf{U}(4.5, 5.0)$. CPU time 1729 seconds, $\varepsilon = 0$.

t	$\widehat{\gamma}_t$	$S_{t,(N)}$	$p\uparrow_{\min}$	$p\downarrow_{\max}$
1	4842820	551289	0.500	0.351
2	6032004	692976	0.500	0.485
3	7114520	805967	0.500	0.498
4	8253897	924785	0.514	0.487
5	9247863	1007224	0.501	0.466
6	9863917	1070735	0.500	0.467
7	1049434	1113882	0.584	0.434
8	1113857	1179822	0.631	0.299
9	1157656	1202118	0.667	0.250
10	1179821	1292903	0.684	0.283
11	1224562	1269969	0.700	0.283
12	1247190	1316007	0.732	0.265
13	1269976	1339294	0.750	0.283
14	1292918	1339277	0.816	0.282
15	1316022	1362703	0.765	0.415
16	1339283	1386292	0.581	0.350
17	1362709	1410035	0.598	0.383
18	1386295	1410042	0.782	0.299
19	1410040	1433943	0.848	0.266
20	1433942	1458000	0.848	0.016

Table 4.8 presents the dynamics of $\widehat{\mathbf{p}}_t = (\widehat{p}_{t,1}, \ldots, \widehat{p}_{t,14})$ for $n = 14,\ m = 7,\ c = 5,\ Z \sim \mathsf{U}(1,5)$ and $N = 3\,n = 42$.

Table 4.8. Dynamics of $\widehat{\mathbf{p}}_t$ for $n = 14,\ m = 7,\ c = 5,\ Z \sim \mathsf{U}(1,5)$ and $N = 3\,n = 42$.

t	$\widehat{\mathbf{p}}_t$													
0	1.00	0.50	0.50	0.50	0.50	0.50	0.50	0.50	0.50	0.50	0.50	0.50	0.50	0.50
1	1.00	0.25	0.75	0.50	0.75	0.75	0.50	0.75	0.00	0.25	0.75	0.25	0.25	0.50
2	1.00	0.75	0.75	0.75	1.00	1.00	1.00	0.25	0.00	0.00	0.25	0.00	0.00	0.00
3	1.00	1.00	1.00	1.00	1.00	1.00	1.00	0.00	0.00	0.00	0.00	0.00	0.00	0.00
4	1.00	1.00	1.00	1.00	1.00	1.00	1.00	0.00	0.00	0.00	0.00	0.00	0.00	0.00
5	1.00	1.00	1.00	1.00	1.00	1.00	1.00	0.00	0.00	0.00	0.00	0.00	0.00	0.00

4.9.2 r-Partition

Here we extend our numerical results to the r-partition problem ($r > 2$), by considering a graph $G = (V, E)$ with the following *symmetric* cost matrix:

$$C = \left(\begin{array}{c|c|c|c} Z_{11} & C_{12} & \cdots & C_{1r} \\ \hline C_{21} & Z_{22} & \cdots & C_{2r} \\ \hline \vdots & \vdots & \ddots & \vdots \\ \hline C_{r1} & C_{r2} & \cdots & Z_{rr} \end{array}\right) . \tag{4.53}$$
$$\qquad \underbrace{}_{V_1^*} \quad \underbrace{}_{V_2^*} \qquad\qquad \underbrace{}_{V_r^*}$$

We choose the submatrices $\{Z_{ii}\}$ and $\{C_{ij}\}$ such that the optimal partition is given by $V^* = \{V_1{}^*, \ldots, V_r{}^*\}$ with

$$\begin{aligned} V_1{}^* &= \{1, \ldots, m_1\} , \\ V_2{}^* &= \{m_1 + 1, \ldots, m_1 + m_2\} , \\ &\vdots \\ V_r{}^* &= \{\sum_{i=1}^{r-1} m_i + 1, \ldots, n\} . \end{aligned}$$

To insure this, we let all elements of the matrices $\{Z_{ii}\}$ be less than all the elements of the matrices $\{C_{ij}\}$. For simplicity, we assume that all elements of the C_{ij} are equal to some constant c and that n integer divisible in r and the optimal partition divides V into r subsets of equal size $m = n/r$. In this case the optimal value γ^* is given by

$$\gamma^* = \binom{r}{2} c m^2 . \tag{4.54}$$

We generated the elements of the Z_{ii} via a random distribution with support on a subset of $[0, b)$, for b small enough. Tables 4.9 and 4.10 below present the evolution of Algorithm 4.5.2 for the r-partition problems with configurations $n = 400$, $r = 4$ and $n = 500$, $r = 5$ respectively. We set $c = 50$ and generate the elements of the Z_{ii} via the $\mathsf{U}(4.5, 5)$-distribution. Note that we only need to generate the elements *above* the diagonal, since C is a symmetric matrix and the diagonal elements of the Z_{ii} are 0.

Table 4.9. Evolution of Algorithm 4.5.2 for the r-partition problem with $n = 400$ nodes, $r = 4$, $\gamma^* = 3000000$, $\varrho = 0.01$, $N = 10000$, $c = 50$, and $Z \sim \mathsf{U}(4.5, 5)$. CPU time is 2097 seconds, $\varepsilon = 0$.

t	$\widehat{\gamma}_t$	$S_{t,(N)}$	$p\uparrow_{\min}$	$p\downarrow_{\max}$
1	2.330840e+06	2.338761e+06	0.259990	0.240017
2	2.330704e+06	2.338354e+06	0.259960	0.240041
3	2.332089e+06	2.341081e+06	0.260016	0.230034
4	2.337166e+06	2.348442e+06	0.259969	0.239941
5	2.349762e+06	2.363804e+06	0.270121	0.230098
6	2.372085e+06	2.389760e+06	0.279922	0.229938
7	2.404083e+06	2.435340e+06	0.289990	0.209948
8	2.445606e+06	2.473909e+06	0.300106	0.209884
9	2.493293e+06	2.528921e+06	0.329936	0.179832
10	2.550295e+06	2.574981e+06	0.299978	0.189747
11	2.611895e+06	2.642337e+06	0.310163	0.150062
12	2.678021e+06	2.713952e+06	0.350111	0.120135
13	2.743622e+06	2.780953e+06	0.460127	0.119892
14	2.806880e+06	2.841077e+06	0.490120	0.079962
15	2.867101e+06	2.895819e+06	0.479967	0.059933
16	2.918022e+06	2.950955e+06	0.669903	0.030030
17	2.960013e+06	2.977818e+06	0.790066	0.019997
18	2.990928e+06	3.000000e+06	0.939996	0.009996
19	3.000000e+06	3.000000e+06	1.000000	0.000000
20	3.000000e+06	3.000000e+06	1.000000	0.000000
21	3.000000e+06	3.000000e+06	1.000000	0.000000

Table 4.10. Evolution of Algorithm 4.5.2 for the r-partition problem with $n = 500$ nodes, $r = 5$, $\gamma^* = 5000000$, $\varrho = 0.01$, $N = 20000$, $c = 50$, and $Z \sim \mathsf{U}(4.5, 5)$. CPU time 9557 seconds, $\varepsilon = 0.00091$.

t	$\widehat{\gamma}_t$	$S_{t,(N)}$	$p \uparrow_{\min}$	$p \downarrow_{\max}$
1	4.105605e+06	4.112836e+06	0.205009	0.194994
2	4.105501e+06	4.115174e+06	0.205005	0.194993
3	4.105646e+06	4.114378e+06	0.209992	0.189999
4	4.106427e+06	4.113426e+06	0.215033	0.184996
...	...		... ...	...
8	4.163015e+06	4.189215e+06	0.215004	0.179973
9	4.197580e+06	4.222975e+06	0.205025	0.185003
10	4.239794e+06	4.268014e+06	0.230000	0.150036
11	4.286359e+06	4.318551e+06	0.244958	0.149990
...	...	...	...	...
14	4.446925e+06	4.485072e+06	0.245036	0.139987
15	4.503736e+06	4.541756e+06	0.259982	0.135044
16	4.558667e+06	4.590573e+06	0.300069	0.104966
17	4.610398e+06	4.638465e+06	0.335027	0.090008
18	4.658267e+06	4.693343e+06	0.349934	0.094989
19	4.699643e+06	4.716176e+06	0.334942	0.054996
...	...	...	...	...
23	4.812588e+06	4.832686e+06	0.494989	0.010006
24	4.848739e+06	4.869789e+06	0.500067	0.005003
25	4.887795e+06	4.911411e+06	0.494996	0.005000
26	4.929371e+06	4.957022e+06	0.579969	0.000000
27	4.965013e+06	4.982266e+06	0.714995	0.000000
28	4.986607e+06	4.995472e+06	0.630120	0.000000
29	4.995472e+06	4.995472e+06	0.995000	0.000000
30	4.995472e+06	4.995472e+06	0.995000	0.000000
31	4.995469e+06	4.995469e+06	0.995000	0.000000
32	4.995469e+06	4.995469e+06	0.995000	0.000000
33	4.995469e+06	4.995469e+06	1.000000	0.000000
34	4.995469e+06	4.995469e+06	1.000000	0.000000

4.9.3 Multiple Solutions

To investigate the convergence behavior of Algorithm 4.5.2 when the max-cut problem has multiple solutions, we ran a modified version of the synthetic max-cut problem. In this problem we select, deterministically or at random, k vertices from both V_1^* and V_2^* and set the weights of *all* edges that are incident to these nodes equal to c. We obtain in this way $\binom{2k}{k}$ different cuts with the same value of objective function.

Tables 4.11–4.12 present the evolution of Algorithm 4.5.2 for a bipartition problem with 70 and 252 multiple solutions, respectively. The problem instances were generated from the same symmetric matrix (4.51) corresponding to a single optimal solution, but using the above modification with edges of equal weights. We set $n = 300$, $c = 50$, $Z \sim \mathsf{U}(4.5, 5.0)$, and $m = 150$.

Table 4.11. Evolution of Algorithm 4.5.2 with 70 multiple solutions. CPU time 956 seconds, $\varepsilon = 0$.

t	$\widehat{\gamma}_t$	$S_{t,(N)}$	$p\uparrow_{\min}$	$p\downarrow_{\max}$
1	653664.2	665978.9	0.500024	0.499776
2	656880.4	670476.3	0.500003	0.499974
3	677162.1	690689.7	0.500017	0.499891
...	...	...	...	...
14	802374.3	821960.2	0.500893	0.498803
15	806397.3	834234.2	0.500335	0.499993
16	813988.4	855626.5	0.500083	0.499733
17	822138.5	846929.7	0.500276	0.499090
...	...	...	...	...
28	873720.7	906628.7	0.501069	0.499286
29	873559.5	911553.8	0.501170	0.499558
30	878122.9	901982.4	0.500080	0.467794
31	882951.0	936865.4	0.531554	0.499926
...	...	...	...	...
42	942160.0	968903.0	0.566601	0.433407
43	947523.2	968873.2	0.500092	0.499161
44	947539.0	968891.8	0.501017	0.499897
45	947540.4	979567.9	0.532519	0.467096
...	...	...	...	...
56	996443.3	1.019435e+06	0.567372	0.499518
57	1.002278e+06	1.049345e+06	0.566760	0.466034
58	1.007850e+06	1.031200e+06	0.533685	0.466328
59	1.013776e+06	1.037309e+06	0.566563	0.499329
...	...	...	...	...
69	1.111974e+06	1.118444e+06	0.599846	0.466461
70	1.118439e+06	1.125000e+06	0.666536	0.466771
71	1.118439e+06	1.125000e+06	0.500000	0.399687
72	1.118441e+06	1.125000e+06	0.566504	0.499707
73	1.118441e+06	1.125000e+06	0.533463	0.499415
74	1.118441e+06	1.125000e+06	0.599805	0.466601
75	1.118441e+06	1.125000e+06	0.599883	0.399922
76	1.118441e+06	1.125000e+06	0.599727	0.333138

Table 4.12. Evolution of Algorithm 4.5.2 with 252 multiple solutions. CPU time 896 seconds, $\varepsilon = 0$.

t	$\widehat{\gamma}_t$	$S_{t,(N)}$	$p\uparrow_{\min}$	$p\downarrow_{\max}$
1	659664.5	671071.7	0.500081	0.499985
2	660285.0	671062.4	0.500001	0.499990
3	659391.4	667080.0	0.500036	0.499938
...	...	...	...	...
14	797840.1	832764.9	0.531047	0.499741
15	808965.4	841014.6	0.533229	0.499118
16	820559.0	849247.3	0.500247	0.499749
17	824540.7	845068.6	0.500101	0.466964
...	...	...	...	...
28	880209.3	918209.4	0.500678	0.499172
29	884915.8	903727.5	0.532711	0.499987
30	889705.4	923102.4	0.532683	0.499826
31	889709.6	923114.4	0.532786	0.497970
...	...	...	...	...
42	953961.2	997393.5	0.500081	0.499887
43	964531.6	991684.7	0.533474	0.466147
44	969800.5	1.002924e+06	0.531970	0.432870
45	970066.6	1.014588e+06	0.532278	0.499721
...	...	...	...	...
56	1.068346e+06	1.092966e+06	0.799371	0.432514
57	1.086535e+06	1.099478e+06	0.733773	0.299485
58	1.099478e+06	1.112058e+06	0.699533	0.033486
59	1.105727e+06	1.112064e+06	0.533270	0.000000
60	1.112058e+06	1.118485e+06	0.599769	0.433359
...	...	...	...	...
65	1.112062e+06	1.118488e+06	0.533679	0.299942
66	1.118483e+06	1.125000e+06	0.566751	0.233288
67	1.118486e+06	1.125000e+06	0.633476	0.200116
68	1.118486e+06	1.125000e+06	0.666279	0.166667
69	1.118486e+06	1.125000e+06	0.666473	0.166861
70	1.118486e+06	1.125000e+06	0.666602	0.166505
71	1.118486e+06	1.125000e+06	0.566150	0.100000
72	1.118486e+06	1.125000e+06	0.666667	0.333333

We see in Tables 4.11 and 4.12 that, although in both cases the optimal value $1.125 \cdot 10^6$ is found, the algorithm takes many iterations to converge and seems to oscillate between multiple optimal solutions, as illustrated by the fluctuating behavior of the vector $\mathbf{p}$. It is interesting to note that eventually Algorithm 4.5.2 settles down in one of the optimal degenerated solutions, provided we increase the parameter d in the stopping rule (4.25). To illustrate this, Table 4.13 presents the evolution of Algorithm 4.5.2 for a bipartition problem with $n = 20$ nodes that has 20 different optimal solutions. We take

$m = 10$, $\varrho = 0.03$, $N = 200$, $c = 50$, $Z \sim \mathsf{U}(4.5, 5.0)$ and increase d until the degeneracy of $\mathbf{p}$ occurs. In this particular example we found that the degeneracy of $\mathbf{p}$ occurs for $d = 13$.

Table 4.13. Evolution of Algorithm 4.5.2 for $d = 13$. CPU time 0.2 seconds, $\varepsilon = 0$.

t	$\widehat{\gamma}_t$	$S_{t,(N)}$	$p \uparrow_{\min}$	$p \downarrow_{\max}$
1	4591.810	4774.041	0.500792	0.497569
2	4548.082	4774.165	0.500033	0.499967
3	4411.859	4773.491	0.657805	0.490885
4	4773.580	4774.041	0.666662	0.499999
5	5000.000	5000.000	0.666667	0.333333
6	5000.000	5000.000	0.666667	0.333333
7	5000.000	5000.000	0.666667	0.333333
8	5000.000	5000.000	0.666667	0.333333
9	5000.000	5000.000	0.666667	0.333333
10	5000.000	5000.000	0.666667	0.166667
11	5000.000	5000.000	0.666667	0.000000
12	5000.000	5000.000	0.833333	0.000000
13	5000.000	5000.000	0.833333	0.166667
14	5000.000	5000.000	1.000000	0.000000
15	5000.000	5000.000	1.000000	0.000000
16	5000.000	5000.000	1.000000	0.000000
17	5000.000	5000.000	1.000000	0.000000
18	5000.000	5000.000	1.000000	0.000000

Table 4.14 presents the dynamics of $\widehat{\mathbf{p}}_t$ for the same data as in Table 4.13. Figures 4.6 and 4.7 present the bar charts based on data of Table 4.14.

Table 4.14. Dynamics of $\widehat{\mathbf{p}}_t$ for the same data as in Table 4.13.

t	$\widehat{\mathbf{p}}_t$																			
0	1.0	0.5	0.5	0.5	0.5	0.5	0.5	0.5	0.5	0.5	0.5	0.5	0.5	0.5	0.5	0.5	0.5	0.5	0.5	0.5
1	1.0	0.7	0.5	0.7	0.5	0.5	0.7	0.7	0.5	0.3	0.5	0.7	0.3	0.5	0.7	0.5	0.5	0.3	0.5	0.2
2	1.0	0.5	0.3	0.5	0.7	0.5	0.7	0.3	0.8	0.5	0.5	0.3	0.5	0.7	0.7	0.5	0.7	0.5	0.5	0.0
3	1.0	0.8	0.8	0.8	0.7	0.8	0.7	0.0	0.7	0.2	0.2	0.3	0.7	0.8	0.7	0.5	0.3	0.3	0.2	0.0
4	1.0	1.0	1.0	0.8	1.0	0.7	1.0	0.0	0.5	0.7	0.3	0.5	0.8	0.5	0.0	0.3	0.0	0.0	0.0	0.0
5	1.0	1.0	1.0	1.0	1.0	1.0	0.7	0.0	0.7	0.3	0.3	0.7	0.8	0.5	0.0	0.0	0.0	0.0	0.0	0.0
6	1.0	1.0	1.0	1.0	1.0	1.0	0.7	0.0	0.8	0.3	0.3	0.7	0.8	0.3	0.0	0.0	0.0	0.0	0.0	0.0
7	1.0	1.0	1.0	1.0	1.0	1.0	0.5	0.0	0.8	0.7	0.3	0.7	0.7	0.3	0.0	0.0	0.0	0.0	0.0	0.0
8	1.0	1.0	1.0	1.0	1.0	1.0	0.5	0.0	0.5	0.5	0.8	0.7	0.7	0.3	0.0	0.0	0.0	0.0	0.0	0.0
9	1.0	1.0	1.0	1.0	1.0	1.0	0.5	0.0	0.7	0.5	1.0	0.8	0.2	0.3	0.0	0.0	0.0	0.0	0.0	0.0
10	1.0	1.0	1.0	1.0	1.0	1.0	0.7	0.0	0.2	0.8	0.7	1.0	0.2	0.5	0.0	0.0	0.0	0.0	0.0	0.0
11	1.0	1.0	1.0	1.0	1.0	1.0	0.7	0.0	0.0	1.0	0.8	1.0	0.0	0.5	0.0	0.0	0.0	0.0	0.0	0.0
12	1.0	1.0	1.0	1.0	1.0	1.0	1.0	0.0	0.0	0.8	0.8	0.8	0.0	0.5	0.0	0.0	0.0	0.0	0.0	0.0
13	1.0	1.0	1.0	1.0	1.0	1.0	1.0	0.0	0.0	0.8	1.0	1.0	0.0	0.2	0.0	0.0	0.0	0.0	0.0	0.0
14	1.0	1.0	1.0	1.0	1.0	1.0	1.0	0.0	0.0	1.0	1.0	1.0	0.0	0.0	0.0	0.0	0.0	0.0	0.0	0.0

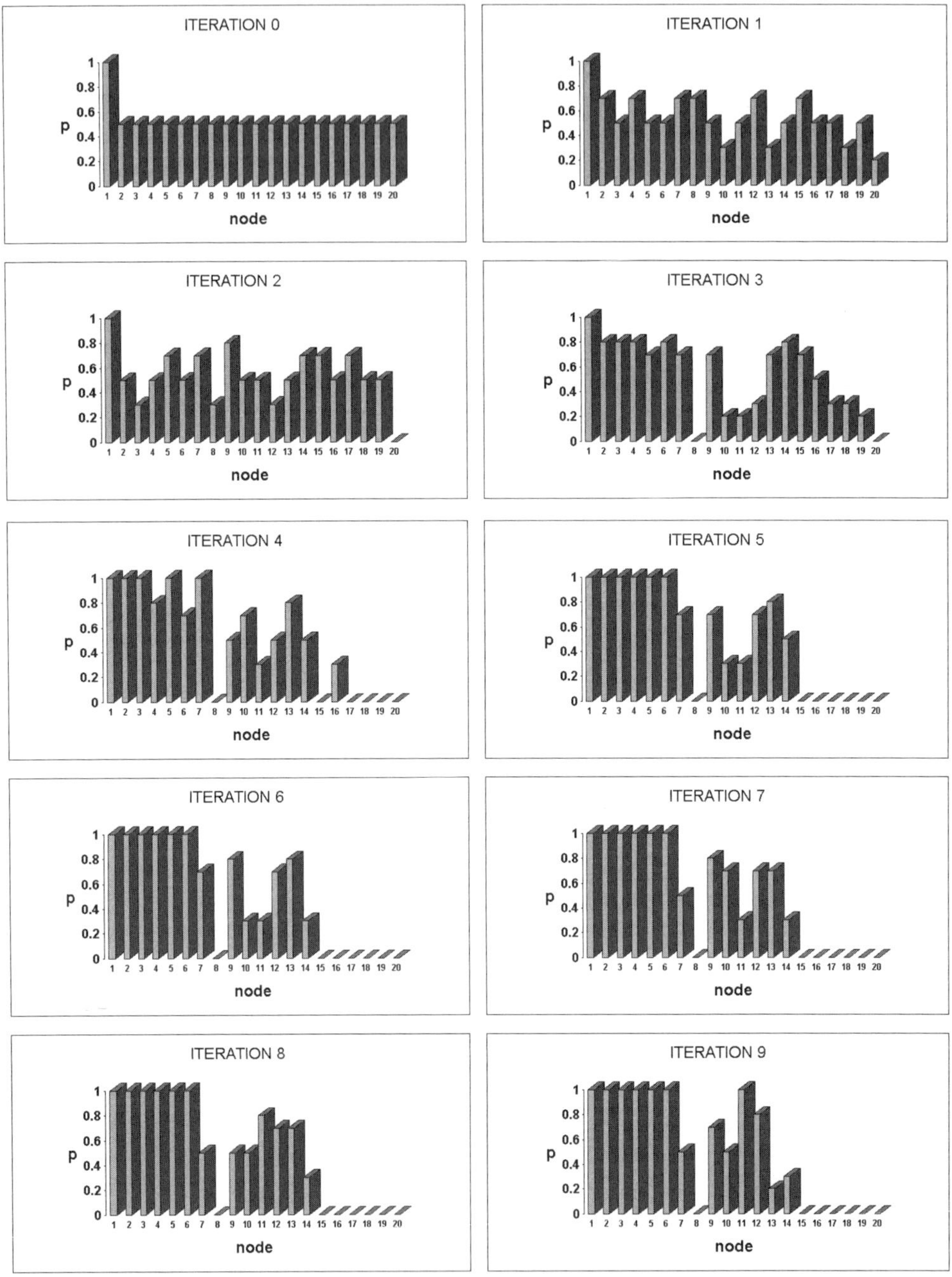

Fig. 4.6. The charts of $\widehat{\mathbf{p}}_t$ as function of t (part a).

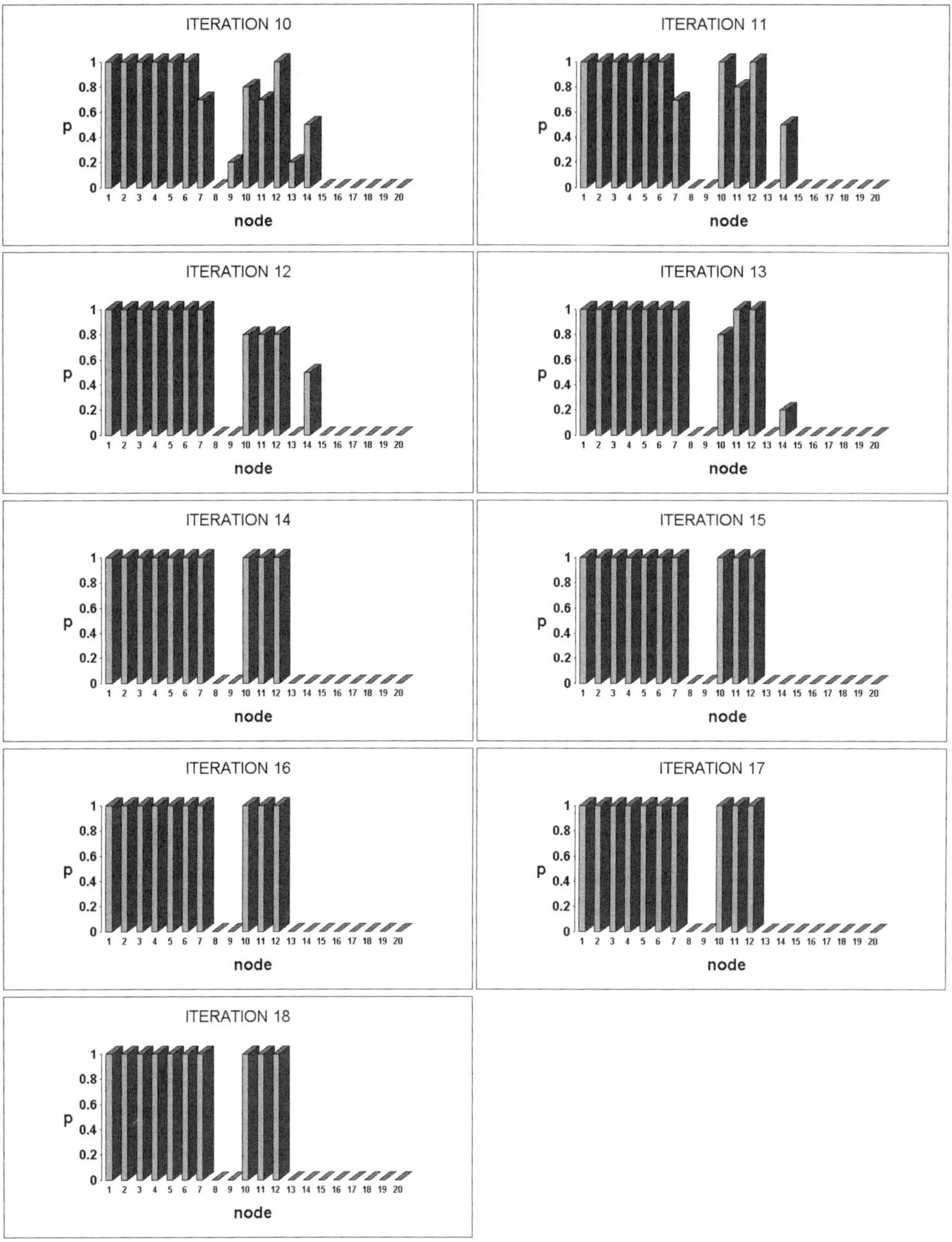

Fig. 4.7. The charts of $\widehat{\mathbf{p}}_t$ as function of t (part b).

It can be readily seen that Algorithm 4.5.2 oscillates between iterations 5 and 13 from one optimal solution to another, but after iteration 13 it settles down in one of the true 20-dimensional (degenerated) solutions. The convergence properties of Algorithm 4.5.2 in the presence of multiple solution is still an open problem.

4.9.4 Empirical Computational Complexity

Let us finally discuss the computational complexity of Algorithm 4.5.2 for the max-cut and the partition problems, which can be defined as

$$\kappa_n = T_n(N_n G_n + U_n) \ . \tag{4.55}$$

Here T_n is the total number of iterations needed before Algorithm 4.5.2 stops; N_n is the sample size, that is, the total number of maximal cuts and partitions generated at each iteration; G_n is the cost of generating the random Bernoulli vectors of size n for Algorithm 4.5.2; $U_n = \mathcal{O}(N_n n^2)$ is the cost of updating the tuple $(\widehat{\gamma}_t, \widehat{\mathbf{p}}_t)$. The latter follows from the fact that computing $S(\mathbf{X})$ in (4.21) is a $\mathcal{O}(n^2)$ operation.

For the model in (4.51) we found empirically that $T_n = \mathcal{O}(\ln n)$, provided $100 \le n \le 1000$. For the max-cut problem, considering that we take $n \le N_n \le 10\,n$ and that G_n is $\mathcal{O}(n)$, we obtain $\kappa_n = \mathcal{O}(n^3 \ln n)$. In our experiments, the complexity we observed was more like

$$\kappa_n = \mathcal{O}(n \ln n) \ .$$

The partition problem has similar computational characteristics. It is important to note that these empirical complexity results are solely for the model with the cost matrix (4.51).

4.10 Numerical Results for SENs

Below we present the results of numerical experiments involving CE Algorithm 4.7.3 for the TSP and QAP. We note that various numerical results for the TSP have already been given in Section 2.5.2, including several benchmark TSP examples. The notation in the tables will be similar to that for SNN experiments in Table 4.3. If not stated otherwise we set the parameters as usual (see also Remark 4.4): $\varrho = 10^{-2}$, $\alpha = 0.7$, $N = 5 \cdot n^2$ and use the stopping rule (4.25) with the parameter $d = 5$. To indicate the convergence of $\widehat{P}_t$ to the ODTM P^* we introduce the following quantities

$$P_t^{mm} = \min_{1\le i\le n} \max_{1\le j\le n} \widehat{p}_{t,ij} \ , \tag{4.56}$$

$t = 1, 2, \ldots$, which corresponds to the min max value of the elements of the matrix $\widehat{P}_{t,ij}$ at iteration t, and, similarly, the max min value

$$P_{t,mm} = \max_{1\le i\le n} \min_{1\le j\le n} \widehat{p}_{t,ij} \ . \tag{4.57}$$

Observe that $P_T^{mm} = 1$ if and only if $\widehat{P}_T = P^*$. In the tables $S_{t,(1)}$ denotes the smallest order statistic of the performances in iteration t; in other words, the length of the smallest tour obtained in iteration t.

4.10.1 Synthetic TSP

Since the TSP is an NP-hard problem, no efficient exact methods are available to verify the accuracy of our method for large n. Total enumeration of all the paths is suitable only for small networks, say with $n = 10$ cities containing $9! = 362{,}880$ trajectories. To still assess the accuracy and speed of the CE algorithm for a large TSP we construct the following artificial problem for which the solution is available in advance: Let the cost matrix C be such that $c_{i,i+1} = 1$, for all $i = 1, 2, \ldots, n-1$ and $c_{n,1} = 1$, while the remaining elements are $c_{ij} \sim \mathsf{U}(a, b),\ j \neq i+1,\ b > a,\ a > 1$ and $b_{ii} \equiv 0$.

It follows that in this case the optimal permutation/tour is given by $\mathbf{x}^* = (1, 2, 3, \ldots, n)$, corresponding to the sequence of visits $1 \to 2 \to 3 \to \cdots \to n \to 1$. The corresponding minimal value of the tour is

$$\gamma^* = \sum_{i=1}^{n-1} c_{i,i+1} + c_{n,1} = n\ . \tag{4.58}$$

Table 4.15 presents a typical evolution of the CE Algorithm 4.7.3 for $n = 30$. The relative experimental error of the final solution was $\varepsilon = 0.0019$. The algorithm was implemented in Matlab and took approximately 45 seconds to converge on a 1.4GHz processor with 256M RAM.

Table 4.15. A typical evolution of Algorithm 4.7.3 for the synthetic TSP with $n = 30$, $d = 5$, $\varrho = 0.01$, $N = 4500$, and $\alpha = 0.7$. $\varepsilon = 0.0019$.

t	$\widehat{\gamma}_t$	$S_{t,(1)}$	P_t^{mm}
1	40.73	39.12	0.0722
2	38.19	36.83	0.0994
3	36.09	34.58	0.1218
4	34.63	32.92	0.1268
5	33.64	32.37	0.1719
6	32.89	31.80	0.1493
7	32.11	31.17	0.2208
8	31.54	30.74	0.3162
9	31.22	30.13	0.2704
10	30.97	30.39	0.2519
11	30.79	30.14	0.3211
12	30.68	30.00	0.2498
13	30.29	30.00	0.2754

The dynamics of this particular problem including an illustration of the convergence can be found in Section 2.5.2.

4.10.2 Multiple Solutions

We consider the behavior of Algorithm 4.7.3 for the TSP in the presence of multiple solutions, similar to Section 4.9.3 for the max-cut problem. In order to generate multiple solutions we first generate our synthetic model of Section 4.10.1. We then select deterministically or at random m different tours and assign to all edges in the tour the cost 1. As a consequence, in addition to the tour $1 \to 2 \to 3 \to \ldots \to n \to 1$ there will be m other tours that have the minimal length $\gamma^* = n$. Performing simulation studies for this model we found that Algorithm 4.7.3 in most cases finds one of the optimal solutions γ^* in a finite number of iterations.

Clearly, in the case of multiple solution one needs to use the stopping rule associated with the level $\widehat{\gamma}_t$ rather than the stopping rule associated with the probability matrix $\widehat{P}_t$. The convergence proof of Algorithm 4.7.3 in this case is still an open problem.

Tables 4.16–4.18 present the evolution of the CE algorithm for a multi-extremal TSP with 40 nodes and 10 optimal solutions.

Table 4.16. Multi-extremal synthetic TSP, $c_{ij} \sim \mathsf{U}(1.0, 10.0)$, 40 nodes, $\gamma^* = 40$, $\varrho = 0.01$, $N = 8000$, $\alpha = 0.90$, 10 optimal tours. CPU time 58 seconds, $\varepsilon = 0.000$.

t	$\widehat{\gamma}_t$	P_t^{mm}	$P_{t,mm}$	$S_{t,(1)}$
1	139.0758	0.058043	0.002564	113.7909
2	103.6459	0.085165	0.000256	81.79360
3	77.32870	0.098468	0.000026	58.98756
4	61.26692	0.122118	0.000003	49.89649
5	52.36412	0.128535	0.000000	45.38383
6	47.44017	0.165937	0.000000	42.24399
7	44.65781	0.151682	0.000000	40.92816
8	42.94660	0.180555	0.000000	40.77399
9	42.22027	0.156351	0.000000	40.00000
10	41.54227	0.159557	0.000000	40.00000
11	41.07858	0.154678	0.000000	40.08350
12	40.89977	0.145107	0.000000	40.00000
13	40.82557	0.147619	0.000000	40.00000
14	40.64345	0.166954	0.000000	40.00000
15	40.62949	0.149387	0.000000	40.00000
16	40.49546	0.144624	0.000000	40.00000
17	40.38665	0.138180	0.000000	40.00000

Table 4.17. Multi-extremal synthetic TSP, $c_{ij} \sim \mathsf{U}(1.0, 10.0)$, 40 nodes, $\gamma^* = 40$, $\varrho = 0.01$, $N = 8000$, $\alpha = 0.90$, 25 optimal tours. CPU time 31 seconds, $\varepsilon = 0.000$.

t	$\widehat{\gamma}_t$	P_t^{mm}	$P_{t,mm}$	$S_{t,(1)}$
1	95.86915	0.049209	0.002564	71.78275
2	65.69504	0.082939	0.000256	53.80846
3	48.94862	0.084247	0.000026	44.08474
4	42.11559	0.101557	0.000003	40.00000
5	40.11133	0.099725	0.000000	40.00000
6	40.00000	0.112419	0.000000	40.00000
7	40.00000	0.112296	0.000000	40.00000
8	40.00000	0.103212	0.000000	40.00000
9	40.00000	0.108986	0.000000	40.00000

Table 4.18. Multi-extremal synthetic TSP, $c_{ij} \sim \mathsf{U}(1.0, 10.0)$, 40 nodes $\gamma^* = 40$, $\varrho = 0.01$, $N = 8000$, $\alpha = 0.90$, 50 optimal tours. CPU time 24 seconds, $\varepsilon = 0.000$.

t	$\widehat{\gamma}_t$	P_t^{mm}	$P_{t,mm}$	$S_{t,(1)}$
1	59.92559	0.050516	0.002564	45.73649
2	43.07207	0.070143	0.000256	40.00000
3	40.00000	0.073504	0.000026	40.00000
4	40.00000	0.075902	0.000003	40.00000
5	40.00000	0.097586	0.000000	40.00000
6	40.00000	0.099463	0.000000	40.00000
7	40.00000	0.097126	0.000000	40.00000

It is interesting to note that before Algorithm 4.7.3 settles down into a particular γ^* it oscillates for a number of iterations from one P^* to another P^*, although it does not appear to result in an increase in number of iterations or in CPU time. Moreover, the larger the subset of optimal solutions, the smaller is the effort to find one of them.

4.10.3 Experiments with Sparse Graphs

Up to now we considered experiments with fully connected graphs. Yet many problems are defined on cost matrices that are not fully connected and even sparse (many entries are set to ∞).

Next, we present numerical results for such problems where the original cost matrix generation is slightly modified to produce "sparse" random cost matrices. We denote by ϱ_{cost} the *density* of a cost matrix, that is, the fraction of active edges relative to the total number of edges of fully connected matrix. To obtain a TSP case with $\varrho_{\text{cost}} < 1$ we first generate our basic cost matrix C as described in Section 4.10.1. Next, for a given $\varrho_{\text{cost}} < 1.0$ we replace at random $(1.0 - \varrho_{\text{cost}}) \times 100$ of the generated c_{ij} with infinities. Thus we obtain a cost matrix with density ϱ_{cost}. Tables 4.19–4.21 present the evolution of the CE algorithm for the deterministic TSP with $\varrho_{\text{cost}} < 1.0$.

Table 4.19. Synthetic TSP, $c_{ij} \sim \mathsf{U}(1.0, 10.0)$, 30 nodes, $\gamma^* = 30$, $\varrho = 0.01$, $N = 4500$, $\alpha = 0.90$, $\varrho_{\text{cost}} = 0.60$. CPU time 20 seconds, $\varepsilon = 0.000$.

t	$\widehat{\gamma}_t$	P_t^{mm}	$P_{t,mm}$	$S_{t,(1)}$
1	730.3994	0.079062	0.003448	440.3003
2	345.7853	0.107328	0.000345	134.3467
3	142.7576	0.114207	0.000034	115.4077
4	116.8807	0.148245	0.000003	94.46148
5	96.78283	0.202041	0.000000	78.22054
6	83.67821	0.221003	0.000000	69.11775
7	73.48995	0.202565	0.000000	57.08287
8	65.36932	0.209768	0.000000	50.82449
9	57.85579	0.290023	0.000000	43.66945
10	51.21792	0.292054	0.000000	41.83288
11	46.08596	0.310255	0.000000	38.35064
12	40.20582	0.386022	0.000000	33.35134
13	35.56522	0.410951	0.000000	30.00000
14	30.00000	0.921320	0.000000	30.00000
15	30.00000	0.992132	0.000000	30.00000
16	30.00000	0.999213	0.000000	30.00000
17	30.00000	0.999921	0.000000	30.00000
18	30.00000	0.999992	0.000000	30.00000

Table 4.20. Synthetic TSP, $c_{ij} \sim \mathsf{U}(1.0, 10.0)$, 40 nodes, $\gamma^* = 40$, $\varrho = 0.01$, $N = 8000$, $\alpha = 0.90$, $\varrho_{\text{cost}} = 0.50$. CPU time 84 seconds, $\varepsilon = 0.000$.

t	$\widehat{\gamma}_t$	P_t^{mm}	$P_{t,mm}$	$S_{t,(1)}$
1	1432.600	0.048886	0.002564	1083.389
2	850.0822	0.071265	0.000256	481.6481
3	453.7181	0.099729	0.000026	276.6252
4	204.5736	0.094775	0.000003	167.9254
5	171.2220	0.120172	0.000000	144.3944
6	148.9051	0.137320	0.000000	123.5032
...	...	...	...	...
11	91.11670	0.182737	0.000000	75.15271
12	85.11291	0.208025	0.000000	63.81974
13	78.54369	0.225161	0.000000	60.25669
14	73.96963	0.214676	0.000000	55.54310
...	...	...	...	...
19	47.37309	0.505358	0.000000	40.00000
20	42.05268	0.499078	0.000000	40.00000
21	40.00000	0.936332	0.000000	40.00000
22	40.00000	0.993633	0.000000	40.00000
23	40.00000	0.999363	0.000000	40.00000
24	40.00000	0.999936	0.000000	40.00000

Table 4.21. Synthetic TSP, $c_{ij} \sim \mathsf{U}(1.0, 10.0)$, 40 nodes, $\gamma^* = 40$, $\varrho = 0.01$, $N = 8000$, $\alpha = 0.90$, $\varrho_{\text{cost}} = 0.30$. CPU time 98 seconds, $\varepsilon = 0.000$.

t	$\widehat{\gamma}_t$	P_t^{mm}	$P_{t,mm}$	$S_{t,(1)}$
1	2115.702	0.058218	0.002564	1630.689
2	1456.354	0.093351	0.000256	1092.379
3	957.4554	0.109995	0.000026	567.8778
4	570.2860	0.124202	0.000003	304.9498
5	365.9991	0.141733	0.000000	178.9949
6	261.9922	0.139430	0.000000	158.5351
7	182.4486	0.128288	0.000000	138.5457
8	168.1792	0.130095	0.000000	135.2442
9	155.0441	0.136634	0.000000	122.6527
10	147.4868	0.164139	0.000000	118.4093
11	137.9609	0.156376	0.000000	106.2428
12	132.3045	0.171164	0.000000	99.46488
13	127.3059	0.187035	0.000000	100.9402
14	122.5778	0.160873	0.000000	96.32555
15	117.6652	0.149974	0.000000	87.01289
16	112.2542	0.155534	0.000000	76.07462
17	104.4893	0.217362	0.000000	80.15214
18	99.34178	0.221301	0.000000	78.23600
19	94.08737	0.173795	0.000000	65.02988
20	88.06668	0.208399	0.000000	61.89260
21	80.64057	0.197207	0.000000	58.10715
22	75.45361	0.276194	0.000000	47.93998
23	68.30934	0.359635	0.000000	40.00000
24	55.84366	0.616132	0.000000	40.00000
25	40.00000	0.961613	0.000000	40.00000
26	40.00000	0.996161	0.000000	40.00000
27	40.00000	0.999616	0.000000	40.00000
28	40.00000	0.999962	0.000000	40.00000

4.10.4 Numerical Results for the QAP

Table 4.22 presents the performance of Algorithm 4.7.3 for $N = 15\,n^2$ for a number of quadratic assignment problem case studies, taken from the URL `www.imm.dtu.dk/~sk/qaplib`. Here we use the same notations as in Table 2.5 for the TSP, with the exception that $\widehat{\gamma}_1^*$ and $\widehat{\gamma}_T^*$ denote the best of the 10 solutions after the first and the final iteration. Note that for $n = 26$ we obtained a solution *better than the best known one.*

Table 4.22. Case studies for QAP.

file	n	$\bar{T}$	$\widehat{\gamma}_1^*$	$\widehat{\gamma}_T^*$	γ^*	$\bar{\varepsilon}$	ε_*	ε^*	CPU
had12	12	33	1772	1666	1652	0.012	0.018	0.008	4.1
had14	14	37	2964	2743	2724	0.014	0.019	0.007	6.5
had16	16	79	4020	3736	3720	0.006	0.009	0.004	15.2
had18	18	112	5748	5412	5358	0.016	0.023	0.010	46.3
had20	20	94	7474	7022	6920	0.020	0.027	0.014	81.8
bur26a	26	172	5547428	5346546	5426670	−0.011	0.002	−0.015	240.5
tho30	30	64	157932	153998	149936	0.045	0.060	0.027	610

4.11 Appendices

4.11.1 Two Tour Generation Algorithm for the TSP

Algorithm 4.7.1 specifies how we can generate a random tour/permutation $\mathbf{X} = (X_1, \ldots, X_n)$ for the TSP. Here is an alternative, but equivalent, representation (see Section 4.4, in particular the first paragraph, and [145]). Define X_{ij} $(i, j \in \{1, \ldots, n\})$ as:

$$X_{ij} = \begin{cases} 1, & \text{if edge } (i,j) \text{ belongs to the current tour,} \\ 0, & \text{otherwise.} \end{cases} \tag{4.59}$$

Obviously, if $X_{ij^*} = 1$, we can conclude that $X_{ij} = 0$, for $j \neq j^*$. The following algorithm, due to Margolin [113], gives a detailed implementation of Algorithm 4.7.1 when tours are represented by the collection $\mathbf{X} = \{X_{ij}\}$.

Let the probability p_{ij} correspond to a transition from node i to node j.

Algorithm 4.11.1 (Node Transition Algorithm)

1. *Define the probability matrix* P:

$$P = \begin{pmatrix} 0 & p_{12} & \ldots & p_{1n} \\ p_{21} & 0 & \ldots & p_{2n} \\ \vdots & \vdots & \vdots & \vdots \\ p_{n1} & p_{n2} & \ldots & 0 \end{pmatrix} . \tag{4.60}$$

   (Assume, that all $p_{ij} \neq 0$, *for* $i \neq j$.*)*
2. *Set counter* $= 1$, $U = \{1\}$, $i = 1$.
3. *Generate a next node* J *from the following distribution:*

$$\mathbb{P}(J = j) = \begin{cases} p_{i1}, & j = 1 , \\ p_{i2}, & j = 2 , \\ p_{i3}, & j = 3 , \\ \vdots & \\ p_{in}, & j = n . \end{cases}$$

Assume that J has received value j^*. Define

$$X_{ij} = \begin{cases} 1, j = j^* , \\ 0, j \neq j^* . \end{cases}$$

4. *Set $U = U \cup \{j^*\}$, counter $=$ counter $+ 1$, $i = j^*$.*
5. *If (counter $< n - 1$) update the i-th row of the matrix:*

```
sum = 0
for j = 1 to n
    if (j ≠ i) and (j ∈ U)
        sum = sum + p_ij
        p_ij = 0
    end if
end for
for j = 1 to n
    if (j ∉ U)
        p_ij = p_ij/(1 − sum)
    end if
end for
```

Go to Step 3.

6. *Set $\{j^*\} = V \setminus U$ (at this step U contains exactly $n - 1$ nodes), $X_{ij^*} = 1$.*

Recall that we can generate trajectories either via *node transitions* (Algorithms 4.7.1 and 4.11.1) or via *node placements*, as in Algorithm 4.7.2. Similar to Algorithm 4.11.1 we next present a more detailed description of Algorithm 4.7.2, using the following representation: Define

$$Y_{ij} = \begin{cases} 1, \text{ the node } i \text{ is arranged to the place } j \text{ in the current tour,} \\ 0, \text{ otherwise.} \end{cases} \tag{4.61}$$

Obviously, if $Y_{ij^*} = 1$, then $Y_{ij} = 0$ (for $j \neq j^*$). Let the probability $p_{(i,j)}$ correspond to the following event: the node i stands on the j-th place in the permutation.

Algorithm 4.11.2 (Node Placement Algorithm)

1. *Define the probability matrix P:*

$$P = \begin{pmatrix} p_{(1,1)} & p_{(1,2)} & \cdots & p_{(1,n)} \\ p_{(2,1)} & p_{(2,2)} & \cdots & p_{(2,n)} \\ \vdots & \vdots & \vdots & \vdots \\ p_{(n,1)} & p_{(n,2)} & \cdots & p_{(n,n)} \end{pmatrix} . \tag{4.62}$$

(Assume, that all $p_{(i,j)} \neq 0$.)

2. *Set $U = \emptyset$, $i = 1$.*
3. *Generate the random variable J from the following distribution:*

$$\mathbb{P}(J = j) = \begin{cases} p_{(i,1)}, & j = 1\ , \\ p_{(i,2)}, & j = 2\ , \\ p_{(i,3)}, & j = 3\ , \\ \quad\vdots & \\ p_{(i,n)}, & j = n\ . \end{cases}$$

Assume that J has received value j^*. Define

$$Y_{ij} = \begin{cases} 1, \ j = j^*\ , \\ 0, \ j \neq j^*\ . \end{cases}$$

4. *Set $U = U \cup \{j^*\}$, $i = i + 1$.*
5. *Update the i-th row of the matrix:*

```
sum = 0
for j = 1 to n
    if (j ∈ U)
        sum = sum + p(i,j)
        p(i,j) = 0
    end if
end for
for j = 1 to n
    if (j ∉ U)
        p(i,j) = p(i,j)/(1 − sum)
    end if
end for
```

If $i < n$ go to Step 3.

6. *Set $\{j^*\} = V \setminus U$ (at this step U contains exactly $n - 1$ nodes), $Y_{n,j^*} = 1$.*

As noted in Remark 4.12 we can use the alias method to speed up trajectory generation. In this case an initial setup and extra storage is required, but since the alias method is very fast it is preferred to the inverse-transform and the acceptance–rejection methods for large dimensions. For the TSP and problems with a similar solution generation we have used a combination of the alias, acceptance–rejection, and inverse-transform methods as described below.

Note that our goal is to generate a permutation of n numbers using the probability matrix P. For simplicity consider Algorithm 4.11.2, that is, the i-th number in a permutation is generated using the i-th row from P. We propose to generate the first x% of the numbers using a combination of the alias method and the acceptance–rejection method. The remaining numbers are generated using the inverse-transform method. Let $Alias(i, P)$ denote a random number generated by the alias method using the i-th row from P. This leads to the following modification of Algorithm 4.11.2.

Algorithm 4.11.3 (Faster Trajectory Generation via Alias)

$U = \emptyset$
for $i = 1$ *to* $[xn]$
    repeat
        $j^* = Alias(i, P)$
    until $j^* \notin U$
    $Y_{ij^*} = 1$
    $U = U \cup \{j^*\}$
end for
for $i = [xn] + 1$ *to* $n - 1$
    Update the i-th row of P as in Step 5 of Algorithm 4.11.2
    Generate j^* *as in Step 3 of Algorithm 4.11.2*
    $Y_{ij^*} = 1$
    $U = U \cup \{j^*\}$
end for
Determine the final j^* *from* $\{j^*\} = V \setminus U$. *Set* $Y_{n,j^*} = 1$.

One should not take x too large, because for large values of x (that is, close to 100%), the algorithm will, as it progresses, generate too many rejections for each acceptance. On the other hand, since the alias method is much faster than the inverse-transform method, x should not be too small either. It has been found numerically that the above algorithm performs best with $x \approx 70-80\%$.

4.11.2 Speeding up Trajectory Generation

Notice that each trajectory of Algorithms 4.7.1 and 4.7.2 requires $n-1$ times the generation from n-point discrete pdf, corresponding to the rows of the matrices $P^{(k)}$. This is clearly time consuming. To overcome this difficulty we shall present next a fast and simple online procedure for trajectory generations for SEN having in mind the TSP. This procedure will employ the well-known *composition technique*; see Section 1.7.3.

Consider Algorithm 4.7.1. Assume as before that $P = (p_{ij})$ is given and that after $k-th$ transitions we arrived at some city (state) X_k, $(k < n-1)$. Let $X_1, \ldots, X_{k-1}$ be the previously selected cities. In Algorithm 4.7.1 we select the $(k+1)$-st city X_{k+1} by sampling from the n-point distribution formed by the X_k-th row of matrix $P^{(k+1)}$, which is simply the X_k-th row of the original P with all probabilities corresponding to the cities $X_1, \ldots, X_k$ set to zero, and the rest properly normalized. Let us denote this probability vector from which we sample by $\mathbf{p} = (p_1, \ldots, p_n)$. Our goal is to show how to sample more efficiently from $\mathbf{p}$ using the new approach.

First, as observed before, k cities have been "marked," corresponding to the already generated part of the tour. We shall call the associated k and $n-k$ elements of $\mathbf{p}$, the *passive* and *active* elements of $\mathbf{p}$.

The main idea of the new approach is to

1. Divide the *unmarked* cities into three groups, denoted $\mathcal{I}(j), j = 1, 2, 3$ such that each p_i with $i \in \mathcal{I}(1)$ is $< \frac{1}{n}$, each p_i with $i \in \mathcal{I}(2)$ is between $\frac{1}{n}$ and $\frac{2}{n}$ and each p_i with $i \in \mathcal{I}(3)$ is $> \frac{2}{n}$.
2. *Approximate* the probabilities for group $\mathcal{I}(1)$ and $\mathcal{I}(2)$ by making them *equal* within each group, while the elements of the third group (with larger values) remain untouched.
3. Generate the desired city X_{k+1} using a simple pdf (see (4.64)), which is based on the approximated distributions.

As we shall see below such a procedure will speed up substantially the trajectory generation while affecting (due to the approximation of probabilities of the cities in the first two groups) very little the accuracy of the CE algorithm.

For $j = 1, 2, 3$, let τ_j denote the number of elements in each group of unmarked cities $\mathcal{I}(j)$. We then proceed as follows.

- Calculate
$$\delta_j = \sum_{i \in \mathcal{I}(j)} p_i, \quad j = 1, 2, 3 \ . \tag{4.63}$$
Note that since the probabilities for the marked cities are 0, we have that $\delta_1 + \delta_2 + \delta_3 = 1$.
Associate the ordered elements in $\mathcal{I}(1)$ with $\{1, \ldots, \tau_1\}$, the ordered elements in $\mathcal{I}(2)$ with $\{\tau_1 + 1, \ldots, \tau_1 + \tau_2\}$, and the ordered elements in $\mathcal{I}(3)$ with $\{\tau_1 + \tau_2 + 1, \ldots, n - k\}$.
- Define a new discrete pdf φ on $\{1, 2, 3\}$ with $\varphi(j) = \delta_j$, $j = 1, 2, 3$.
- Define two *new* discrete pdfs, $h_1(x)$ and $h_2(x)$, which are uniformly distributed on $\{1, \ldots, \tau_1\}$ and on $\{\tau_1 + 1, \tau_1 + \tau_2\}$, respectively. These are the pdfs of the approximated probabilities in the first and second group.
- Define a *new* τ_3-point discrete pdf $h_3(x)$, on $\{\tau_1 + \tau_2 + 1, \ldots, n - k\}$ that is associated with the original probabilities of the cities in the third group.
- Based on the four discrete pdfs above ($h_i(x)$, $i = 1, 2, 3$, and $\varphi(y)$) define the following pdf

$$\phi(x) = \sum_{i=1}^{3} h_i(x)\varphi(i) = \sum_{i=1}^{3} h_i(x)\delta_i \ . \tag{4.64}$$

The pdf $\phi(x)$ will be used to generate the next city X_{k+1} using the composition method; see Section 1.7.3. The associated algorithm is as follows:

Algorithm 4.11.4 (Fast Algorithm for Trajectory Generation)

1. *Generate a random variate* Y *with the outcomes* $1, 2, 3$ *from* $\varphi(y)$.
2. *Given* $Y = i$, *generate a random variate from the corresponding pdf* $h_i(x)$. *Denote the resulting outcome by* r, $r = 1, \dots, n - s$.
3. *Let* X_{k+1} *be the city that is matched to the outcome* r *above.*

Remark 4.17. After several iterations with the CE Algorithm (4.7.3) we will typically have that $\delta_3 > \delta_j$, $j = 1, 2$, and that (for large n) the number of elements τ_3 in the third group will be much smaller than in the first two. This means that we will be sampling mainly from $h_3(x)$ rather than from $h_j(x), j = 1, 2$, which is beneficial since for small τ_3 matching the outcome r to X_{k+1} is fast.

Remark 4.18. Note that for $\tau_1 = \tau_2 = 0$, we obtain that $\tau_3 = n - s$ and thus arrive at conventional sampling. When $\tau_2 = \tau_1$ we obtain only two groups — the second group being empty. In such case it is again desirable to choose the elements of the first group $< \frac{1}{n}$, while the elements of the third group $\geq \frac{1}{n}$.

4.11.3 An Analysis of Algorithm 4.5.2 for a Partition Problem

In this section we show that for a special case of the partition problem (or the max-cut problem) the CE algorithm can be evaluated exactly. Specifically, we can employ the deterministic rather than the stochastic version of Algorithm 4.5.2. This can be used to examine the behavior of the deterministic CE algorithm, and enables us to compare the stochastic and deterministic versions of the algorithm.

Consider a fully connected network with $n = 2\,m$. Assume that

$$c_{ij} = \begin{cases} a, & 1 \leq i, j \leq m \quad \text{or} \quad m + 1 \leq i, j \leq n; \\ b, & \text{otherwise}, \end{cases} \tag{4.65}$$

where $a < b$. Thus, (4.65) defines the symmetric distance matrix

$$C = \begin{pmatrix} A & B \\ B & A \end{pmatrix}, \tag{4.66}$$

where all components of the upper left-hand and lower right-hand quadrants equal a and the remaining components equal b.

We want to partition the graph into two equal parts (sets), such that the sum of distances (weights of the edges running from one subset to the other) is maximized.

Clearly, the optimal partition is $V^* = \{V_1^*, V_2^*\}$, with $V_1^* = \{1, \dots, m\}$. We assume $1 \in V_1$. Moreover, by symmetry, we can assume that the distribution of cut vector $\mathbf{X} = (X_1, X_2, \dots, X_n)$ at stage t of the CE algorithm is of the form $\mathbf{p}_t = (1, p_{t,1}, \ldots, p_{t,1}, p_{t,2}, \ldots, p_{t,2})$, so at each iteration we only have two parameters to update.

Denote by $\mathcal{A}_i \subseteq \mathcal{X}$, $i = 0, \ldots, m-1$ the subsets of $\mathcal{X}$ of all possible partitions obtained from the optimal partition V^* by replacing i nodes from V_1^* with i nodes from V_2^*. It is readily seen that

$$\mathcal{X} = \bigcup_{i=0}^{m-1} \mathcal{A}_i \ . \tag{4.67}$$

The cardinality of $\mathcal{A}_i$ represents the total number of ways in which we can choose i nodes from $V_1^* \setminus \{1\}$ and i nodes from V_2^*, which is given by

$$|\mathcal{A}_i| = \binom{m-1}{i}\binom{m}{i} \ . \tag{4.68}$$

The cardinality of the set of all possible bipartitions is

$$|\mathcal{X}| = \sum_{i=0}^{m-1} |\mathcal{A}_i| = \frac{n!}{2m!\,m!} \ .$$

Denote by S_i, $i = 0, \ldots, m-1$, the cost of a bipartition associated with the subset $\mathcal{A}_i$. Then, the optimal bipartition V^* corresponds to the cost $S_0 \equiv \gamma^* = b\,m^2$, and

$$S_i = S_0 - (b-a)i(n-i) = bm^2 - (b-a)i(n-i) \ . \tag{4.69}$$

We shall calculate now the deterministic sequence of triplets $\{(\gamma_t, p_{t,1}, p_{t,2})\}$ $(t = 1, 2, \ldots)$ using (4.11), (4.12),(4.67) and (4.68).

First, note that

$$\mathbb{P}_{\mathbf{p}_t}(\mathbf{X} \in \mathcal{A}_i) = \binom{m-1}{i} p_{t,1}^{m-1-i}(1-p_{t,1})^i \times \binom{m}{i} p_{t,2}^i (1-p_{t,2})^{m-i} \ . \tag{4.70}$$

We assume that $\mathbf{p}_0 = (1, \frac{1}{2}, \ldots, \frac{1}{2})$. That is, with probability $\left(\frac{1}{2}\right)^{m-1}$ we generate any of $|\mathcal{X}|$ possible bipartitions.

Second, for a given $\mathbf{p}_{t-1}$ and ϱ, say $\varrho = 0.01$, we determine γ_t from (4.12). Since $S(\mathbf{X})$ can only take the values $S_0 > S_1 > \ldots > S_{m-1}$ on the sets $\mathcal{A}_0, \ldots, \mathcal{A}_{m-1}$, it follows that γ_t is given by

$$\gamma_t = S_{m-k} \ ,$$

where k is the largest integer k $(1 \leq k \leq m)$ such that

$$\mathbb{P}_{\mathbf{p}_{t-1}}(S(\mathbf{X}) \geqslant S_{m-k}) = \sum_{i=m-k}^{m-1} \mathbb{P}_{\mathbf{p}_{t-1}}(\mathbf{X} \in \mathcal{A}_i) \geqslant \varrho \ ,$$

where $\mathbb{P}_{\mathbf{p}_{t-1}}(\mathbf{X} \in \mathcal{A}_i)$ is given in (4.70). Of course k can be different at each iteration.

Next, in the usual way we find that the deterministic updating formula (4.12) for $p_{t,1}$ is

$$p_{t,1} = \frac{\mathbb{E}_{\mathbf{p}_{t-1}} I_{\{S(\mathbf{X}) \geqslant \gamma_t\}} (m-1)^{-1} \sum_{i=2}^{m} X_i}{\mathbb{E}_{\mathbf{p}_{t-1}} I_{\{S(\mathbf{X}) \geqslant \gamma_t\}}} , \tag{4.71}$$

and similar for $p_{t,2}$ (replacing $(m-1)^{-1} \sum_{i=2}^{m} X_i$ with $m^{-1} \sum_{i=m+1}^{n} X_i$). The denominator of (4.71), is given by

$$\mathbb{P}_{\mathbf{p}_{t-1}}(S(\mathbf{X}) \geqslant S_{m-k}) = \sum_{i=0}^{m-k} \mathbb{P}_{\mathbf{p}_{t-1}}(\mathbf{X} \in \mathcal{A}_i)$$

and the numerator by

$$\begin{aligned}
&\mathbb{E}_{\mathbf{p}_{t-1}} I_{\{S(\mathbf{X}) \geqslant \gamma_t\}} \frac{\sum_{i=2}^{m} X_i}{m-1} \\
&= \sum_{i=0}^{m-k} \mathbb{E}_{\mathbf{p}_{t-1}} I_{\{\mathbf{X} \in \mathcal{A}_i\}} \frac{\sum_{i=2}^{m} X_i}{m-1} \\
&= \sum_{i=0}^{m-k} \mathbb{E}_{\mathbf{p}_{t-1}} I_{\{\mathbf{X} \in \mathcal{A}_i\}} \frac{m-1-i}{m-1} \\
&= \sum_{i=0}^{m-k} \frac{m-1-i}{m-1} \, \mathbb{P}_{\mathbf{p}_{t-1}}(\mathbf{X} \in \mathcal{A}_i) .
\end{aligned}$$

Similar explicit expressions can be found for $p_{t,2}$. Hence, we can explicitly calculate the sequence of triplets $\{(\gamma_t, p_{t,1}, p_{t,2})\}$ as a function of t, n, ϱ, a, b and also the stopping time T of Algorithm 4.5.2, as a function of n, ϱ, a, b.

We employed Algorithm 4.5.2 for the model (4.65)–(4.66) using both simulation and the analytical formulas above. For the former, Algorithm 4.5.2 generated the standard stochastic sequences of tuples $\{(\widehat{\gamma}_t, \widehat{\mathbf{p}}_t)\}$, while for the latter we replaced $\{(\widehat{\gamma}_t, \widehat{\mathbf{p}}_t)\}$ by the deterministic sequences of tuples $\{(\gamma_t, \mathbf{p}_t)\}$. We obtained that the theoretical and the simulation result are very close provided that $N \geq 10\,n$ in the latter. Consider, for example, the simulation results in Table 4.5 for the cases $Z_1 = 4$ and $Z_2 = 4.9$, which correspond to $\varepsilon_1(T_1)$ and $\varepsilon_2(T_2)$. For this case we found that for $N \geq 10\,n$ the deterministic $\{(\gamma_t, \mathbf{p}_t)\}$ and the stochastic $\{(\widehat{\gamma}_t, \widehat{\mathbf{p}}_t)\}$ sequences are, in fact, indistinguishable in the sense that the stopping time T is the same for both cases, and after several iterations with Algorithm 4.5.2 both sequences $\{\gamma_t\}$ and $\{\widehat{\gamma}_t\}$ differ by no more than 10^{-2}.

4.11.4 Convergence of CE Using the Best Sample

The purpose of this section is to provide insight in the convergence properties of the CE algorithm. We consider thereto a very simple, but also general,

CE-like algorithm in which the updating is performed on the basis of a *single* elite (best) sample. In contrast, Algorithm 4.2.1 updates the parameters on the basis of $\lceil \varrho N \rceil$ elite samples. We investigate under what general conditions the algorithm will converge to the optimal solution. We show that convergence is guaranteed, provided that the probability of obtaining a better value in the next iteration increases fast enough and in a smooth way. For alternative convergence proofs of the CE method, see Chapter 3 and [85, 113, 145], respectively.

Let $\mathcal{X}$ be a finite set and $S(x)$ be a real-valued function on $\mathcal{X}$. We are interested in finding the maximum of S on $\mathcal{X}$, and to this end use the following general randomized algorithm:

Algorithm 4.11.5

1. *Pick a point $X_t \in \mathcal{X}$ according to a probability density f_t on $\mathcal{X}$.*
2. *Compute the quantity $S(X_t)$ and the best found value so far of S*

$$S_t = \max_{1 \leqslant \tau \leqslant t} S(X_\tau) .$$

3. *Update the probability density f_t, thus getting a new probability density f_{t+1}, replace t with $t+1$, and loop.*

The above description, of course, is generic — we have not specified how the initial density f_1 is chosen and what is the updating $f_t \mapsto f_{t+1}$. Our goal is to point out general conditions under which we can guarantee the following two desirable properties of the algorithm:

A. Convergence with probability 1 as $t \to \infty$, of the best found value so far S_t of S to the maximum value $\gamma^* = \max_{x \in \mathcal{X}} S(x)$;
B. Convergence, with probability 1 as $t \to \infty$, of the density f_t to a density supported by the set $\mathcal{X}^*$ of maxima of S on $\mathcal{X}$.

A natural sufficient condition for Property **A** is given by the following proposition.

Proposition 4.19. *Assume that*

(i) *The initial density is positive everywhere on $\mathcal{X}$:*

$$f_1(x) > 0 \quad \textit{for all } x \in \mathcal{X} ;$$

(ii) *The updating rules $f_t \mapsto f_{t+1}$ are "safe" in the following sense:*

$$f_{t+1}(x) \geqslant e^{-a_t} f_t(x) \quad \textit{for all } x \in \mathcal{X} ,$$

where $a_t \geqslant 0$ are such that

$$\sum_{t=1}^{T-1} a_t \leqslant \ln(T) + C \tag{4.72}$$

for all $T \geq 2$ and a certain $C \in (0, \infty)$.

Then,

$$\mathbb{P}(S_t = \gamma^* \textit{ for all large enough values of } t) = 1 . \tag{4.73}$$

Proof. Let x^* be one of the maximizers of S on $\mathcal{X}$. From the description of our generic algorithm it is clear that

$$1 - \mathbb{P}(S_t = \gamma^* \text{ for all large enough values of } t) \leqslant \mathbb{P}(X_t \neq x^* \text{ for all } t)$$

so that to prove (4.73) it suffices to verify that $\mathbb{P}(X_t \neq x^* \text{ for all } t) = 0$, or, (which is the same) that

$$\lim_{T\to\infty} \mathbb{P}(X_t \neq x^*,\, t = 1, \ldots, T) = 0 . \tag{4.74}$$

Setting $p = f_1(x^*)$, we have

$$f_t(x^*) \geqslant p\,\mathrm{e}^{-a_1}\mathrm{e}^{-a_2}\cdots\mathrm{e}^{-a_{t-1}} \geqslant \underbrace{p\exp\{-C\}}_{\beta\in(0,1)}\, t^{-1}$$

where the concluding inequality is implied by (4.72). Consequently

$$\begin{aligned}
\mathbb{P}(X_t \neq x^*,\, t = 1, ..., T) &= \prod_{t=1}^{T} (1 - f_t(x^*)) \\
&\leqslant \prod_{t=1}^{T} \left(1 - \beta t^{-1}\right) \\
&\leqslant \prod_{t=1}^{T} \exp\{-\beta t^{-1}\} \\
&\text{[since } \exp\{-s\} \geqslant 1 - s \geqslant 0 \text{ for } 0 \leqslant s \leqslant 1] \\
&= \exp\{-\beta \sum_{t=1}^{T} t^{-1}\} \to 0,\ T \to \infty ,
\end{aligned} \tag{4.75}$$

and (4.74) follows. □

Now let us give a sufficient condition for Property **B**. Consider a realization $x_1, x_2, ...$ of algorithm's trajectory. We call τ the *record time* for the sequence $S(x_i)$, if $\tau = 1$ or $\tau > 1$ and $S(x_t) < S(x_\tau)$ for all $t < \tau$ (that is, at time instant τ we pick a point which is better, in terms of S, than all points we have seen before). For $t \geqslant 1$, let $\tau(t)$ be the last record time which is $\leqslant t$, and let $x^t = x_{\tau(t)}$ be the *record* at instant t (that is, the best, in terms of S, among the points $x_1, ..., x_t$; if there are several best points among $x_1, ..., x_t$, the record x^t is the first of them). Let us make the following natural assumption on the updating rules $f_t \mapsto f_{t+1}$:

> $(*)$ Assume that a point $x \in \mathcal{X}$ becomes a record at a certain time T and remains record till time $U \geqslant T$: $x = x^T = x^{T+1} = ... = x^U$ (and $S(x^{U+1}) > S(x)$). There exists a sequence $\{\beta_t \geqslant 0\}_{t=1}^{\infty}$ (perhaps depending on x and T as on parameters) such that $\sum_{t=1}^{\infty} \beta_t = \infty$ and
>
> $$1 - f_{t+1}(x^T) \leqslant (1 - \beta_t)(1 - f_t(x^T)),\ t = T, T+1, ..., U .$$

(∗) says that the rule for updating the current probability density increases the probability of the current record, and this increase is, in a sense, not too small.

Proposition 4.20. *Under condition (∗), the density f_t converges, with probability 1 as $t \to \infty$, to a single-point density. If, in addition to (∗), the conditions of Proposition 4.19 take place, the limiting distribution with probability 1 corresponds to a maximizer of S on $\mathcal{X}$.*

Proof. Since $\mathcal{X}$ is finite, the sequence of records, along every trajectory of the algorithm, stabilizes: starting from a time instant T (depending on the trajectory), we have $x^T = x^{T+1} = \dots$. Let $\{\beta_t\}$ be the sequence corresponding to (x^T, T) in view of (∗). Then by (∗) for $t > T$ one has

$$1 - f_t(x^T) \leqslant (1 - \beta_T) \cdots (1 - \beta_{t-1})(1 - f_T(x^T))$$

and the left-hand side in this inequality tends to 0 as $t \to \infty$ due to $\sum_t \beta_t = \infty$. Thus $f_t(x^T) \to 1$ as $t \to \infty$, and therefore the density f_t along the trajectory in question converges to the single-point density corresponding to the point x^T — the last record in the realization. Under conditions of Proposition 4.19, this point, by Proposition 4.19, is, with probability 1, a maximizer of S on $\mathcal{X}$.

Example 4.21 (Bernoulli Case). Consider the case where

- $\mathcal{X} = \{0, 1\}^n$ is the collection of all 2^n n-dimensional vectors $\mathbf{x}$ with coordinates x_i taking values 0 and 1;
- f_t, for every t, is the pdf of $\mathbf{X} = (X_1, \dots, X_n)$ where $\mathbf{X}$ has independent components with $X_i \sim \mathsf{Ber}(p_{t,i})$; in other words,

$$f_t(\mathbf{x}) = \prod_{i=1}^{n} p_{t,i}^{x_i} (1 - p_{t,i})^{1 - x_i} .$$

Let $\mathbf{p}_t = (p_{t,1}, \dots, p_{t,n})$. Assume, for the sake of simplicity, that $\mathbf{p}_1$ of f_1 has all coordinates equal to 1/2 and that the rules for updating $f_t \to f_{t+1}$ are as follows:

> At step t, after the corresponding record $\mathbf{x}^t$ is defined, we convert $\mathbf{p}_t$ into $\mathbf{p}_{t+1}$ (which induces the transformation $f_t \mapsto f_{t+1}$) by shifting $p_{t,i}$ "in the directions of x_i^t." Specifically, for certain given real numbers $\beta_{t,i} > 0$, we decrease $p_{t,i}$ for i with $x_i^t = 0$ and increase $p_{t,i}$ for i with $x_i^t = 1$ according to the following rules:
>
> - *In the case of* $x_i^t = 0$:
>
> $$p_{t+1,i} = (1 - \beta_{t,i})\, p_{t,i} .$$
>
> - *In the case of* $x_i^t = 1$:
>
> $$1 - p_{t+1,i} = (1 - \beta_{t,i})(1 - p_{t,i}) .$$

It can be easily verified that if

$$\frac{\varepsilon_-}{t} \leqslant \min_{1 \leqslant i \leqslant n} \beta_{t,i} \leqslant \max_{1 \leqslant i \leqslant n} \beta_{t,i} \leqslant \frac{\varepsilon_+}{t} ,$$

with $0 \leqslant \varepsilon_- \leqslant \varepsilon_+ \leqslant \frac{1}{2n}$, then the conditions of Proposition 4.19 as well as condition $(*)$ are satisfied.

4.12 Exercises

Max-Cut

1. Consider the deterministic max-cut algorithm in Example 4.7.
   a) Repeat the steps of the example, using $\varrho = 1/3$ instead of $\varrho = 0.1$. Show that $\mathbf{p}_t$ converges to $\mathbf{p}^* = (1, 1, 0, 0, 0)$.
   b) Repeat the same, but now with the indicator $I_{\{S(\mathbf{X}) \geqslant \gamma_t\}}$ in Step 3 of Algorithm 4.2.2 replaced with $I_{\{S(\mathbf{X}) \geqslant \gamma_t\}} S(\mathbf{X})$. Show that $\mathbf{p}_t$ again converges to $\mathbf{p}^*$. Does the deterministic version of Algorithm 4.5.2 converge with the same number of iterations as in (a)?
   c) Run Algorithm 4.2.1 on this problem and obtain a table similar to Table 2.3.

TSP

2. Consider the 4-node TSP with distance matrix

$$C = \begin{pmatrix} \infty & 1 & 2 & 3 \\ 2 & \infty & 1 & 3 \\ 2 & 1 & \infty & 3 \\ 1 & 3 & 2 & \infty \end{pmatrix} .$$

   Carry out, by hand, the deterministic CE Algorithm 4.2.2 for this problem, using node transitions (see also Algorithm 4.7.3 for more details). Take the following transition matrix

$$P = \begin{pmatrix} 0 & p_{12} & p_{13} & p_{14} \\ p_{21} & 0 & p_{23} & p_{24} \\ p_{31} & p_{32} & 0 & p_{34} \\ p_{41} & p_{42} & p_{43} & 0 \end{pmatrix} ,$$

   set $\varrho = 1/3$. Start with the initial P_0 where all off-diagonal elements are 1/3.

3. Run Algorithm 4.2.1 on the data from the URL

`http://www.iwr.uni-heidelberg.de/groups/comopt/software/TSPLIB95/atsp/`

   and obtain a table similar to Table 2.5.

4. Choose a benchmark TSP of approximately 100 nodes. Select 15 nodes at random, and construct a miniature TSP as described in Remark 4.13. Find a good choice for (α, C, ϱ) by trial and error, and apply this to the original problem. Check if this procedure produces better results than the "standard" parameter choice $(0.7, 5, 0.01)$.

5. Select a TSP of your choice. Verify the following statements about the choice of CE parameters:
   a) Reducing ϱ or increasing α the convergence is faster but we can be trapped in a local minimum.
   b) Reducing ϱ one needs to decrease simultaneously α and visa versa in order to avoid convergence to a local minimum.
   c) Increasing the sample size N one can simultaneously reduce ϱ or (and) increase α.

6. The longest path problem (LPP) is a celebrated problem in combinatorics. Consider a complete graph with n nodes. With each edge from node i to j is associated a cost c_{ij}. For simplicity we assume that all costs are nonnegative and finite. Some edges can have zero cost. The goal is to find the *longest* self-avoiding path from a certain *source* node to a *sink* node. The complexity of any known LPP algorithm grows geometrically in the number of nodes. This is in sharp contrast with the *shortest* path problem (SPP) for which efficient (i.e., polynomially bounded) algorithms exist, such as Dijkstra's algorithm [48] or the Bellman-Ford algorithm [21, 56], which can also be applied to negative-weight graphs.
   a) Assuming the source node is 1 and the sink node is n, formulate the LPP similar to the TSP in (4.32). (Note that the main difference with the TSP is that the vectors, or paths, in the LPP can have different lengths.)
   b) Specify a path generation mechanism and the corresponding updating rules for the CE algorithm.
   c) Verify that the ODTM P^* in general does not have only ones and zeros, in contrast to the ODTM of the TSP.

5

Continuous Optimization and Modifications

In this chapter we discuss how the CE method can be used for continuous multi-extremal optimization, and provide a number of modifications of the basic CE algorithm, applied to both combinatorial and continuous multi-extremal optimization. Specific modifications include the use of alternative "reward/loss" functions (Section 5.2), and the formulation of a fully automated version of the CE algorithm (Section 5.3). This modified version allows automatic tuning of all the parameters of the CE algorithm. Numerical results for continuous multi-extremal optimization problems and the FACE modifications are given in Sections 5.4, and 5.5, respectively.

5.1 Continuous Multi-Extremal Optimization

In this section we apply the CE Algorithm 4.2.1 to continuous multi-extremal optimization problems. We consider optimization problems of the form $\max_{\mathbf{x}\in\mathscr{X}} S(\mathbf{x})$ or $\min_{\mathbf{x}\in\mathscr{X}} S(\mathbf{x})$, where the objective function $S : \mathbb{R}^n \to \mathbb{R}$ is either unconstrained ($\mathscr{X} = \mathbb{R}^n$) or the constraints are very simple (for example, when $\mathscr{X}$ is an n-dimensional rectangle $[\mathbf{a}, \mathbf{b}]^n$). Generation of a random vector $\mathbf{X} = (X_1, \ldots, X_n) \in \mathscr{X}$ in such cases is straightforward. The easiest way is to generate the coordinates independently from an arbitrary 2-parameter distribution, such that by applying Algorithm 4.2.1 the joint distribution converges to the degenerated distribution at the point $\mathbf{x}^*$ where the global extremum is attained. Examples of such distributions are the normal, double-exponential, and beta distributions.

Example 5.1 (Normal Updating). We wish to optimize the function S given by

$$S(x) = \mathrm{e}^{-(x-2)^2} + 0.8\,\mathrm{e}^{-(x+2)^2}, \quad x \in \mathbb{R}\,.$$

Note that S has a local maximum at point -2.00 (approximately) and a global maximum at 2.00. At each stage t of the CE procedure we simulate a sample $X_1, \ldots, X_N$ from a $\mathsf{N}(\widehat{\mu}_{t-1}, \widehat{\sigma^2}_{t-1})$ distribution, determine $\widehat{\gamma}_t$ in the usual way, and update $\widehat{\mu}_t$ and $\widehat{\sigma}_t$ as the mean and standard deviation of all samples X_i that exceed level $\widehat{\gamma}_t$; see (3.66) and (3.67). Note that the likelihood ratio term is not used. A simple Matlab implementation may be found in Appendix A.3. The CE procedure is illustrated in Figure 5.1, using starting values $\widehat{\mu}_0 = -6, \widehat{\sigma}_0 = 100$ and CE parameters $\alpha = 0.7, \varrho = 0.1$ and $N = 100$. The algorithm is stopped when the standard deviation becomes smaller than 0.05. We observe that the vector $(\widehat{\mu}_t, \widehat{\sigma}_t)$ quickly converges to the optimal $(\mu^*, \sigma^*) = (2.00, 0)$, easily avoiding the local maximum.

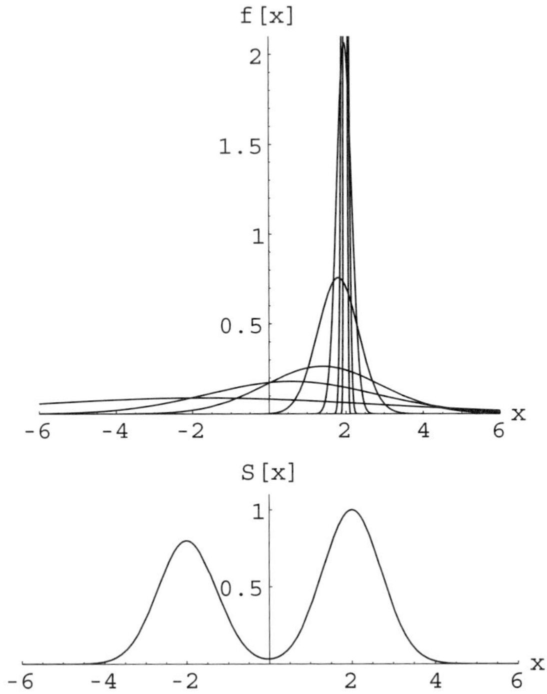

Fig. 5.1. Continuous Multi-Extremal Optimization.

The normal updating procedure above is easily adapted to the multidimensional case. Specifically, when the components of the random vectors $\mathbf{X}_1, \ldots, \mathbf{X}_N$ are chosen independently, updating of the parameters can be done separately for each component. While applying Algorithm 4.2.1, the mean vector $\widehat{\boldsymbol{\mu}}_t$ should converge to $\mathbf{x}^*$ and the vector of standard deviations $\widehat{\boldsymbol{\sigma}}$ to the zero vector. In short, we should obtain a degenerated pdf with all mass concentrated at the vicinity of the point $\mathbf{x}^*$. When using $\mathsf{Beta}(\mathbf{a}, \mathbf{b})$ random variables the updating of the parameters should be obtained from the numerical (rather than analytical) solution of (4.8), but again, after stopping Algorithm 4.2.1, the resulting parameter vectors $\widehat{\mathbf{a}}$ and $\widehat{\mathbf{b}}$ should define a degenerated pdf at the vicinity of the point $\mathbf{x}^*$.

Remark 5.2 (Smoothing Scheme). During the course of Algorithm 4.2.1 the sampling distribution "shrinks" to a degenerate distribution. When the smoothing parameter α is large, say 0.9, this convergence to a degenerate distribution may happen too quickly, which would "freeze" the algorithm in a suboptimal solution. To prevent this from happening the following dynamic smoothing scheme is suggested (considering the case where the sampling distribution is normal): At iteration t update the variance σ^2 using a smoothing parameter

$$\beta_t = \beta - \beta\left(1 - \frac{1}{t}\right)^q , \tag{5.1}$$

where q is a small integer (typically between 5 and 10), and β is a large smoothing constant (typically between 0.8 and 0.99). The parameter μ can be updated in the conventional way, with constant smoothing parameter α, say between 0.7 and 1. By using β_t instead of α the convergence to the degenerate case has polynomial speed instead of exponential. We have used (5.1) so far only for the Rosenbrock case; see below.

In Section 5.4 we demonstrate the performance of Algorithm 4.2.1, where we will consider the *minimization* of two well-known test functions: the *Rosenbrock* function

$$S(\mathbf{x}) = \sum_{i=1}^{n-1} 100\,(x_{i+1} - x_i^2)^2 + (x_i - 1)^2 \tag{5.2}$$

and the *trigonometric* function

$$S(\mathbf{x}) = 1 + \sum_{i=1}^{n} 8\sin^2(\eta(x_i - x_i^*)^2) + 6\sin^2(2\eta(x_i - x_i^*)^2) + \mu(x_i - x_i^*)^2 \,. \tag{5.3}$$

The graphical representations of Rosenbrock's and the trigonometric functions for $\eta = 7$, $\mu = 1$, $x_i^* = x^* = 0.9$ in the two-dimensional case are given in Figures 5.2 and 5.3, respectively. It is not difficult to see that in the n-dimensional case the global minima for the Rosenbrock and the trigonometric function are attained at points $\mathbf{x}^* = (1, 1, \ldots, 1)$ and $\mathbf{x}^* = (0.9, 0.9, \ldots, 0.9)$, respectively. The corresponding minimal function values are $S(\mathbf{x}^*) = 0$ and $S(\mathbf{x}^*) = 1$, respectively. If not stated otherwise we assume (as in [145]) for the trigonometric function that $\eta = 7$, $\mu = 1$.

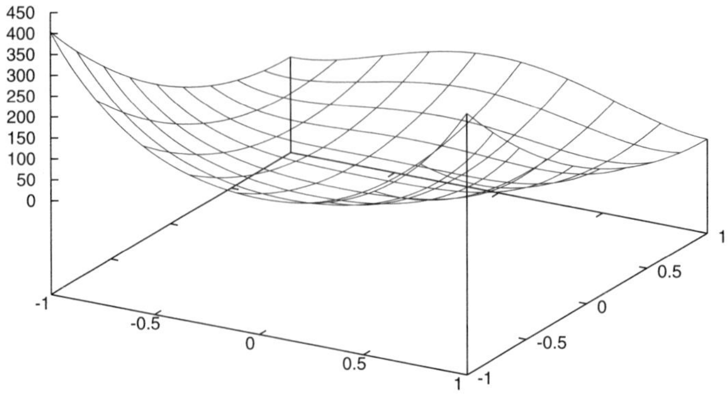

Fig. 5.2. Rosenbrock's function in $\mathbb{R}^2$ for $-1 \leqslant x_i \leqslant 1$.

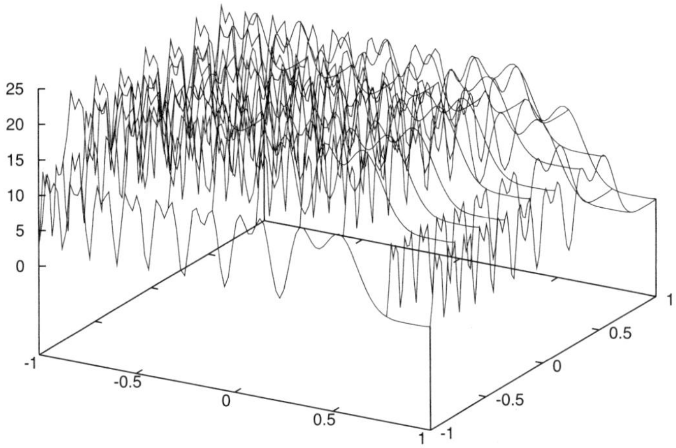

Fig. 5.3. The trigonometric function in $\mathbb{R}^2$ with $\eta = 7$, $\mu = 1$, $x_i^* = x^* = 0.9$ and $-1 \leqslant x_i \leqslant 1$.

5.2 Alternative Reward Functions

Consider the maximization problem (4.2) where $S(\mathbf{x})$ is some positive objective function defined on $\mathcal{X}$. In the associated stochastic problem (4.3) we consider instead estimating the rare-event probability

$$\ell(\gamma) = \mathbb{E}_{\mathbf{u}} I_{\{S(\mathbf{X}) \geqslant \gamma\}} \; .$$

Note that this can be written as

$$\ell(\gamma) = \mathbb{E}_{\mathbf{u}} \varphi(S(\mathbf{X}); \gamma) ,$$

where $\varphi(s;\gamma)$ is the indicator $I_{\{s \geqslant \gamma\}}$, that is,

$$\varphi(s;\gamma) = \begin{cases} 1 & \text{if } s \geqslant \gamma , \\ 0 & \text{if } s < \gamma , \end{cases} \tag{5.4}$$

for $s \geqslant 0$. The standard CE algorithm contains the now well-known steps (a) updating $\widehat{\gamma}_t$ via (4.6), and (b) updating $\widehat{\mathbf{v}}_t$ via the (analytic) solution of (4.8). A natural modification of Algorithm 4.2.1 would be to update $\mathbf{v}_t$ using an alternative function $\varphi(s;\gamma)$. For a maximization problem such a function should be increasing in s for each fixed $\gamma \geqslant 0$, and decreasing in γ for each fixed $s \geqslant 0$. In particular one could use

$$\varphi(s;\gamma) = I_{\{s \geqslant \gamma\}} \, \psi(s) ,$$

for some increasing function $\psi(s)$. Using such a $\varphi(s;\gamma)$ instead of the indicator (5.4) we now proceed similarly to how we went before. Specifically, the updating step (a) of $\widehat{\gamma}_t$ remains *exactly the same*, and the updating step (b) of $\widehat{\mathbf{v}}_t$ now reduces to the solution of the following program

$$\max_{\mathbf{v}} \widehat{D}(\mathbf{v}) = \max_{\mathbf{v}} \frac{1}{N} \sum_{i=1}^{N} I_{\{S(\mathbf{X}_i) \geqslant \widehat{\gamma}_t\}} \, \psi(S(\mathbf{X}_i)) \ln f(\mathbf{X}_i; \mathbf{v}) . \tag{5.5}$$

Numerical evidence suggests that $\psi(s) = s$ can lead to some speed-up of the algorithm. As an example, for the max-cut problem we obtain in this case (see (4.24)) the analytic updating formulas

$$\widehat{p}_{t,i} = \frac{\sum_{k=1}^{N} I_{\{S(\mathbf{X}_k) \geqslant \widehat{\gamma}_t\}} \, S(\mathbf{X}_i) \, X_{ki}}{\sum_{k=1}^{N} I_{\{S(\mathbf{X}_k) \geqslant \widehat{\gamma}_t\}} \, S(\mathbf{X}_i)} , \tag{5.6}$$

$i = 2, \ldots, n$.

However, other numerical experiments suggest that high-power polynomials, $\psi(s) = s^{\beta}$ with large β, and the Boltzmann function $\psi(s) = \mathrm{e}^{-s/\beta}$ are not advisable, as they may lead more easily to local minima.

5.3 Fully Adaptive CE Algorithm

We present a modification of Algorithm 4.2.1, called the *fully adaptive* (or *automated*) *CE* (FACE) algorithm in which the sample size is updated adaptively at each iteration t of the algorithm, that is, $N = N_t$. In addition, this modification is able to identify some "difficult" and pathological problems.

Consider a *maximization* problem and let

$$S_{t,(1)} \leqslant \ldots \leqslant S_{t,(N_t)}$$

denote the *ordered* sample performances of the N_t samples at the t-th iterations. To make notation easier, we denote $S_{t,(N_t)}$ by S_t^*.

The main assumption in the FACE Algorithm is that at the end of each iteration t the updating of the parameters is done on the basis of a *fixed* number, N^{elite}, say, of the best performing samples, the so-called *elite* samples. Thus, the set of elite samples $\mathcal{E}_t$ consist of those N^{elite} samples in $\{\mathbf{X}_1, \ldots, \mathbf{X}_{N_t}\}$ for which the performances $S(\mathbf{X}_1), \ldots, S(\mathbf{X}_{N_t})$ are highest. The updating Steps 2 and 3 of Algorithm 4.2.1 are modified such that

$$\widehat{\gamma}_t = S_{(N_t - N^{\text{elite}}+1)} ,$$

and

$$\widehat{\mathbf{v}}_t = \operatorname*{argmax}_{\mathbf{v}} \sum_{\mathbf{X}_i \in \mathcal{E}_t} \ln f(\mathbf{X}_i; \mathbf{v}) .$$

Note that $\widehat{\gamma}_t$ is equal to the worst sample performance of the best N^{elite} sample performances, and S_t^* is the best of the elite performances (indeed, of all performances).

In the FACE algorithm the parameters ϱ and N of Algorithm 4.2.1 are updated adaptively. Specifically, they are "replaced" by a single parameter: the number of elite samples N^{elite}. The above updating rules are consistent with Algorithm 4.2.1 — provided we view ϱ in Algorithm 4.2.1 as the parameter which changes inversely proportional to N_t: $\varrho_t = N^{\text{elite}}/N_t$.

It was found experimentally that a sound choice is $N^{\text{elite}} = c_0\, n$ and $N^{\text{elite}} = c_0\, n^2$ for SNNs and SENs, respectively, where c_0 is a fixed positive constant (usually in the interval $0.01 \leq c_0 \leq 0.1$). The easiest way to explain FACE is via the flow chart in Figure 5.4.

For each iteration t of the FACE algorithm we design a sampling plan which ensures with high probability that

$$S_t^* > S_{t-1}^* . \tag{5.7}$$

Note that (5.7) implies *improvement* of the maximal order statistics (best elite performance) at each iteration. To ensure (5.7) with high probability, we allow N_t to vary at each iteration t in a quite wide range

$$N^{\min} \leq N_t \leq N^{\max},$$

where, say, for SNN-type problems $N^{\min} = n$ (or $N^{\min} = N^{\text{elite}}$), and $N^{\max} = 20\, n$. Note that we always start each iteration by generating $N^{\min}$ samples. If at any iteration t, while increasing N_t we obtain that $N_t = N^{\max}$ and (5.7) is violated, then we directly proceed with updating $(\widehat{\gamma}_t, \widehat{\mathbf{v}}_t)$ according to Algorithm 4.2.1 before proceeding with the next iteration of the FACE Algorithm. However, if FACE keeps generating samples of size $N^{\max}$ for several iterations in turn, say, for $c = 3$ iterations, then we stop, and announce

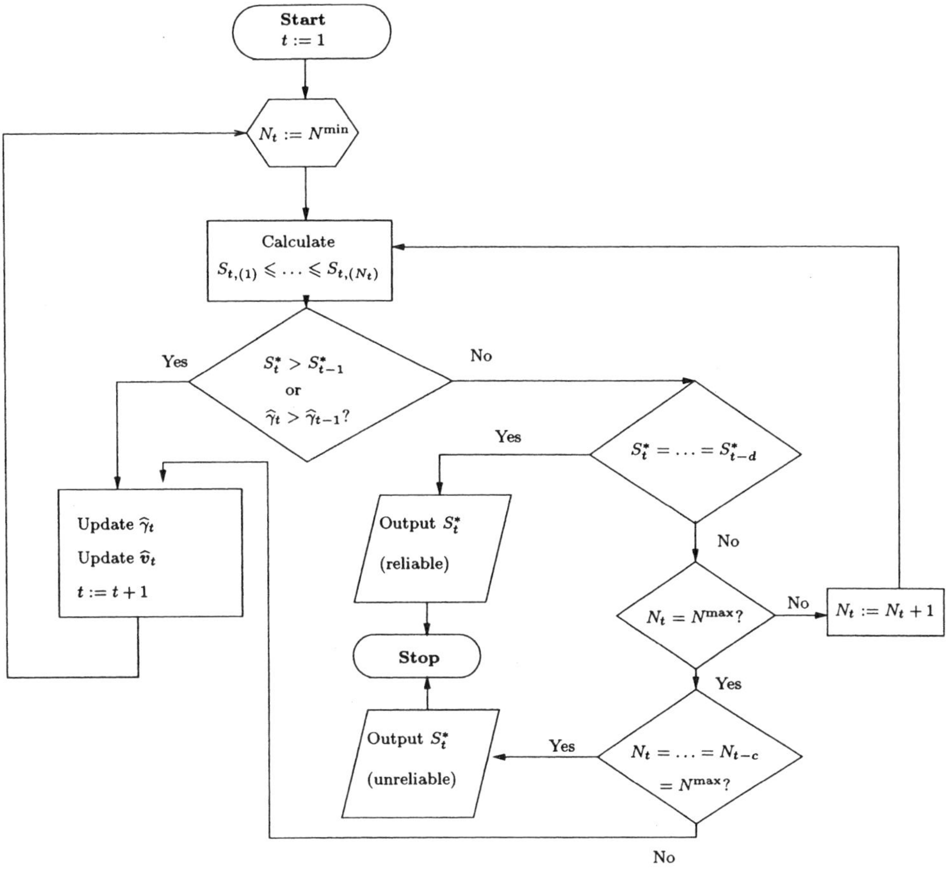

Fig. 5.4. The flowchart for the FACE algorithm.

that FACE identified a "hard" problem for which the estimate of the optimal solution is unreliable. Parallel to (5.7) we will require at each iteration that

$$\widehat{\gamma}_t > \widehat{\gamma}_{t-1} . \tag{5.8}$$

Note that (5.8) implies *improvement* of the worst elite sample performance $\widehat{\gamma}_t$ at each iteration.

Similar to Algorithm 4.2.1 we initialize by choosing some $\widehat{\mathbf{v}}_0$. For example, for the max-cut problem we choose (using $\mathbf{p}$ instead of $\mathbf{v}$), $\widehat{\mathbf{p}}_0 = \mathbf{p}_0 = (1, 1/2, \ldots, 1/2)$. We assume that the FACE parameters $N^{\min}$, $N^{\max}$, α and the stopping constants c and d are chosen in advance. For example, for the max-cut problem we take $N^{\min} = n$. We let $t = 1$ and $N_1 = N^{\min}$ and proceed as follows:

Algorithm 5.3.1 (FACE Algorithm)

1. *At iteration* t, $t = 1, 2, \ldots$ *take an initial sample of size* N_t, *with* $N^{\min} \leq N_t \leq N^{\max}$ *from* $f(\cdot; \widehat{\mathbf{v}}_{t-1})$. *Denote the corresponding ordered sample performances by* $S_{t,(1)} \leqslant \cdots S_{t,(N_t)}$.
2. *If (5.7) or (5.8) holds, proceed with the updating steps (4.6) and (4.8) using the* N_t *samples in Step 1.*
3. *If (5.7) and (5.8) are violated, check whether or not*

$$S_t^* = \cdots = S_{t-d}^* \,. \tag{5.9}$$

If so, stop and deliver S_t^* *as an estimate of the optimal solution. Call such* S_t^* *a* **reliable** *estimate of the optimal solution. If* (5.9) *does not hold, and if* $N_t < N^{\max}$, *increase* N_t *by 1, recalculate* S_t^* *and* $\widehat{\gamma}_t$, *and repeat Step 3.*

4. *If* $N_t = N^{\max}$ *and each of (5.7), (5.8) and (5.9) is violated, proceed with (4.6) and (4.8) using the* $N^{\max}$ *samples mentioned in Step 1 and go to Step 3.*
5. *If* $N_t = N^{\max}$ *for several iterations in turn, say for* $c = 3$ *iterations, and each of (5.7), (5.8) and (5.9) is violated, stop and announce that FACE identified a "hard" problem. Call* S_t^* *an* **unreliable** *estimate of the optimal solution.*

The stopping criterion (5.9) means that the best samples in the last d iterations are the same. Note that if (5.7) holds for *all* $t \geq 1$ we automatically obtain that $N_t = N^{\min}$ for all t. In such case FACE reduces to the original Algorithm 4.2.1. In Section 5.5 we illustrate a number of numerical experiments with the FACE algorithm.

5.4 Numerical Results for Continuous Optimization

We apply Algorithm 4.2.1 to the Rosenbrock and trigonometric functions of Section 5.1. For both cases we use the normal sampling distribution with independent components. For an n-dimensional function we thus need to update at each iteration n means and n variances. We observed that the algorithm finds the global optimum flawlessly, even for highly irregular cases such as the Rosenbrock function with $n = 50$. For the Rosenbrock function it is important that the modified smoothing scheme of Remark 5.2 is applied, in order to obtain convergence to the global minimum. For the trigonometric function the standard (fixed) smoothed updating scheme suffices. The updating of the parameters is carried out on the basis of a fixed number of samples $N^{\text{elite}} = \lceil \varrho N \rceil$. The Matlab implementation can be found in Appendix A.5.

Table 5.1 presents the evolution of Algorithm 4.2.1 for the minimization of the Rosenbrock function on the region $[-\mathbf{2}, \mathbf{2}]^{10}$. We deal with this constraint by imposing a high *penalty* on the original Rosenbrock function outside this region. Specifically, for each coordinate x_i outside $[-2, 2]$ we add $10^9\, d_i$ to the original Rosenbrock function, where d_i is the distance from x_i to the interval $[-2, 2]$. We take $N = 1000$, $\alpha = 0.8$ and $N^{\text{elite}} = 20$. For the modified smoothing scheme of (5.1) we use $\beta = 0.9$ and $q = 6$. The initial means are chosen independently from a uniform distribution on [-2,2]; the initial variances are 10^4. In the table $S_{t,(1)}$ denotes the best (that is, smallest) function value found in iteration t, $\widehat{\gamma}_t$ the worst of the elite performances, and $\widehat{\boldsymbol{\mu}}_t$ the vector of means at the t-th iteration. In repeated experiments the global minimum was consistently found in less than 9 seconds on a 1.8GHz computer. It is interesting to note that the first component always converges first, then the second, then third, etc.

Table 5.1. The evolution of Algorithm 4.2.1 for the Rosenbrock function with $n = 10$.

t	$\widehat{\gamma}_t$	$S_{t,(1)}$	$\widehat{\boldsymbol{\mu}}_t$									
100	9.379	7.851	0.64	0.41	0.18	0.04	0.01	0.00	0.01	0.01	0.01	0.01
200	4.142	3.891	0.95	0.91	0.84	0.70	0.49	0.24	0.07	0.02	0.01	0.00
300	1.686	1.585	0.99	0.98	0.96	0.93	0.86	0.75	0.57	0.32	0.11	0.01
400	0.526	0.495	1.00	0.99	0.99	0.98	0.95	0.90	0.82	0.67	0.44	0.19
500	0.224	0.207	1.00	1.00	0.99	0.99	0.97	0.94	0.89	0.79	0.63	0.39
600	0.108	0.098	1.00	1.00	1.00	0.99	0.98	0.96	0.93	0.86	0.74	0.55
700	0.051	0.048	1.00	1.00	1.00	0.99	0.99	0.98	0.95	0.91	0.82	0.68
800	0.026	0.022	1.00	1.00	1.00	1.00	0.99	0.98	0.97	0.94	0.88	0.77
900	0.013	0.011	1.00	1.00	1.00	1.00	0.99	0.99	0.98	0.96	0.92	0.84
1000	0.007	0.005	1.00	1.00	1.00	1.00	1.00	0.99	0.99	0.97	0.94	0.89
1100	0.004	0.003	1.00	1.00	1.00	1.00	1.00	1.00	0.99	0.98	0.96	0.92
1200	0.002	0.001	1.00	1.00	1.00	1.00	1.00	1.00	0.99	0.99	0.97	0.95
1300	0.001	0.001	1.00	1.00	1.00	1.00	1.00	1.00	1.00	0.99	0.98	0.97
1400	0.001	0.001	1.00	1.00	1.00	1.00	1.00	1.00	1.00	1.00	0.99	0.98
1500	0.001	0.000	1.00	1.00	1.00	1.00	1.00	1.00	1.00	1.00	0.99	0.99
1600	0.000	0.000	1.00	1.00	1.00	1.00	1.00	1.00	1.00	1.00	1.00	0.99
1700	0.000	0.000	1.00	1.00	1.00	1.00	1.00	1.00	1.00	1.00	1.00	1.00
1800	0.000	0.000	1.00	1.00	1.00	1.00	1.00	1.00	1.00	1.00	1.00	1.00
1900	0.000	0.000	1.00	1.00	1.00	1.00	1.00	1.00	1.00	1.00	1.00	1.00
2000	0.000	0.000	1.00	1.00	1.00	1.00	1.00	1.00	1.00	1.00	1.00	1.00

Table 5.2 presents the evolution of Algorithm 4.2.1 for the trigonometric function with $\eta = 7$, $\mu = 1$ and $n = 10$. The same notation is used as in Table 5.1. The CE parameters are exactly the same as for the Rosenbrock case. In repeated experiments the global maximum was consistently found in less than 2 seconds on a 1.8GHz computer. Different values for the parameters μ and ν had little effect on the excellent accuracy of the method.

Table 5.2. The evolution of Algorithm 4.2.1 for trigonometric function with $\eta = 7$, $\mu = 1$ and $n = 10$.

t	$\widehat{\gamma}_t$	$S_{t,(1)}$	$\widehat{\boldsymbol{\mu}}_t$									
10	145.890	100.990	0.29	1.06	0.42	1.36	0.99	0.32	1.51	1.69	2.32	1.13
20	57.356	43.581	1.25	0.93	0.79	1.03	0.77	0.70	1.38	1.08	1.40	0.93
30	44.173	28.687	1.26	0.83	1.16	0.93	0.67	0.80	0.87	1.11	1.00	1.00
40	35.233	23.276	0.95	1.01	0.71	0.86	0.86	1.08	0.81	0.69	1.40	0.96
50	31.960	18.373	0.77	0.94	0.77	0.97	0.88	0.93	0.70	0.77	1.02	0.79
60	27.461	10.693	0.89	1.02	1.07	0.85	0.85	0.99	0.94	0.87	1.11	0.94
70	21.804	10.954	0.94	0.91	0.86	0.83	0.97	0.95	0.84	0.94	0.92	0.94
80	17.044	10.617	0.85	0.81	0.79	0.89	0.95	0.89	0.94	0.84	0.80	0.88
90	10.247	4.550	0.89	0.88	0.89	0.89	0.85	0.90	0.90	0.90	0.87	0.89
100	6.667	2.646	0.91	0.91	0.91	0.87	0.90	0.90	0.88	0.95	0.95	0.86
110	3.459	1.706	0.92	0.87	0.92	0.91	0.85	0.90	0.88	0.89	0.90	0.92
120	2.226	1.155	0.89	0.94	0.90	0.90	0.89	0.90	0.90	0.90	0.88	0.90
130	1.808	1.137	0.92	0.90	0.90	0.90	0.90	0.88	0.89	0.91	0.90	0.90
140	1.535	1.080	0.91	0.88	0.88	0.91	0.91	0.87	0.90	0.89	0.87	0.89
150	1.310	1.053	0.90	0.91	0.90	0.90	0.90	0.91	0.88	0.92	0.89	0.89
160	1.209	1.051	0.90	0.90	0.88	0.90	0.90	0.91	0.91	0.89	0.92	0.90
170	1.115	1.007	0.89	0.89	0.90	0.91	0.90	0.90	0.90	0.89	0.89	0.91
180	1.076	1.033	0.91	0.89	0.92	0.89	0.90	0.91	0.90	0.90	0.89	0.90
190	1.046	1.010	0.90	0.89	0.90	0.91	0.89	0.90	0.90	0.90	0.90	0.91
200	1.041	1.017	0.89	0.91	0.89	0.91	0.91	0.91	0.90	0.90	0.90	0.90
210	1.026	1.004	0.90	0.90	0.90	0.90	0.90	0.91	0.89	0.90	0.89	0.91
220	1.018	1.008	0.89	0.90	0.90	0.90	0.89	0.90	0.90	0.90	0.90	0.90
230	1.013	1.004	0.90	0.90	0.91	0.90	0.89	0.90	0.90	0.89	0.90	0.90
240	1.010	1.003	0.90	0.90	0.90	0.90	0.91	0.90	0.90	0.91	0.91	0.90

Remark 5.3 (Noisy Optimization). The experiments for the continuous optimization experiments above were repeated for their *noisy* counterparts, that is, when the objective functions are corrupted with noise. The CE algorithm worked well, provided the parameters were suitably altered (e.g., by increasing the sample size or decreasing β). Noisy optimization will be considered in detail in Chapter 6.

5.5 Numerical Results for FACE

Table 5.3 represents the results of the FACE Algorithm 5.3.1 applied to the artificial max-cut problem of Table 4.6. The parameter setup is almost the same, except for the fact that the sample size N_t is variable. In particular we set $c_0 = 0.1$, $N^{\min} = 200$, $\alpha = 0.7$. We obtained relative experimental error $\varepsilon = 0.01$. Comparing the results of Tables 4.6 and 5.3 indicates that the FACE algorithm behaves very similarly to the basic CE algorithm 4.5.2. But the big advantage is that the parameters ϱ and N need not be specified in advance and that sample size N_t for FACE is usually much smaller than the sample size 1000 in Table 4.6.

Table 5.3. Evolution of the FACE algorithm for the artificial max-cut problem (4.51) of Table 4.6, with $n = 200,\ \ m = 100$, $b = 1$, and $Z = 0$.

t	$\widehat{\gamma}_t$	$S_{t,(N_t)}$	N_t	$p \uparrow$max	$p \downarrow$min
1	5050	5162	200	0.5000	0.0000
2	5088	5234	207	0.5000	0.0000
3	5132	5260	200	0.5000	0.0031
4	5286	5560	200	0.5000	0.0031
5	5510	5840	200	0.5000	0.0035
6	5760	6254	200	0.5000	0.0015
7	6064	6560	200	0.5000	0.0042
8	6404	7046	200	0.5000	0.0017
9	6768	7516	200	0.5000	0.0005
10	7146	7800	200	0.5000	0.0138
11	7584	8040	200	0.5000	0.0095
12	7964	8280	200	0.5000	0.0028
13	8432	9048	200	0.5000	0.0018
14	8864	9140	312	0.5000	0.0522
15	9136	9324	200	0.5000	0.0051
16	9324	9606	200	0.5000	0.0549
17	9606	9900	200	0.5000	0.1615
18	9900	9900	4000	0.5000	0.3984

Table 5.4 represents the results of the FACE Algorithm 5.3.1 applied to the artificial TSP of Table 4.15. Again, the parameter setup is the same. The FACE parameters are $c_0 = 0.1$, $N^{\min} = 900$, $\alpha = 0.7$. We now obtained a relative experimental error $\varepsilon = 0$, that is, the optimal solution is found, which is an improvement on what was achieved with the basic method. We see that roughly 80% more iterations are needed, but that most sample sizes are less than the original 4500.

Table 5.5 illustrates the evolution of the FACE algorithm applied to the benchmark `ft53` TSP of Table 2.4.

Figure 5.5 and Table 5.6 present a more detailed comparison between the standard and FACE algorithm for the same `ft53` TSP. The results are averaged over 10 separate runs for each algorithm. In particular, Figure 5.5 depicts the mean and the standard deviation (vertical bars) of the best trajectory values produced by the algorithms in each iteration. In addition, we have measured the relative experimental error

$$\varepsilon = \frac{\gamma^* - \gamma_T^b}{\gamma^*} , \tag{5.10}$$

where $\gamma^* = 6905$ is the best trajectory value (known to exist) and γ_T^b is the resulting best trajectory value at the end of the current run; in other words $\gamma_T^b = S_{T,(1)}$. Another important parameter is the total number of samples N^{tot}

Table 5.4. Evolution of the FACE algorithm for the synthetic TSP (4.10.1) of Table 4.15, with $n = 30$. Relative experimental error $\varepsilon = 0.0000$.

t	$\widehat{\gamma}_t$	$S_{t,(1)}$	N_t	P_t^{mm}	t	$\widehat{\gamma}_t$	$S_{t,(1)}$	N_t	P_t^{mm}
1	42.52	38.39	900	0.0567	12	32.78	31.68	900	0.1762
2	40.88	37.61	900	0.0769	13	32.54	31.33	900	0.1540
3	39.27	36.85	900	0.0978	14	32.27	31.17	900	0.1953
4	38.02	35.37	900	0.1031	15	32.08	31.12	900	0.2215
5	36.87	34.55	900	0.1063	16	31.92	30.83	900	0.2432
6	35.87	34.38	900	0.1252	17	31.72	30.43	1111	0.2323
7	35.14	33.44	900	0.1537	18	31.15	30.43	18000	0.2686
8	34.40	32.68	900	0.1475	19	30.92	30.41	4949	0.2900
9	33.77	32.40	1511	0.1220	20	30.88	30.32	900	0.2737
10	33.39	32.07	900	0.1517	21	30.63	30.00	3219	0.3018
11	33.00	31.90	900	0.1933	22	30.43	30.00	18000	0.3782

Table 5.5. Evolution of the FACE algorithm for the `ft53` TSP of Table 2.4. Relative experimental error $\varepsilon = 0.0284$.

t	$\widehat{\gamma}_t$	$S_{t,(1)}$	N_t	P_t^{mm}	t	$\widehat{\gamma}_t$	$S_{t,(1)}$	N_t	P_t^{mm}
1	24528	21549	2810	0.0281	29	8943	8051	8288	0.1863
2	23025	19778	2810	0.0361	30	8822	7856	2810	0.1829
3	21424	18523	2810	0.0406	31	8680	7742	3794	0.1894
18	11352	9527	2810	0.1015	32	8342	7732	21759	0.2088
19	11054	9439	5425	0.1102	33	8301	7304	13943	0.2070
20	10661	9076	31020	0.1252	34	8130	7224	7531	0.2091
21	10271	9076	56200	0.1283	35	7743	7224	56200	0.2336
22	10078	9001	8497	0.1334	36	7666	7209	11591	0.2514
23	9989	8997	7326	0.1247	37	7609	7129	2810	0.2672
24	9887	8904	2810	0.1395	38	7534	7128	24968	0.2745
25	9711	8658	10219	0.1589	39	7392	7123	5864	0.2741
26	9550	8509	2810	0.1722	40	7246	7119	40641	0.2518
27	9423	8307	2810	0.1688	41	7159	7101	9080	0.6110
28	9251	8229	2810	0.1727	42	7129	7101	56200	0.5046

produced by the algorithm before stopping. For CE algorithm N^{tot} equals $T \cdot N$ ($N = 5\,n^2$ - constant) and for FACE algorithm $N^{\text{tot}} = \sum_{t=1}^{T} N_t$, where T is the number of iterations in a particular run and N_t is the sample size in iteration t. The comparison of minimum, maximum and average values of these parameters is given in Table 5.6. These data indicate that the FACE algorithm is only slightly better than CE. Indeed, the FACE algorithm requires less sample size in each iteration. However, the number of iterations (see Figure 5.5) is greater.

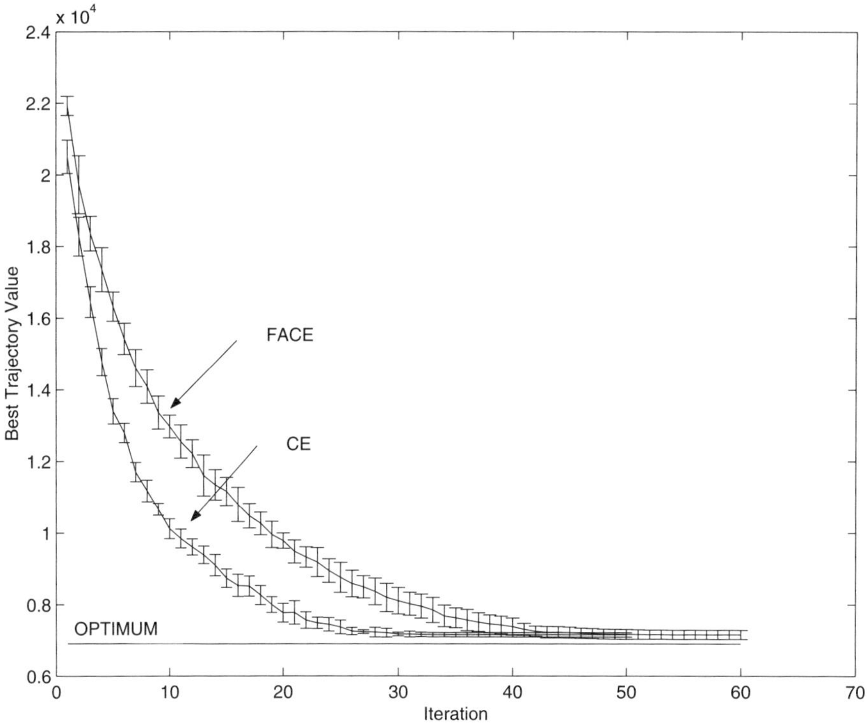

Fig. 5.5. Convergence of the CE and FACE algorithms applied to `ft53` TSP.

Table 5.6. Comparison of the performance of the CE and FACE algorithms applied to `ft53` TSP for 10 different runs.

	CE		FACE	
	ε	N^{tot}	ε	N^{tot}
min	0.0245	379215	0.0012	243210
max	0.0468	533710	0.0647	539168
mean	0.0368	457867	0.0373	376346

5.6 Exercises

Continuous Optimization

1. Verify Example 5.1, for example by using the program `cecont.m` from Appendix A.3. Which choice of parameters gives the fastest convergence?

2. The most simple sampling distribution for continuous optimization is the $\mathsf{U}(a,b)$ distribution.
   a) Derive the updating formulas for a and b from (4.8).
   b) The advantage of uniform updating is that it is very fast. What is the main disadvantage?

3. Suppose we use the $\mathsf{Beta}(a,b)$ distribution in the CE procedure for continuous optimization. The updating formulas for a and b follow from (3.41), that is

$$\max_{a,b} \widehat{D}(a,b) = \max_{a,b} \frac{1}{N} \sum_{i=1}^{N} I_{\{S(\mathbf{X}_i) \geqslant \gamma\}} \ln f(X_i; a, b) \,,$$

   where $f(\cdot; a, b)$ is the pdf of the $\mathsf{Beta}(a,b)$ distribution.
   Show that the optimal a and b satisfy

$$\frac{\partial \ln(\Gamma(a+b))}{\partial a} - \frac{\partial \ln(\Gamma(a))}{\partial a} + \frac{\sum_{i=1}^{N} I_{\{S(X_i) \geqslant \gamma\}} \ln(X_i)}{\sum_{i=1}^{N} I_{\{S(X_i) \geqslant \gamma\}}} = 0 \qquad (5.11)$$

   and

$$\frac{\partial \ln(\Gamma(a+b))}{\partial b} - \frac{\partial \ln(\Gamma(a))}{\partial b} + \frac{\sum_{i=1}^{N} I_{\{S(X_i) \geqslant \gamma\}} \ln(1-X_i)}{\sum_{i=1}^{N} I_{\{S(X_i) \geqslant \gamma\}}} = 0\,. \qquad (5.12)$$

   Although there is no closed-form solution to this system of equations, efficient algorithms such as the *ellipsoid method* [22] can be used to find the solution numerically.

4. Instead of using normal updating, apply the CE method to Example 5.1 using beta updating. Do this by modifying the program `cecont.m` of Appendix A.3. You may wish to use the programs of Appendix A.6; in that case you will also need the standard Matlab functions `betapdf.m`, `betarnd.m`, `gamrnd.m`, and `rndcheck.m` from the *statistics toolbox*.

5. Run Algorithm 4.2.1 for the Rosenbrock and trigonometric functions in Section 5.1 and obtain tables similar to Tables 5.1 and 5.2. Use both the normal and the $\mathsf{Beta}(a,b)$ pdfs for generating sample trajectories. Observe the effect of the modified smoothing scheme (5.1).

6. Consider the constrained optimization of the Rosenbrock function as in Table 5.1; but now generate the components from a *truncated normal*

distribution on $[-2, 2]$, using the acceptance–rejection method. Note that the updating formulas for the mean and variance remain the same. Why?

7. Although we typically assume that the components in a CE procedure for continuous optimization are *independent*, it is possible to formulate a CE procedure in which the components are *dependent*. In particular, consider the CE algorithm where at each stage t we draw $\mathbf{X}_1, \ldots, \mathbf{X}_N$ from a multivariate normal distribution with mean vector $\boldsymbol{\mu}_{t-1}$ and covariance matrix Σ_{t-1}.
   a) Give the updating formulas for $\boldsymbol{\mu}_t$ and Σ_t in terms of $\mathbf{X}_1, \ldots, \mathbf{X}_N$.
   b) Implement this procedure and explore under which conditions the CE method with dependent component is better than the simpler (and faster) method with independent components.

8. Although the beta distribution is defined only on [0,1] it can be used to optimize functions on any finite interval $[c, d]$. Explain how.

Modifications

9. Implement a FACE algorithm for the n-queen problem of Exercise 2.6. Compare your implementation with that of Appendix A.4.

10. Run the FACE Algorithm 5.3.1 on the data from the URL

`http://www.iwr.uni-heidelberg.de/groups/comopt/software/TSPLIB95/atsp/`

and obtain a table similar to Table 2.5. Compare the results with those obtained using the main CE Algorithm 4.2.1.

11. Sometimes it is not feasible — due to memory problems — to store all vectors $\mathbf{X}_1, \ldots, \mathbf{X}_N$ at Step 2 of the CE Algorithm 4.2.1. A memory-efficient alternative is to determine $\widehat{\gamma}_t$ in Step 2 using a relatively *small* sample size. Once $\widehat{\gamma}_t$ is determined, Step 3 can be carried out using a *new* sample, updating the parameters "online" while ignoring each $\mathbf{X}_i$ for which the performance is less than $\widehat{\gamma}_t$. Implement and apply this modification to a large max-cut problem.

12.* To obtain a larger increment $\Delta\widehat{\gamma}_t = \widehat{\gamma}_t - \widehat{\gamma}_{t-1}$ in the objective function S at iteration t we can modify Algorithm 4.2.1 via a *big-step* procedure similar to Algorithm 3.11.1.
   a) Formulate a big-step modification of Algorithm 4.2.1.
   b) Implement the algorithm and apply it to a problem of your choice.
   c) Compare the results with the original CE algorithm.

6

Noisy Optimization with CE

6.1 Introduction

In this chapter we show how the CE method optimizes *noisy* objective functions, that is, objective functions corrupted with noise. Noisy optimization, also called stochastic optimization, can arise in various ways. For example, in the analysis of data networks [24], a well-known problem is to find a routing table that minimizes the congestion in the network. Typically this problem is solved under deterministic assumptions. In particular, the stream of data is represented by a constant "flow," and the congestion at a link is measured in terms of a constant arrival rate to the link, which is assumed to depend on the routing table only. These assumptions are reasonable only when the arrival process changes very slowly relative to the average time required to empty the queues in the network. In a more realistic setting however, using random arrival and queueing processes, the optimization problem becomes noisy. Other examples of noisy optimization include stochastic scheduling and stochastic shortest/longest path problems. References on noisy optimization include [9, 26, 36, 37, 60, 78].

A classical example of noisy optimization is simulation-based optimization [148]. A typical instance is the *buffer allocation problem* (BAP). The objective in the BAP is to allocate n buffer spaces amongst the $m-1$ "niches" (storage areas) between m machines in a serial production line, so as to optimize some performance measure, such as the steady-state throughput. This performance measure is typically not available analytically and thus must be estimated via simulation. A more detailed description of the BAP will be given in Section 6.3, but for general references on production lines we refer to [5, 41, 63, 116, 155]. Buzacott and Shanthikumar [33] provide a good reference on stochastic modeling of manufacturing systems, while Gershwin and Schor [61] present a comprehensive summary paper on optimization of buffer allocation models.

The rest of the chapter is organized as follows: In Section 6.2 we explain how the standard CE Algorithm 4.2.1 can be easily modified to handle general noisy optimization problems. In Section 6.3 we apply the noisy version of

Algorithm 4.2.1 to the BAP. The material in this section closely follows [9]. In Sections 6.4 and 6.5 we provide various numerical experiments for the BAP and the noisy TSP, respectively.

6.2 The CE Algorithm for Noisy Optimization

Consider the maximization problem (4.2) and assume that the performance function $S(\mathbf{x})$ is noisy, that is, corrupted with noise. We denote such a noisy function as $\widehat{S}(\mathbf{x})$. We will assume that for each $\mathbf{x}$ we can readily obtain an outcome of $\widehat{S}(\mathbf{x})$, for example via generation of some additional random vector $\mathbf{Y}$, whose distribution may depend on $\mathbf{x}$. In that case we write symbolically $\widehat{S}(\mathbf{x}) = \widehat{S}(\mathbf{x}, \mathbf{Y})$. We introduce noisy optimization via the noisy (stochastic) TSP.

Example 6.1 (Noisy TSP). Consider the deterministic version of the TSP in (4.32). Suppose that the matrix (c_{ij}), replaced now by $\mathbf{Y} = (Y_{ij})$, is random. Thus, Y_{ij}, $i, j = 1, \ldots, n$ is a random variable representing the "random" time to travel from city i to city j. The total time of a tour $\mathbf{x} = (x_1, \ldots, x_n)$ is given by

$$\widehat{S}(\mathbf{x}, \mathbf{Y}) = \sum_{i=1}^{n-1} Y_{x_i, x_{i+1}} + Y_{x_n, x_1} \; . \tag{6.1}$$

Let us assume that $\mathbb{E}Y_{ij} = c_{ij}$. For example, if $Y_{ij} \sim \mathsf{Po}(c_{ij})$, then $\mathbb{E}Y_{ij} = c_{ij}$. For another example, if

$$Y_{ij} = \frac{1}{p_{ij}}\, c_{ij}\, \widetilde{Y}_{ij} \; ,$$

where $\widetilde{Y}_{ij} \sim \mathsf{Ber}(p_{ij})$, $(0 < p_{ij} < 1)$, then clearly again, $\mathbb{E}Y_{ij} = c_{ij}$.

We shall associate with $\widehat{S}(\mathbf{x}, \mathbf{Y})$ two different types of problems. The first type deals with the random variable $S^*(\mathbf{Y}) = \max_{\mathbf{x} \in \mathscr{X}} \widehat{S}(\mathbf{x}, \mathbf{Y})$. For example we may wish to estimate either

$$\mathbb{E}S^*(\mathbf{Y}) = \mathbb{E} \max_{x \in \mathscr{X}} \widehat{S}(\mathbf{x}, \mathbf{Y}) \tag{6.2}$$

or $\mathbb{P}(S^*(\mathbf{Y}) \geqslant \gamma)$, that is, to estimate the expected longest tour or the probability that the longest tour $S^*(\mathbf{Y})$ exceeds some length γ. If $S^*(\mathbf{Y})$ can be easily computed for each $\mathbf{Y}$, we are basically back in the situation discussed in Chapter 3.

The second type of problem deals with the maximization problem

$$\max_{\mathbf{x}} \mathbb{E}\widehat{S}(\mathbf{x}, \mathbf{Y}) \; . \tag{6.3}$$

Note that if either the distribution of $\mathbf{Y}$ is known, or if $S(\mathbf{x}) = \mathbb{E}\widehat{S}(\mathbf{x}, \mathbf{Y}) = \mathbb{E}\widehat{S}(\mathbf{x})$ is easily computable for each $\mathbf{x}$, then (6.3) reduces to the deterministic optimization problem (4.2). Note also that (6.3) and (6.2) are different, since the expected value of a maximum of a random function is not equal to the maximum of the expected value of the same random function.

For the rest of this section we deal with the maximization problem (6.3) alone. We assume that $S(\mathbf{x}) = \mathbb{E}\widehat{S}(\mathbf{x})$ is *not* available, but the sample value $\widehat{S}(\mathbf{x})$ (unbiased estimate of $\mathbb{E}\widehat{S}(\mathbf{x})$) is available, for example via simulation. For this type of problem we now present the modified (noisy) versions of our basic CE formulas (4.6)–(4.8). These are given for completeness only, since we replace only $S(\mathbf{x})$ by $\widehat{S}(\mathbf{x})$, while all other data remain the same.

1. **Adaptive updating of γ_t.** For a fixed $\mathbf{v}_{t-1}$, let γ_t be a $(1-\varrho)$-quantile of $\widehat{S}(\mathbf{X})$ under $\mathbf{v}_{t-1}$. A simple estimator $\widehat{\gamma}_t$ of γ_t is

$$\widehat{\gamma}_t = \widehat{S}_{(\lceil (1-\varrho)N \rceil)}\ , \tag{6.4}$$

   where, for a random sample $\mathbf{X}_1, \ldots, \mathbf{X}_N$ from $f(\cdot; \mathbf{v}_{t-1})$, $\widehat{S}_{(i)}$ is the i-th order statistic of the performances $\widehat{S}(\mathbf{X}_1), \ldots, \widehat{S}(\mathbf{X}_N)$, similar to (4.6).
2. **Adaptive updating of $\mathbf{v}_t$.** For fixed γ_t and $\mathbf{v}_{t-1}$, derive $\mathbf{v}_t$ from the solution of the CE program

$$\max_{\mathbf{v}} D(\mathbf{v}) = \max_{\mathbf{v}} \mathbb{E}_{\mathbf{v}_{t-1}} I_{\{\widehat{S}(\mathbf{X}) \geqslant \gamma_t\}} \ln f(\mathbf{X}; \mathbf{v})\ . \tag{6.5}$$

   The stochastic counterpart of (6.5) is as follows: for fixed $\widehat{\gamma}_t$ and $\widehat{\mathbf{v}}_{t-1}$, derive $\widehat{\mathbf{v}}_t$ from the following program

$$\max_{\mathbf{v}} \widehat{D}(\mathbf{v}) = \max_{\mathbf{v}} \frac{1}{N} \sum_{i=1}^{N} I_{\{\widehat{S}(\mathbf{X}_i) \geqslant \widehat{\gamma}_t\}} \ln f(\mathbf{X}_i; \mathbf{v})\ . \tag{6.6}$$

Clearly, the main CE optimization Algorithm 4.2.1 with noisy functions remains the same as for the deterministic one, provided again that $S(\mathbf{x})$ is replaced by $\widehat{S}(\mathbf{x})$. It is not difficult to understand that such noisy Algorithm 4.2.1 will work nicely as well since it will *filter out* the noise component $\mathbf{Y}$ from $\widehat{S}(\mathbf{x}) = \widehat{S}(\mathbf{x}, \mathbf{Y})$.

To provide more insight into the noisy version of Algorithm 4.2.1 assume, as before, that $\mathbb{E}\widehat{S}(\mathbf{x}) = \mathbb{E}\widehat{S}(\mathbf{x}, \mathbf{Y}) = S(\mathbf{x})$ for all $\mathbf{x}$ and that $S(\mathbf{x})$ has a unique maximum γ^* at $\mathbf{x}^*$. The first thing to observe is that γ_t in the noisy and deterministic case — let us denote them by γ_{1t} and γ_{2t} — are generally *different*. Namely, γ_{1t} is the $(1-\varrho)$-quantile of the random variable $S(\mathbf{X})$, with $\mathbf{X} \sim f(\cdot; \mathbf{v}_{t-1})$ and γ_{2t} is the $(1-\varrho)$-quantile of the random variable $\widehat{S}(\mathbf{X})$, with $\mathbf{X} \sim f(\cdot; \mathbf{v}_{t-1})$. As a result, the sequences of estimators $\{\widehat{\gamma}_{1t}\}$ and $\{\widehat{\gamma}_{2t}\}$ will generally converge to different values as well.

Suppose next, as in (4.4), that the theoretically optimal pdf $f(\cdot; \mathbf{v}^*)$ is the delta function with mass at $\mathbf{x}^*$. For the deterministic CE algorithm the sequence of reference parameters $\{\widehat{\mathbf{v}}_{1t}\}$ is steered towards the optimal (degenerate) $\mathbf{v}^*$, and since $\mathbb{E}\widehat{S}(\mathbf{x}) = S(\mathbf{x})$, it is conceivable that the sequence $\{\widehat{\mathbf{v}}_{2t}\}$ also tends to $\mathbf{v}^*$. This is indeed what we have observed numerically, although no general proof of this result is yet available. Observe also that $\gamma^* = \mathbb{E}\widehat{S}(\mathbf{x}^*, \mathbf{Y})$, so one can estimate γ^* based on the final noisy reference parameter $\widehat{\mathbf{v}}_{2T}$ as

$$\widehat{\gamma}_T = \frac{1}{K} \sum_{i=1}^{K} \widehat{S}(\mathbf{X}_i) , \tag{6.7}$$

where $\mathbf{X}_1, \ldots, \mathbf{X}_K$ a random sample from $f(\cdot; \widehat{\mathbf{v}}_{2T})$.

Note finally that trajectory generation for noisy problems is similar to their deterministic counterparts. As an example consider again the noisy TSP. Here at each iteration t we first generate a sample $\mathbf{X}_1, \ldots, \mathbf{X}_N$ according to a MCWR with transition matrix $\widehat{P}_t$, and then, independently, generate $\mathbf{Y}_k$, $k = 1, \ldots, N$. In fact, for each $\mathbf{Y}_k$ we only need to generate the random costs on the links traversed by the path $\mathbf{X}_k$. This produces a sequence of tuples $\{(\widehat{\gamma}_{2t}, \widehat{P}_t)\}$ approximating (γ_2^*, P^*), where P^* is the ODTM for the deterministic case and γ_2^* the $(1 - \varrho)$-quantile of $\widehat{S}(\mathbf{X}, \mathbf{Y})$, with $\mathbf{X} \sim P^*$. Finally, to estimate the maximum $\gamma^* = \gamma_1^*$ for the deterministic case we apply (6.7).

Stopping Rules

The standard stopping criterion (4.10) does not apply to noisy optimization problems, since typically $\widehat{\gamma}_t$ will not converge to a fixed value. However, we can still use alternative stopping criteria such as (4.30) and (4.43), which are applicable to both the deterministic and the noisy case and are based on the convergence of a sequence such as $\{\widehat{P}_t\}$ to the associated degenerated case.

Another possibility is to stop the algorithm when $\{\widehat{\gamma}_{2t}\}$ has reached stationarity. There exist an extensive literature on stationarity detection of a stochastic process. We present here two simple rules of thumb, involving moving averages of $\{\widehat{\gamma}_{2t}\}$ and the fact that for some $t \geq T$ the first two moments of $\{\widehat{\gamma}_{2t}\}$ settle down to some constant values, say C_1 and C_2, respectively, that is $\mathbb{E}\widehat{\gamma}_{2t} = C_1$, and $\mathrm{Var}(\widehat{\gamma}_{2t}) = C_2$. Our first stopping criterion is based on $\mathbb{E}\widehat{\gamma}_{2t} = C_1$, while our second stopping criterion is based on both $\mathbb{E}\widehat{\gamma}_{2t} = C_1$ and $\mathrm{Var}(\widehat{\gamma}_{2t}) = C_2$.

First Stopping Criterion

To identify T, we consider the following moving-average process

$$B_t(k) = \frac{1}{k} \sum_{s=t-k+1}^{t} \widehat{\gamma}_{2s}, \; t = s, s+1, \dots, \; s \geq k \,, \tag{6.8}$$

where k is fixed, say $k = 10$. Define next

$$B_t^-(k, r) = \min_{j=1,\dots,r} B_{t+j}(k) \tag{6.9}$$

and

$$B_t^+(k, r) = \max_{j=1,\dots,r} B_{t+j}(k) \,, \tag{6.10}$$

respectively, where r is fixed, say $r = 5$. Clearly, for $t \geq T$ and moderate k and r, we may expect that $B_t^+(k,r) \approx B_t^-(k,r)$, provided the variance of $\{\widehat{\gamma}_{2t}\}$ is bounded.

With this at hand, we define our first stopping criterion as

$$T = \min \left\{ t \; : \; \frac{B_t^+(k,r) - B_t^-(k,r)}{B_t^-(k,r)} \leq \varepsilon \right\} \,, \tag{6.11}$$

where k and r are fixed and ε is a small number, say $\varepsilon \leq 0.01$.

Second Stopping Criterion

Our second stopping criterion is the same as (6.11) with $B_t(k, r)$ replaced by

$$C_t(k) = \frac{\frac{1}{k-1} \left\{ \sum_{s=t-k+1}^{t} [\widehat{\gamma}_{2s} - B_t(k)]^2 \right\}}{B_t(k)^2} \,. \tag{6.12}$$

Note that $C_t(k)$ in (6.12) and $B_t(k)$ in (6.11) represent in fact the sample SCV and the sample mean of moving averages, respectively. Although the second criterion is more time consuming, it is typically more accurate than the first one.

Once the algorithm stops at time T, say for the noisy TSP, we take the matrix $\widehat{P}_T$ in Algorithm 4.7.3 as an estimate of the ODTM P^*. An alternative is to first approximate $\widehat{P}_T$ by a degenerated one, say $\widehat{P}^d$, and then take the latter as an estimate of P^*.

6.3 Optimal Buffer Allocation in a Simulation-Based Environment [9]

Here we apply the noisy version of Algorithm 4.2.1 to the *buffer allocation problem* (BAP), a well-known difficult problem in the design of production lines.

The basic setting of the BAP is the following (see Figure 6.1): Consider a production line consisting of m machines in series, numbered $1, 2, \ldots, m$. Jobs are processed by all machines in consecutive order. The processing time at machine i has a fixed distribution with rate μ_i (hence the mean processing time is $1/\mu_i$), $i = 1, \ldots, m$. The machines are assumed to be unreliable, with exponential life and repair times. Specifically, machine i has failure rate β_i and repair rate r_i, $i = 1, \ldots, m$. All life, repair and processing times are assumed to be independent of each other.

The machines are separated by $m-1$ storage areas, or *niches*, in which jobs can be stored. However, the total number of storage places, or *buffer* places, is limited to n. When a machine breaks down, this can have consequences for other machines up or down the production line. In particular, an upstream machine could become *blocked* (when it cannot pass a processed job on to the next machine or buffer) and a downstream machine could become *starved* (when no jobs are offered to this machine). We assume that a starved or blocked machine has the same failure rate as a "busy" machine. The first machine in the line is never starved and the last machine is never blocked.

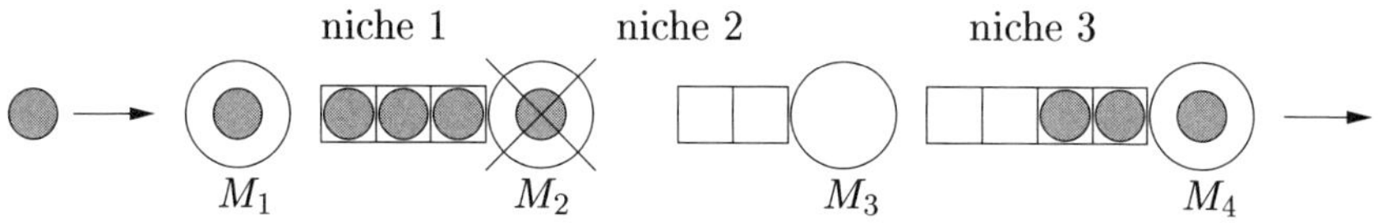

Fig. 6.1. A production line with $m = 4$ machines. The total available buffer space is $n = 9$. The current buffer allocation is (3,2,4). Machine 1 has an infinite supply, but is currently blocked. Machine 2 has failed and is under repair. Machine 3 is starved. Machine 4 is never blocked.

The BAP deals with the optimal allocation of n buffers among $m-1$ niches. Here "optimal" refers to some performance measure of the flowline. Typical performance measures are the steady-state *throughput* and the expected amount of *work-in-process* (WIP). We shall only deal with the steady-state throughput. Note that there are

$$\binom{n+m-2}{m-2}$$

possible buffer allocations.

There are two reasons why the BAP might be considered to be a difficult optimization problem. The first reason is that BAP presents a combinatorial optimization problem with $\binom{n+m-2}{m-2}$ elements which even for moderate values of n and m is quite large. The second reason is that, for a given buffer allocation, the exact value of the objective function — the steady-state throughput

— is often difficult or impossible to calculate analytically. In fact, the objective function is only available for relatively simple production lines in which the processing times have exponential or (simple) phase-type distributions [61, 81]. So, in a more general setting the BAP is a *noisy* or *simulation-based* optimization problem, where the objective function needs to be estimated via simulation [126, 148].

We will use the following mathematical formulation of the BAP. Each possible *buffer allocation* (BA) will be represented by a vector $\mathbf{x} = (x_1, \ldots, x_{m-1})$ in the set

$$\mathcal{X} := \left\{ (x_1, \ldots, x_{m-1}) : x_i \in \{0, 1, \ldots, n\}, i = 1, \ldots, m-1, \sum_{i=1}^{m-1} x_i = n \right\}.$$

Here x_i represents the number of buffer spaces allocated to niche i, $i = 1, \ldots, m-1$.

For each buffer allocation $\mathbf{x}$ let $S(\mathbf{x})$ denote the expected steady-state throughput of the production line. Thus the BAP can be formulated as the optimization problem (6.3):

$$\text{maximize } S(\mathbf{x}) \text{ over } \mathbf{x} \in \mathcal{X}. \tag{6.13}$$

As mentioned before, we assume that $S(\mathbf{x})$ is not available analytically and thus must be estimated via simulation. We denote by $\widehat{S}(\mathbf{x})$ the estimate of $S(\mathbf{x})$.

In order to apply the CE algorithm we need to specify (a) how to generate random buffer allocations, and (b) how to update the parameters at each iteration.

We present below two algorithms for random buffer allocation generation: the first is based on a multivariate discrete distributions with dependent marginals, and the second is based on generating random partitions.

First Random Buffer Allocation Algorithm

A simple method to generate a random vector $\mathbf{X} = (X_1, \ldots, X_{m-1})$ in $\mathcal{X}$ is to first draw $X_1, X_2, \ldots, X_{m-1}$ independently, each X_i according to an $(n+1)$-point discrete distribution $(p_{i0}, \ldots, p_{in}), i = 1, \ldots, m-1$, and accept the sample only if $X_1 + \cdots + X_{m-1} = n$. From Sections 4.3 and 6.2 we see immediately that the (noisy) updating formulas are given by

$$\widehat{p}_{t,ij} = \frac{\sum_{k=1}^{N} I_{\{\widehat{S}(\mathbf{X}_t) \geqslant \widehat{\gamma}_t\}} I_{\{X_{ki}=j\}}}{\sum_{k=1}^{N} I_{\{\widehat{S}(\mathbf{X}_t) \geqslant \widehat{\gamma}_t\}}}, \tag{6.14}$$

where $X_{ki} = j$ means that j space is allocated to niche i at replication (sample) k. Note that (6.14) presents the fraction of times that $\mathbf{X}_t$ has an estimated performance greater than $\widehat{\gamma}_t$ and have their i-th coordinate equal to j. For convenience we amalgamate for each t the $\widehat{p}_{t,ij}$ into a matrix $\widehat{P}_t$. The following algorithm generates random buffer allocations according to the conditional distribution of independent $\mathbf{X}$ above, given that $\sum_i X_i = n$.

Algorithm 6.3.1 (Generation of Buffer Allocations)

For given $P = (p_{ij})$, *generate a random permutation* $(\pi_1, \ldots, \pi_{m-1})$ *of* $\{1, \ldots, m-1\}$.
Let $k = 0$
For $i = 1, \ldots, m-1$
  Let $t = \sum_{j=0}^{n-k} p_{\pi_i, j}$
  For $j = 0, \ldots, n-k$ *let* $p_{\pi_i, j} = p_{\pi_i, j}/t$
  For $j = n-k+1, \ldots, n$ *let* $p_{\pi_i, j} = 0$
  Generate X_{π_i} *according to* $(p_{\pi_i, 0}, \ldots, p_{\pi_i, n})$.
  Let $k = k + X_{\pi_i}$
End

Algorithm 6.3.1 is further illustrated in Figure 6.2.

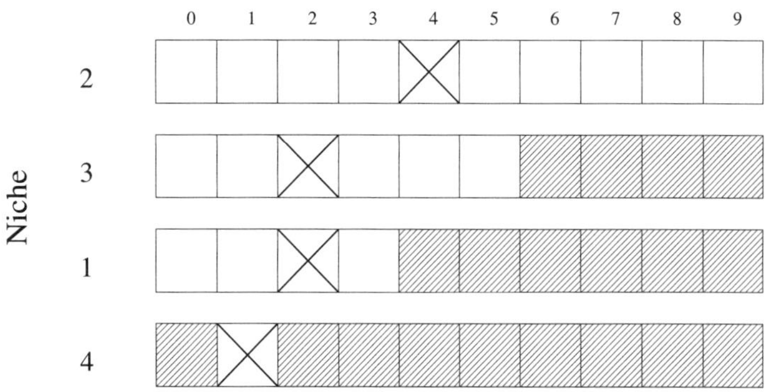

Fig. 6.2. Generation of the BA vector (2, 4, 2, 1), for the case $m = 5$, $n = 9$ and the permutation $\pi = (2, 3, 1, 4)$. For the second niche there are initially 9 possible buffer places; 4 buffer places are allocated. This reduces the number of available buffer places for the third niche to 5; 2 buffer places are allocated, etc.

To complete the algorithm, we need to specify the initialization and stopping conditions. For the initial matrix $\widehat{P}_0$ we simply take all elements equal to $1/(n+1)$. The stopping criterion is based on the convergence of the sequence of matrices $\widehat{P}_0, \widehat{P}_1, \ldots$ (see also Sections 4.2 and 4.3.2), which is found to converge to a degenerate matrix P^*, that is, a matrix in which each row has exactly 1 one and n zeros. Specifically, the algorithm is terminated if for some integer d, for example, $d = 5$,

$$\xi_t(i) = \xi_{t-1}(i) = \cdots = \xi_{t-d}(i), \quad \text{for all } i = 1, \ldots, m-1 , \tag{6.15}$$

where $\xi_t(i)$ denotes the index of the maximal element of the i-th row of $\widehat{P}_t$.

Second Buffer Allocation Algorithm

An alternative method for random buffer allocations is based on the bipartition problem in Section 4.6. Specifically, one can view the BAP as a "network" containing $n+m-2$ nodes, which must be partitioned into two mutually disjoint sets, where the first and the second set contain exactly $m-2$ and n nodes, respectively. Similarly, in the urn model terminology, it could be viewed as the problem of allocating (distributing) $m-2$ balls into to $n+m-2$ different cells. This leads to Algorithm 4.6.1 with the corresponding updating rules given in (4.24).

Consider for example the case $m = 5$ (4 niches) and $n = 9$ buffer spaces as in Figure 6.2. Any partition of $\{1, 2, \ldots, 12\}$ into two groups of 3 and 9 nodes corresponds to exactly one partition. Think of 3 balls, which we can place at any of 12 positions. The partition described by the binary vector $(0, 0, 1, 0, 0, 0, 0, 1, 0, 0, 1, 0)$ corresponds to the buffer allocation (2,4,2,1).

If not stated otherwise we shall use the first algorithm below. For easy reference we summarize the main Algorithm 4.2.1 for the noisy BAP as follows:

Algorithm 6.3.2 (Main Algorithm for the noisy BAP)

1. *Choose* $\widehat{P}_0$ *such that all elements are equal to* $1/(n+1)$. *Set* $t = 1$.
2. *Generate a sample of buffer allocations* $\mathbf{X}_1, \ldots, \mathbf{X}_N$ *according to Algorithm 6.3.1, with* $P = P_{t-1}$, *and compute the* $(1-\varrho)$*-quantile* $\widehat{\gamma}_t$ *of the estimated throughputs according to* (6.4).
3. *Using the same sample, update* $\widehat{P}_t$ *according to* (6.14) *and then smooth it out using* (4.9).
4. *If for some* $t \geqslant d$ *convergence condition* (6.15) *is met, then stop; otherwise set* $t = t + 1$ *and reiterate from Step 2.*

For the deterministic BAP, that is, the problem (6.13), the only change in Algorithm 6.3.2 is that Step 2 is replaced by

2'. ... *the actual throughputs according to* (6.4).

It is intuitively clear that the noisy BAP converges to the deterministic BAP if we decrease the relative error of the estimated throughputs to 0.

Simulation Issues

Recall that our approach involves two types of randomness: the *natural* one associated with steady-state simulation of the production line and the *artificial* one associated with the probability matrix P, which we purposely introduced to generate random allocation using Algorithm 6.3.1. The following discussion deals with the first type. Note that although steady-state simulation of BAP is time consuming and it takes up, in fact, most of the CPU time of Algorithm 6.3.2, our main goal here is to show how to incorporate efficiently the steady-state output data in a simulation-based optimization framework in BAP using CE rather than deal with efficient steady-state simulation techniques by themselves. Clearly, use of efficient variance reduction techniques, such as importance sampling, control variables, etc., can speed up Algorithm 6.3.2 dramatically. These issues will be addressed elsewhere. It is worth mentioning that the simulation model presents a "black box" for the CE optimization procedure. Any other method which calculates or approximates the expected steady-state objective function for a given buffer allocation can be used as well.

At this juncture we mention several standard rules for simulation-based systems [148] that should be considered when implementing the CE method. Since for any buffer allocation (with a total of $\binom{n+m-2}{m-2}$) the expected steady-state performance (like WIP and throughput) is unknown, one has to allow some warmup (terminating simulation) before collecting the statistical data.

One typically starts simulating a vacant system, that is, when all the buffers in the production line are empty and all machines are idle. After the first job output the buffers are still relatively empty, thus the throughput is high. As time progresses the WIP and the throughput converge to their steady state. As soon as the process has reached steady state, we start collecting the statistical data disregarding those from the warmup period. We run a single (long) simulation, called the *batch mean* [148], in order to estimate the unknown expected performance. To identify (approximate) the steady state we collect the output transient data in blocks (each of, say, 50 jobs) and perform a statistical test for each such block, using the means and variances of all the subsequent blocks in the same simulation until the relative error changes very little from block to block. We found empirically that for a configuration not exceeding 12 machines and 45 buffer spaces a warmup period of 100 jobs is quite sufficient. Clearly, with the system at steady state for an output process, such as throughput, one can start collecting the output data simultaneously for *any* other performance, such as WIP etc [148]. In our numerical studies we selected the size $\mathcal{B}$ of the batches as a function of the relative error of the underlying output process.

More specifically, for the percentage relative error $\kappa \times 100\% = 0.3, 0.5, 1.0$ we took $\mathcal{B} = 300, 600, 2000$, respectively. A documented Matlab source code file is available at

`http://iew3.technion.ac.il/~talraviv/Publications/buffer.zip`

6.4 Numerical results for the BAP

To evaluate the effectiveness of Algorithm 6.3.2 we applied it to various test problems. Specifically, we applied Algorithm 6.3.2 to a suite of 70 test cases in Vouros and Papadopolous [168]. In all these cases the machine processing times have exponential or Erlang_2 distributions. Since, in addition, the life and repair times are assumed to be exponentially distributed, we can in principle calculate the exact optimal buffer allocation and corresponding steady-state throughput for these systems, using Markov chain theory, as described in [81]. It should be noted, however, that the solutions are in practice only obtainable for relatively small n and m. In addition to the 70 test cases, we applied Algorithm 6.3.2 to various relatively large systems for which the solutions were not available from [168]. In this section we summarize the results on a selection of these test problems.

In all test cases below we set $\varrho = 0.1$, $\alpha = 0.7$ (smoothing parameter), took $d = 5$ in our stopping rule (6.15) and generated at each iteration $N = 2\,m\,n$ random buffer allocations (we need to estimate the components of the $(m-1) \times (n+1)$ matrices $\widehat{P}_t$, so we need at least $m\,n$ replications). Similar results where obtained with $0.05 \leq \varrho \leq 0.2$ and $0.5 \leq \alpha \leq 0.95$. The algorithm was implemented in Matlab 5.2 and ran on an Intel Pentium III 500MHz processor. For a given buffer allocation we used the *batch means* method [148] to estimate the steady-state throughput, each simulation run starting with a sufficiently long warmup period.

For each test case we generated 10 independent solutions via Algorithm 6.3.2, say $\gamma_T^{(i)}$, $i = 1, \ldots, 10$. These were compared with either the optimal solution (steady-state output) γ^*, or with the best known solution $\gamma^\dagger$. In the tables below, we use the following notation: The average relative experimental error of the 10 solutions is defined either as

$$\bar{\varepsilon} = \frac{1}{10} \sum_{i=1}^{10} \frac{|\gamma^* - \gamma_T^{(i)}|}{\gamma^*} \qquad \text{or as} \qquad \bar{\varepsilon} = \frac{1}{10} \sum_{i=1}^{10} \frac{|\gamma^\dagger - \gamma_T^{(i)}|}{\gamma^\dagger}\,, \tag{6.16}$$

depending on whether the true optimal solution is known or not. Also, below $\bar{\gamma}_T$ denotes the average of the 10 generated solutions, and ε_* and ε^* denote the worst and the best relative experimental error among the 10 generated solutions. Here, we take again $\gamma^\dagger$ instead of γ^* when the optimal solution is not known. Finally, BA denotes the optimal buffer allocation vector, and $\bar{T}$ and CPU denote the average total number of iterations needed before stopping and the average CPU time in seconds, respectively.

Tables 6.1, 6.2 and 6.3 present the results for a number of test cases in Vouros and Papadopolous [168]. In particular, in Tables 6.1 and 6.2 we consider systems with exponential processing times with rates $\mu_i, i = 1, \ldots, m$, and in Table 6.3 we consider systems with Erlang_2 processing times, with rates $\mu_i, i = 1, \ldots, m$; thus, for each machine i the processing time consists of two exponential phases with rates $2\mu_i$. We recall that the machine life and repair times are assumed to be exponential with rates $\beta_i, i = 1, \ldots, m$ and $r_i, i = 1, \ldots, m$, respectively. We see that the allocations found by the CE method are very close to the exact optimal ones (γ^*) in [168].

Tables 6.4 and 6.5 present the performance of Algorithm 6.3.2 for $m = 6$ and $m = 10$, respectively, with exponential processing times and different values of n. We could not compare the results of Tables 6.4 and 6.5 with any alternatives since to the best of our knowledge no case studies are available yet for such *relatively large systems*. We argue, however, that our results are accurate and reliable and could serve as case studies to compare different algorithms. Note also that $\gamma^\dagger$ in Tables 6.4 and 6.5 corresponds to our best solution obtained (on the basis of 10 different runs) for each fixed n.

We obtained similar accuracies for different processing time distributions (that is, exponential, normal, Erlang, uniform and deterministic), provided $0.05 \le \varrho \le 0.2$ and $0.5 \le \alpha \le 0.95$.

Table 6.1. Performance of Algorithm 6.3.2 for BAPs with $m - 1 = 2$ niches and different values of n, exponential processing times with rates $\mu_1 = 1$, $\mu_2 = 1.2$, $\mu_3 = 1.4$, failure rates $\beta_i = 0.05$, and repair rates $r_i = 0.5, i = 1, \ldots, 3$.

n	$\bar{T}$	BA	$\bar{\gamma}_T$	γ^*	$\bar{\varepsilon}$	ε_*	ε^*	CPU
1	2.0	(1,0)	.6341	.6341	0.0000	0.0000	0	7
2	2.0	(1,1)	.6715	.6744	0.0044	0.0088	0	8
3	2.6	(2,1)	.6998	.7113	0.0164	0.0590	0	9
4	3.5	(3,1)	.7349	.7361	0.0016	0.0054	0	14
5	3.8	(3,2)	.7574	.7587	0.0018	0.0059	0	14
6	4.3	(4,2)	.7688	.7777	0.0037	0.0211	0	16
7	6.2	(5,2)	.7811	.7922	0.0052	0.0171	0	22
8	5.1	(5,3)	.8040	.8060	0.0025	0.0084	0	20
9	9.1	(6,3)	.8142	.8178	0.0044	0.0163	0	32
10	8.3	(7,3)	.8255	.8274	0.0024	0.0095	0	30

Table 6.2. Performance of Algorithm 6.3.2 for $m - 1 = 4$ niches, different values of n, exponential processing times with rates $\mu_1 = 1$, $\mu_2 = 1.1$, $\mu_3 = 1.2$, $\mu_4 = 1.3$, $\mu_5 = 1.5$, failure rates $\beta_i = 0.05$, and repair rates $r_i = 0.5, i = 1, \ldots, 5$.

n	$\bar{T}$	BA	$\bar{\gamma}_T$	γ^*	$\bar{\varepsilon}$	ε_*	ε^*	CPU
1	2.6	(0,1,0,0)	.5213	.5213	0.0000	0.0000	0	12
2	4.6	(1,1,0,0)	.5479	.5514	0.0060	0.0110	0	32
3	3.6	(1,1,1,0)	.5824	.5824	0.0000	0.0000	0	39
4	6.4	(1,2,1,0)	.6015	.6027	0.0020	0.0085	0	67
5	9.0	(2,2,1,0)	.6202	.6213	0.0018	0.0032	0	103
6	5.7	(2,2,1,1)	.6420	.6422	0.0003	0.0031	0	89
7	7.7	(2,2,2,1)	.6572	.6585	0.0020	0.0087	0	116
8	7.2	(3,2,2,1)	.6731	.6744	0.0020	0.0120	0	132
9	9.1	(3,3,2,1)	.6885	.6894	0.0013	0.0067	0	166
10	10.7	(3,3,3,1)	.7004	.7005	0.0002	0.0003	0	197

Table 6.3. Performance of Algorithm 6.3.2 for $m - 1 = 4$ niches, with Erlang$_2$ processing times with rates $\mu_1 = 1$, $\mu_2 = 1.1$, $\mu_3 = 1.2$, $\mu_4 = 1.3$, $\mu_5 = 1.5$, failure rates $\beta_i = 0.05$, and repair rates $r_i = 0.5$, $i = 1, \ldots, 5$.

n	T	BA	$\bar{\gamma}_T$	γ^*	$\bar{\varepsilon}$	ε_*	ε^*	CPU
1	2.8	(0,1,0,0)	.5968	.5968	0.0000	0.0000	0	23
2	3.5	(1,1,0,0)	.6331	.6338	0.0011	0.0114	0	39
3	3.9	(1,1,1,0)	.5824	.5824	0.0000	0.0000 0	0	55
4	5.8	(2,1,1,0)	.6802	.6808	0.0009	0.0073	0	86
5	8.3	(2,2,1,0)	.6985	.6996	0.0016	0.0028	0	159
6	6.9	(2,2,1,1)	.7180	.7195	0.0114	0.0020	0	187
7	12.5	(3,2,2,1)	.7335	.7341	0.0018	0.0007	0	202
8	9.8	(3,2,2,1)	.7496	.7501	0.0043	0.0007	0	181
9	9.7	(3,3,2,1)	.7620	.7627	0.0068	0.0009	0	177
10	13.6	(4,3,2,1)	.7714	7740	0.0124	0.0330	0	261

Table 6.4. Performance of Algorithm 6.3.2 for $m - 1 = 5$ niches and various n, exponential processing times with rates $\mu_1 = 8$, $\mu_2 = 11$, $\mu_3 = 14$, $\mu_4 = 14$, $\mu_5 = 11$, $\mu_6 = 8$, failure rates $\beta_i = 0.05$, and repair rates $r_i = 0.5$, $i = 1, \ldots, 6$.

n	$\bar{T}$	BA	$\bar{\gamma}_T$	$\gamma^\dagger$	$\bar{\varepsilon}$	ε_*	ε^*	CPU
2	4.2	(1,0,0,0,1)	5.4935	5.5027	0.0017	0.0084	0	33
4	5.4	(1,0,0,0,1)	5.9245	5.9334	0.0015	0.0076	0	66
6	12	(1,1,0,1,1)	6.2443	6.2555	0.0018	0.0050	0	156
8	13.4	(2,1,0,1,2)	6.5197	6.5253	0.0009	0.0022	0	199
10	25.6	(3,1,1,1,2)	6.7510	6.7589	0.0012	0.0057	0	386
12	49	(4,2,1,2,3)	6.9316	6.9360	0.0006	0.0011	0	766
14	28.2	(4,2,1,2,5)	7.0684	7.0934	0.0035	0.0079	0	604
16	59.2	(5,2,2,2,5)	7.1783	7.1846	0.0009	0.0026	0	1128
18	92.6	(6,2,2,3,5)	7.4149	7.4291	0.0019	0.0036	0	2048

Table 6.5. Performance of Algorithm 6.3.2 for $m - 1 = 9$ niches, exponential processing times with rates $\mu_1 = 8$, $\mu_2 = 8$, $\mu_3 = 11$, $\mu_4 = 14$, $\mu_5 = 14$, $\mu_6 = 11$, $\mu_7 = 8$, $\mu_8 = 8$, $\mu_9 = 6$, $\mu_{10} = 6$, failure rates $\beta_i = 0.05$, and repair rates $r_i = 0.5$, $i = 1, \ldots, 10$.

n	$\bar{T}$	BA	$\bar{\gamma}_T$	$\gamma^\dagger$	$\bar{\varepsilon}$	ε_*	ε^*	CPU
2	4.00	(0,0,0,0,0,0,0,1,1)	3.8281	3.8749	0.0122	0.0376	0	110
4	11.67	(0,0,0,0,0,0,1,1,2)	4.1160	4.1236	0.0018	0.0028	0	402
6	19.67	(0,1,0,0,0,0,1,2,2)	4.3220	4.3289	0.0016	0.0024	0	964
8	23.33	(0,1,0,0,0,1,1,2,3)	4.5325	4.5420	0.0021	0.0058	0	1200
10	12.67	(1,1,0,0,0,1,2,2,3)	4.6146	4.6426	0.0060	0.0184	0	1165
12	14.33	(1,1,0,0,0,1,2,3,4)	4.7814	4.7946	0.0028	0.0084	0	1719
14	37.00	(1,1,0,0,1,1,2,3,5)	4.8852	4.8895	0.0008	0.0020	0	3325
16	49.33	(1,1,0,1,0,2,2,4,5)	4.9832	4.9891	0.0018	0.0033	0	5117
18	186.67	(2,1,1,0,0,1,3,4,6)	5.0414	5.0638	0.0045	0.0117	0	20714

Dynamics

We illustrate the dynamics of the matrices $\widehat{P}_t$ for a benchmark problem with 4 niches, 10 buffer spaces, normally distributed processing times with $\mu = 6, \sigma = 2$, and $N = 80$.

$$\widehat{P}_0 = \begin{pmatrix} 0.0909 & 0.0909 & 0.0909 & 0.0909 & 0.0909 & 0.0909 & 0.0909 & 0.0909 & 0.0909 & 0.0909 & 0.0909 \\ 0.0909 & 0.0909 & 0.0909 & 0.0909 & 0.0909 & 0.0909 & 0.0909 & 0.0909 & 0.0909 & 0.0909 & 0.0909 \\ 0.0909 & 0.0909 & 0.0909 & 0.0909 & 0.0909 & 0.0909 & 0.0909 & 0.0909 & 0.0909 & 0.0909 & 0.0909 \\ 0.0909 & 0.0909 & 0.0909 & 0.0909 & 0.0909 & 0.0909 & 0.0909 & 0.0909 & 0.0909 & 0.0909 & 0.0909 \end{pmatrix} .$$

$$\vdots$$

$$\widehat{P}_4 = \begin{pmatrix} 0.0002 & 0.0013 & 0.0139 & 0.4484 & 0.5349 & 0.0002 & 0.0002 & 0.0002 & 0.0002 & 0.0002 & 0.0002 \\ 0.0002 & 0.0002 & 0.0303 & 0.8226 & 0.1432 & 0.0014 & 0.0013 & 0.0002 & 0.0002 & 0.0002 & 0.0002 \\ 0.0002 & 0.0483 & 0.9410 & 0.0089 & 0.0002 & 0.0002 & 0.0002 & 0.0002 & 0.0002 & 0.0002 & 0.0002 \\ 0.0014 & 0.6007 & 0.3913 & 0.0051 & 0.0002 & 0.0002 & 0.0002 & 0.0002 & 0.0002 & 0.0002 & 0.0002 \end{pmatrix} .$$

$$\vdots$$

$$\widehat{P}_9 = \begin{pmatrix} 0.0000 & 0.0000 & 0.0038 & 0.0179 & 0.9783 & 0.0000 & 0.0000 & 0.0000 & 0.0000 & 0.0000 & 0.0000 \\ 0.0000 & 0.0000 & 0.0001 & 0.9996 & 0.0003 & 0.0000 & 0.0000 & 0.0000 & 0.0000 & 0.0000 & 0.0000 \\ 0.0000 & 0.0001 & 0.9445 & 0.0554 & 0.0000 & 0.0000 & 0.0000 & 0.0000 & 0.0000 & 0.0000 & 0.0000 \\ 0.0000 & 0.9801 & 0.0199 & 0.0000 & 0.0000 & 0.0000 & 0.0000 & 0.0000 & 0.0000 & 0.0000 & 0.0000 \end{pmatrix} .$$

It follows from the results above that starting from $\widehat{P}_0$ with the elements $1/(n+1) = 1/11 = 0.0909$ Algorithm 6.3.2 stopped after 9 iterations allocating 4, 3, 2, and 1 buffer spaces to niches 1, 2, 3, and 4, respectively.

Our numerical studies suggest that the proposed algorithm is fast and typically performs well, in the sense that in approximately 99% of the cases the relative experimental error ε does not exceed 0.01.

Further topics for investigation include (a) establishing convergence of Algorithm 6.3.2 for finite sampling (that is, $N < \infty$) with emphasis on the complexity and the speed of convergence under the suggested stopping rules; (b) establishing confidence intervals (regions) for the optimal solution; (c) application of parallel optimization techniques to the proposed methodology; and (d) investigations regarding a further speed-up of the algorithm. With respect to (d), we note that initially the throughputs do not need to be estimated very accurately, since the procedure just needs a rough idea which buffer allocations are good or not. However, furtheron in the procedure the accuracy needs to be increased to distinguish between competing "good" solutions. In the present test cases the same accuracy was used for all iterations, since the goal of this section was to show that for the "noisy" BAP high accuracy could be achieved within a reasonable time.

6.5 Numerical Results for the Noisy TSP

In this section we present the performance of Algorithm 4.7.3 for both deterministic and stochastic (noisy) versions of the travelling salesman problem

(TSP) respectively, using the formalism of Section 6.2. If not stated otherwise we use the indicator reward function $I_{\{S(\mathbf{X})\geqslant\gamma\}}$ (or $I_{\{\widehat{S}(\mathbf{X})\geqslant\gamma\}}$ in the noisy case) in our traditional two-stage procedure for updating the sequence $\{(\widehat{\gamma}_t, \widehat{P}_t)\}$, (for alternative reward functions see Section 5.2).

To generate different case studies we choose the elements c_{ij} of the cost matrix $C = (c_{ij})$ as i.i.d. random variables distributed according to some specified distribution. Once the matrix C is fixed we generate at each stage of the CE algorithm the Y_{ij} according to another specified distribution, with expectation c_{ij}. In all examples, we set $\varrho = 0.01$ and, depending on whether the deterministic or stochastic (noisy) version is considered, we stopped Algorithm 4.7.3 either according to the stopping rule (4.10) or according to the stopping rule (6.11). We also used stopping rule (4.43), which is suitable for both the deterministic and the stochastic (noisy) problems. In all cases (see (4.10), (6.11) and (4.43)) we took $d = k = 5$. We also present the CPU time of Algorithm 4.7.3 implemented in Matlab on 1.4GHz processor with 256M RAM along with the relative experimental error, defined as

$$\varepsilon = \frac{|\gamma^* - \widehat{\gamma}_T|}{\gamma^*}\,, \tag{6.17}$$

provided $\gamma^* \neq 0$, where $\widehat{\gamma}_T$ is the estimate of γ^* obtained after stopping. As estimates of the unknown parameters γ_{1t} and γ_{2t} (see Section 6.2) we take $\widehat{\gamma}_{1t}$ and $\widehat{\gamma}_{2t}$ respectively, where indices 1 and 2 correspond to the deterministic and stochastic versions, respectively. To indicate the convergence of $\widehat{P}_t$ to ODTM P^* we also present the following quantities:

$$P_t^{mm} = \min_{1\leq i\leq n}\max_{1\leq j\leq n} \widehat{p}_{t,ij}\,, \tag{6.18}$$

and

$$P_{t,mm} = \max_{1\leq i\leq n}\min_{1\leq j\leq n} \widehat{p}_{t,ij}\,. \tag{6.19}$$

They correspond to the min max and max min values of the elements of the matrix $\widehat{P}_t = (\widehat{p}_{t,ij})$ at iteration t, respectively. It is readily seen that $P_T^{mm} = 1$ if and only if $\widehat{P}_T = P^*$. We use the notations P_{1t}^{mm} and P_{2t}^{mm} for the deterministic and the stochastic versions, respectively.

Consider the synthetic TSP of Section 4.10.1 with optimal tour $1 \to 2 \to 3 \to \cdots \to n \to 1$ with total cost $\gamma^* = n$.

We consider separately (a) deterministic and (b) stochastic TSP.

(a) Deterministic TSP

Table 6.6 presents the relative experimental errors $\varepsilon_k,\ k = 1,\ldots,4$ (and the associated stopping times $T_k,\ k = 1,\ldots,4$) as a function of the sample size $N = r\,n^2 = r\,2500$ for $n = 50$ and the following four cases of c_{ij}: $c_{1,ij} \sim \mathsf{U}(2,10)$; $c_{2,ij} = 2$; $c_{3,ij} \sim \mathsf{U}(9,10)$, and $c_{4,ij} = 9$.

Table 6.6. The relative experimental errors ε_i, $i = 1, \ldots, 4$ (and the associated stopping times T_i, $i = 1, \ldots, 4$) as a function of the sample size $N = r\,n^2 = r\,2500$ for the four cases of c_{ij} above.

N	$\varepsilon_1(T_1)$	$\varepsilon_2(T_2)$	$\varepsilon_3(T_3)$	$\varepsilon_4(T_4)$
$n^2 = 2500$	0.104 (29.3)	0.0 (16.3)	0.0 (15.8)	0.0 (16)
$3n^2 = 7500$	0.0 (18)	0.0 (13.3)	0.0 (13.5)	0.0 (13.8)
$5n^2 = 12500$	0.0 (17.5)	0.0 (13.8)	0.0 (13.3)	0.0 (13)
$10n^2 = 25000$	0.0 (17)	0.0 (13)	0.0 (13)	0.0 (13)

Table 6.7 presents the associated CPU times, denoted $\text{CPU}_1, \ldots, \text{CPU}_4$, as a function of the sample size N.

Table 6.7. The CPU times as a function of the sample size N.

N	CPU_1	CPU_2	CPU_3	CPU_4
$n^2 = 2500$	48.5	27.0	26.5	26.8
$3n^2 = 7500$	92.0	64.3	66.5	67.8
$5n^2 = 12500$	143.0	111.3	108.3	105.5
$10n^2 = 25000$	279.5	205.5	205.0	204.8

It is seen from Table 6.6 that for fixed N, $T_1 \geq T_3$ and $T_2 \geq T_4$. The reason is that the lengths of all tours in cases 1 and 2 are "closer" each to other than in the alternative ones 3 and 4, respectively. As result, convergence of $(\widehat{\gamma}_t, \widehat{P}_t)$ to (γ^*, P^*) is slower in cases 1 and 2 as compared to the cases 3 and 4, respectively.

(b) Stochastic TSP

The test cases for the noisy TSP were constructed by first generating the deterministic matrix (c_{ij}) exactly as in the deterministic case above, and then generate the Y_{ij} from a predescribed distribution with $\mathbb{E}Y_{ij} = c_{ij}$. Below we present numerical results for the same TSP of Section 4.10.1, with $\gamma^* = n = 40$ and for the following six cases of Y_{ij}:

1. $Y_{ij} = c_{ij}$ (deterministic case).
2. $Y_{ij} \sim \mathsf{Beta}(10, 10, c_{ij} - 1, c_{ij} + 1)$.
3. $Y_{ij} \sim \mathsf{U}(c_{ij} - 0.5, c_{ij} + 0.5)$.
4. $Y_{ij} \sim \mathsf{U}(c_{ij} - 1, c_{ij} + 1)$.
5. $Y_{ij} \sim \mathsf{Po}(c_{ij})$.
6. $Y_{ij} \sim \mathsf{Exp}(c_{ij}^{-1})$.

In each of the six experiments we averaged the output statistics for the relative experimental errors ε_i, $i = 1, \ldots, 6$ and the associated stopping times T_i, $i = 1, \ldots, 6$ on the basis of five independent replications, each of size $N = r\,n^2 = r\,1600$, $r > 1$. We found that as $r \geqslant 10$ the (average) relative experimental error equals zero, that is Algorithm 4.7.3 is exact.

Table 6.8 presents the average relative experimental errors $\bar{\varepsilon}_k$, $k = 1, \ldots, 6$ (and the associated stopping times $\bar{T}_k$, $k = 1, \ldots, 6$) as a function of the sample size $N = r\,n^2 = r\,1600$ for $n = 40$ and the six cases of Y_{ij} above. Here, for the deterministic network ($Y_{ij} = c_{ij}$) we used the stopping rule (4.10), while for the remaining five stochastic networks we used the stopping rule (6.11).

Table 6.8. The average relative experimental errors $\bar{\varepsilon}_i$, $i = 1, \ldots, 6$ (and the associated stopping times $\bar{T}_i$, $i = 1, \ldots, 6$) as a function of the sample size $N = r\,n^2 = r\,1600$ for the six cases of Y_{ij} above.

N	$\bar{\varepsilon}_1(\bar{T}_1)$	$\bar{\varepsilon}_2(\bar{T}_2)$	$\bar{\varepsilon}_3(\bar{T}_3)$	$\bar{\varepsilon}_4(\bar{T}_4)$	$\bar{\varepsilon}_5(\bar{T}_5)$	$\bar{\varepsilon}_6(\bar{T}_6)$
8000	0.006 (24.3)	0.029(38)	0.037 (36.5)	0.040 (42.5)	0.053 (62.8)	0.058 (89.5)
16000	0.000 (21.8)	0.007(35)	0.007 (35)	0.019 (42)	0.025 (61.5)	0.034 (90)
32000	0.000 (20.5)	0.000(34)	0.000 (35.3)	0.000 (40)	0.013 (57)	0.013 (88.8)

Table 6.9 presents the associated average CPU times, denoted $\mathrm{CPU}_1, \ldots, \mathrm{CPU}_6$, as a function of the sample size N.

Table 6.9. The average CPU times as a function of the sample size N.

N	CPU_1	CPU_2	CPU_3	CPU_4	CPU_5	CPU_6
8000	81.3	266.8	157.3	176.0	286.0	326.8
16000	138.5	478.0	294.8	347.8	575.8	839.0
32000	262.5	960.5	581.3	667.5	1081.3	1341.3

It is seen in Table 6.8 that for fixed N the accuracy of Algorithm 4.7.3 decreases as the variance (noise) of Y_{ij} increases. This is in agreement with the fact that $\bar{\varepsilon}_k$, $k = 1, \ldots, 6$ in Table 6.8 are arranged purposely in the nonincreasing order, that is, $\bar{\varepsilon}_1 \leqslant \bar{\varepsilon}_2 \leqslant \cdots \leqslant \bar{\varepsilon}_6$ and this arrangement matches with the corresponding variances of $Y_{k,ij} = 1, \ldots, 6$ in the sense that

$$0 = \mathrm{Var}(Y_{1,ij}) \leqslant \mathrm{Var}(Y_{2,ij}) \leqslant \cdots \leqslant \mathrm{Var}(Y_{6,ij})\,.$$

We also found that for small samples ($r \leqslant 2$), Algorithm 4.7.3 does not converge for most cases.

Tables 6.10 and 6.11 present data similar to Tables 6.8 and 6.9, respectively using the same stopping rule (4.43) for both the deterministic ($Y_{ij} = c_{ij}$) and stochastic networks. We see that the results of Tables 6.10 and 6.11, and Tables 6.8 and 6.9 are nearly the same.

Table 6.10. Data similar to Table 6.8 using stopping rule (4.43).

N	$\bar{\varepsilon}_1(T_1)$	$\bar{\varepsilon}_2(T_2)$	$\bar{\varepsilon}_3(T_3)$	$\bar{\varepsilon}_4(T_4)$	$\bar{\varepsilon}_5(T_5)$	$\bar{\varepsilon}_6(T_6)$
8000	0.008 (23.7)	0.011 (23)	0.016 (24.5)	0.037 (29)	0.038 (45.8)	0.045 (71.8)
16000	0.000 (23.5)	0.007 (22.5)	0.015 (23.5)	0.023 (29.8)	0.030 (43.3)	0.031 (67)
32000	0.000 (21.3)	0.000 (21.5)	0.000(23.3)	0.007 (25.8)	0.009 (43.8)	0.013 (67.5)

Table 6.11. Data similar to Table 6.9 using stopping rule (4.43).

N	CPU_1	CPU_2	CPU_3	CPU_4	CPU_5	CPU_6
8000	79.5	160.0	101.8	122.0	214.5	263.5
16000	150.0	314.5	196.0	247.8	409.5	488.8
32000	272.5	605.5	431.8	431.8	830.5	990.3

Table 6.12 presents the dynamics of Algorithm 4.7.3 for cases 1 and 4 associated with Table 6.10, using the same stopping rule (4.43) with $n = 40$. More precisely, Table 6.12 presents the comparative evolution of the shortest tour estimates $\widehat{\gamma}_{1t}$ for the deterministic case ($Y_{ij} = c_{ij}$) (using $N_1 = 8000$), and $\widehat{\gamma}_{2t}$ for the noisy case with $Y_{ij} \sim \mathsf{U}(c_{ij} - 1, c_{ij} + 1)$ (using $N_2 = 16000$). It also displays the evolution of P_{1t}^{mm} and P_{2t}^{mm}. The deterministic and noisy versions of the algorithm stopped after 22 and 44 iterations and both found the shortest tour (with $\gamma^* = 40$) exactly. The CPU times were 125 and 383 seconds for the deterministic and noisy case, respectively. Figure 6.3 presents $\widehat{\gamma}_{1t}$ and $\widehat{\gamma}_{2t}$ as functions of iteration t based on the data of Table 6.12.

Table 6.12. Comparative evolution of Algorithm 4.7.3 for the deterministic and the noisy (U(−1, +1)) case with $n = 40$ and with sizes of the sample equal to $N_1 = 8000$ and $N_2 = 16000$, respectively.

t	$\widehat{\gamma}_{1t}$	$\widehat{\gamma}_{2t}$	P_{1t}^{mm}	P_{2t}^{mm}
1	179.0520	179.4279	0.058578	0.048116
2	144.0247	145.0523	0.076042	0.065749
3	116.1935	118.6788	0.096642	0.090130
4	97.05315	99.11970	0.129601	0.098680
5	83.49589	84.54807	0.125402	0.121932
13	49.87714	53.80331	0.256380	0.192082
14	48.47305	52.34546	0.315090	0.204799
15	45.63236	51.81999	0.381699	0.208865
16	43.48742	50.07529	0.414624	0.192094
17	40.00000	48.48727	0.925295	0.212014
18	40.00000	46.81393	0.992530	0.193051
23	40.00000	40.55133	0.999925	0.324640
24	40.00000	39.42279	0.999925	0.334615
25	40.00000	38.12740	0.999925	0.397181
26	40.00000	35.99771	0.999925	0.540564
27	40.00000	34.15617	0.999925	0.821609
28	40.00000	32.38640	0.999925	0.967815
41	40.00000	31.46442	0.999925	1.000000
42	40.00000	31.46441	0.999925	1.000000
43	40.00000	31.57093	0.999925	1.000000
44	40.00000	31.38555	0.999925	1.000000

Fig. 6.3. $\widehat{\gamma}_{1t}$ and $\widehat{\gamma}_{2t}$ as function of t based on the data of Table 6.12.

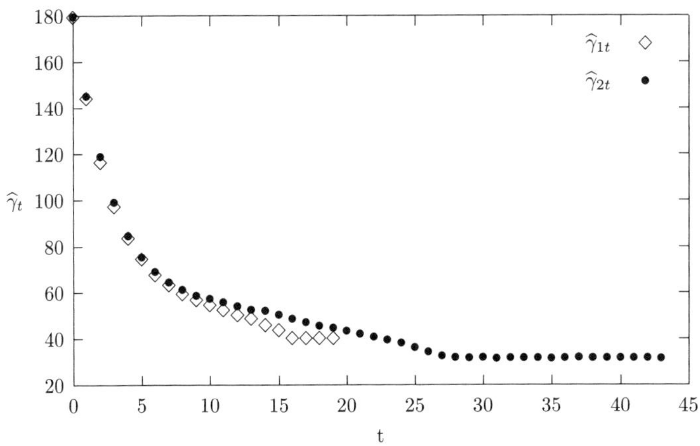

Table 6.13 represents similar data as in Table 6.12, but now for the noisy case $Y_{ij} \sim \mathsf{Exp}(c_{ij}^{-1})$. The sample sizes were $N_1 = 8000$ and $N_2 = 32000$. The CPU times were 125 and 614 seconds for the deterministic and noisy cases, respectively. For the deterministic case, the shortest tour was found exactly. Note that Table 6.13 presents a situation when the variance of Y_{ij} is quite large. For this reason, Algorithm 4.7.3 stopped for the noisy case after 74 iterations compared with 25 iterations for the deterministic case.

Table 6.13. Comparative evolution of Algorithm 4.7.3 for the deterministic and the exponential noisy case with $n = 40$ and with sizes of the sample equal to $N_1 = 8000$ and $N_2 = 32000$, respectively.

t	$\widehat{\gamma}_{1t}$	$\widehat{\gamma}_{2t}$	P_{1t}^{mm}	P_{2t}^{mm}
1	178.9196	133.5925	0.048438	0.039064
2	145.1071	121.1037	0.073678	0.047518
3	119.1800	112.4635	0.100546	0.055369
4	98.29205	103.1123	0.130158	0.060051
5	84.25539	95.44152	0.158822	0.069481
...	...	...	...	...
17	46.01890	57.40773	0.321939	0.131134
18	44.11950	56.34756	0.364391	0.123450
19	42.54073	55.62273	0.351199	0.136384
20	42.49261	55.18718	0.331856	0.137117
21	41.39288	54.24865	0.542151	0.146697
22	41.00371	54.52926	0.475954	0.147014
23	40.00000	53.50527	0.926640	0.150714
24	40.00000	52.93636	0.992664	0.147023
25	40.00000	51.94572	0.999266	0.149987
26	40.00000	50.90347	0.999266	0.157964
...	...	...	...	...
54	40.00000	35.40847	0.999266	0.500419
55	40.00000	33.79186	0.999266	0.648272
56	40.00000	31.33194	0.999266	0.807651
57	40.00000	28.92427	0.999266	0.944733
58	40.00000	27.44452	0.999266	0.977367
59	40.00000	26.91028	0.999266	0.995102
60	40.00000	26.94722	0.999266	0.999510
61	40.00000	26.85958	0.999266	0.999951
62	40.00000	26.72758	0.999266	0.999995
63	40.00000	26.78825	0.999266	1.000000
...	...	...	...	...
72	40.00000	26.76143	0.999266	1.000000
73	40.00000	26.81044	0.999266	1.000000
74	40.00000	26.66721	0.999266	1.000000

We also ran Algorithm 4.7.3 for the case studies in Table 2.5 in situations of noise, namely, we used the noisy cost matrices instead of the original deterministic ones. The noise was generated in a fashion similar to that generated in Tables 6.8 and 6.10. Tables 6.14 and 6.15 present such case studies for the the problems `ftv47` and `ftv55` in Table 2.5.

Table 6.14. Case studies for the noisy TSP `ftv47` with $n = 48$ and $N = 10\,n^2$.

Y_{ij}	$\bar{T}$	$\bar{\gamma}_1$	$\bar{\gamma}_T$	$\widehat{\gamma}$	$\bar{\varepsilon}$	ε_*	ε^*	CPU
$Y_{ij} = c_{ij}$	61	5622	1787	1776	0.018	0.043	0.006	284.0
$Y_{ij} \sim c_{ij} + \mathsf{U}(-7,7)$	61	5723	1814	1776	0.035	0.044	0.021	459.8
$Y_{ij} \sim \mathsf{Po}(c_{ij})$	60	5732	1817	1776	0.045	0.068	0.023	2287.2

Table 6.15. Case studies for the noisy TSP `ftv55` with $n = 56$ and $N = 10\,n^2$.

Y_{ij}	T	$\widehat{\gamma}_1$	$\widehat{\gamma}_T$	$\widehat{\gamma}$	$\widehat{\varepsilon}$	ε_*	ε^*	CPU
$Y_{ij} = c_{ij}$	67	6256	1614	1608	0.006	0.014	0.004	543.3
$Y_{ij} \sim c_{ij} + \mathsf{U}(-5,5)$	75	6402	1615	1608	0.029	0.040	0.004	9809.9
$Y_{ij} \sim \mathsf{Po}(c_{ij})$	78	6255	1646	1608	0.064	0.087	0.024	4412.2

Similar to the results in Tables 6.8 and 6.10 we found that the relative accuracy remains approximately the same as for the deterministic case, provided the sample size is increased from $5n^2$ to $15\,n^2$, respectively.

Finally, we compare the dynamics of the deterministic TSP at the end of Section 2.5.2 with its noisy counterpart. Thus, below we present $\widehat{P}_t$ as function of the iteration number t for the synthetic noisy TSP with n=10 cities. We take $Y_{ij} \sim \mathsf{U}(c_{ij} - 1, c_{ij} + 1)$ and a sample size of N=2000. Note that for the deterministic case a sample size of $N = 500$ was taken.

Dynamics of $\widehat{P}_t$

$$\widehat{P}_1 = \begin{pmatrix} 0.00 & 0.12 & 0.12 & 0.08 & 0.17 & 0.12 & 0.08 & 0.08 & 0.17 & 0.04 \\ 0.17 & 0.00 & 0.29 & 0.08 & 0.08 & 0.08 & 0.04 & 0.08 & 0.08 & 0.08 \\ 0.07 & 0.07 & 0.00 & 0.32 & 0.07 & 0.04 & 0.07 & 0.14 & 0.07 & 0.14 \\ 0.04 & 0.21 & 0.08 & 0.00 & 0.12 & 0.12 & 0.12 & 0.08 & 0.12 & 0.08 \\ 0.09 & 0.05 & 0.09 & 0.09 & 0.00 & 0.27 & 0.09 & 0.09 & 0.18 & 0.05 \\ 0.08 & 0.08 & 0.08 & 0.04 & 0.04 & 0.00 & 0.42 & 0.08 & 0.12 & 0.04 \\ 0.15 & 0.23 & 0.04 & 0.08 & 0.08 & 0.04 & 0.00 & 0.15 & 0.08 & 0.15 \\ 0.05 & 0.10 & 0.05 & 0.15 & 0.25 & 0.05 & 0.05 & 0.00 & 0.20 & 0.10 \\ 0.09 & 0.05 & 0.14 & 0.14 & 0.14 & 0.09 & 0.09 & 0.05 & 0.00 & 0.23 \\ 0.25 & 0.08 & 0.04 & 0.08 & 0.08 & 0.04 & 0.04 & 0.29 & 0.08 & 0.00 \end{pmatrix}.$$

$$\widehat{P}_2 = \begin{pmatrix}
0.00 & 0.32 & 0.08 & 0.06 & 0.12 & 0.16 & 0.06 & 0.06 & 0.12 & 0.03 \\
0.11 & 0.00 & 0.60 & 0.06 & 0.06 & 0.04 & 0.03 & 0.04 & 0.04 & 0.04 \\
0.05 & 0.05 & 0.00 & 0.52 & 0.05 & 0.02 & 0.05 & 0.07 & 0.05 & 0.15 \\
0.03 & 0.13 & 0.04 & 0.00 & 0.13 & 0.09 & 0.09 & 0.22 & 0.18 & 0.07 \\
0.04 & 0.03 & 0.06 & 0.06 & 0.00 & 0.47 & 0.06 & 0.08 & 0.16 & 0.04 \\
0.06 & 0.04 & 0.06 & 0.03 & 0.03 & 0.00 & 0.67 & 0.06 & 0.04 & 0.03 \\
0.21 & 0.29 & 0.03 & 0.06 & 0.04 & 0.03 & 0.00 & 0.21 & 0.06 & 0.08 \\
0.04 & 0.04 & 0.04 & 0.09 & 0.31 & 0.04 & 0.04 & 0.00 & 0.31 & 0.09 \\
0.04 & 0.03 & 0.04 & 0.08 & 0.24 & 0.07 & 0.07 & 0.03 & 0.00 & 0.40 \\
0.46 & 0.06 & 0.03 & 0.08 & 0.06 & 0.04 & 0.03 & 0.19 & 0.06 & 0.00
\end{pmatrix} .$$

$$\widehat{P}_3 = \begin{pmatrix}
0.00 & 0.62 & 0.05 & 0.04 & 0.08 & 0.07 & 0.04 & 0.04 & 0.04 & 0.02 \\
0.08 & 0.00 & 0.72 & 0.04 & 0.04 & 0.03 & 0.02 & 0.03 & 0.03 & 0.03 \\
0.03 & 0.03 & 0.00 & 0.65 & 0.03 & 0.02 & 0.03 & 0.05 & 0.03 & 0.12 \\
0.08 & 0.09 & 0.03 & 0.00 & 0.23 & 0.15 & 0.06 & 0.08 & 0.23 & 0.05 \\
0.04 & 0.02 & 0.05 & 0.05 & 0.00 & 0.54 & 0.05 & 0.04 & 0.17 & 0.04 \\
0.04 & 0.03 & 0.04 & 0.02 & 0.02 & 0.00 & 0.77 & 0.04 & 0.03 & 0.02 \\
0.17 & 0.13 & 0.02 & 0.04 & 0.03 & 0.02 & 0.00 & 0.42 & 0.04 & 0.13 \\
0.03 & 0.03 & 0.04 & 0.06 & 0.37 & 0.03 & 0.03 & 0.00 & 0.37 & 0.04 \\
0.03 & 0.03 & 0.03 & 0.09 & 0.17 & 0.04 & 0.05 & 0.04 & 0.00 & 0.52 \\
0.52 & 0.04 & 0.02 & 0.04 & 0.04 & 0.03 & 0.02 & 0.24 & 0.04 & 0.00
\end{pmatrix} .$$

$$\vdots$$

$$\widehat{P}_9 = \begin{pmatrix}
0.00 & 0.92 & 0.01 & 0.01 & 0.02 & 0.01 & 0.01 & 0.01 & 0.01 & 0.00 \\
0.02 & 0.00 & 0.93 & 0.01 & 0.01 & 0.01 & 0.00 & 0.01 & 0.01 & 0.01 \\
0.01 & 0.01 & 0.00 & 0.89 & 0.01 & 0.00 & 0.01 & 0.06 & 0.01 & 0.01 \\
0.01 & 0.01 & 0.00 & 0.00 & 0.88 & 0.04 & 0.01 & 0.02 & 0.02 & 0.01 \\
0.01 & 0.00 & 0.01 & 0.01 & 0.00 & 0.90 & 0.01 & 0.01 & 0.04 & 0.01 \\
0.01 & 0.01 & 0.01 & 0.01 & 0.01 & 0.00 & 0.94 & 0.01 & 0.01 & 0.01 \\
0.02 & 0.01 & 0.00 & 0.01 & 0.01 & 0.00 & 0.00 & 0.88 & 0.01 & 0.05 \\
0.01 & 0.01 & 0.01 & 0.01 & 0.04 & 0.00 & 0.00 & 0.00 & 0.92 & 0.01 \\
0.01 & 0.00 & 0.01 & 0.06 & 0.01 & 0.01 & 0.01 & 0.01 & 0.00 & 0.89 \\
0.92 & 0.01 & 0.00 & 0.01 & 0.01 & 0.01 & 0.00 & 0.02 & 0.01 & 0.00
\end{pmatrix} .$$

$$\widehat{P}_{10} = \begin{pmatrix}
0.00 & 0.93 & 0.01 & 0.01 & 0.01 & 0.01 & 0.01 & 0.01 & 0.01 & 0.00 \\
0.02 & 0.00 & 0.90 & 0.01 & 0.05 & 0.01 & 0.00 & 0.01 & 0.01 & 0.01 \\
0.01 & 0.01 & 0.00 & 0.91 & 0.01 & 0.00 & 0.01 & 0.05 & 0.01 & 0.01 \\
0.01 & 0.01 & 0.00 & 0.00 & 0.88 & 0.05 & 0.01 & 0.01 & 0.02 & 0.01 \\
0.01 & 0.00 & 0.05 & 0.01 & 0.00 & 0.88 & 0.01 & 0.01 & 0.03 & 0.01 \\
0.01 & 0.01 & 0.01 & 0.00 & 0.00 & 0.00 & 0.95 & 0.01 & 0.01 & 0.00 \\
0.02 & 0.01 & 0.00 & 0.01 & 0.00 & 0.00 & 0.00 & 0.90 & 0.01 & 0.04 \\
0.00 & 0.00 & 0.01 & 0.01 & 0.03 & 0.00 & 0.00 & 0.00 & 0.93 & 0.01 \\
0.00 & 0.00 & 0.00 & 0.05 & 0.01 & 0.01 & 0.01 & 0.01 & 0.00 & 0.91 \\
0.93 & 0.01 & 0.00 & 0.01 & 0.01 & 0.01 & 0.00 & 0.02 & 0.01 & 0.00
\end{pmatrix} .$$

$$\widehat{P}_{11} = \begin{pmatrix}
0.00 & 0.94 & 0.01 & 0.01 & 0.01 & 0.01 & 0.01 & 0.01 & 0.01 & 0.00 \\
0.01 & 0.00 & 0.92 & 0.01 & 0.04 & 0.00 & 0.00 & 0.00 & 0.00 & 0.00 \\
0.00 & 0.00 & 0.00 & 0.93 & 0.00 & 0.00 & 0.00 & 0.04 & 0.00 & 0.01 \\
0.01 & 0.01 & 0.00 & 0.00 & 0.91 & 0.04 & 0.01 & 0.01 & 0.02 & 0.00 \\
0.01 & 0.00 & 0.04 & 0.01 & 0.00 & 0.90 & 0.01 & 0.01 & 0.03 & 0.01 \\
0.01 & 0.01 & 0.01 & 0.00 & 0.00 & 0.00 & 0.95 & 0.01 & 0.01 & 0.00 \\
0.01 & 0.01 & 0.00 & 0.01 & 0.00 & 0.00 & 0.00 & 0.92 & 0.01 & 0.04 \\
0.00 & 0.00 & 0.00 & 0.01 & 0.03 & 0.00 & 0.00 & 0.00 & 0.94 & 0.00 \\
0.00 & 0.00 & 0.00 & 0.04 & 0.01 & 0.01 & 0.01 & 0.01 & 0.00 & 0.92 \\
0.94 & 0.01 & 0.00 & 0.01 & 0.01 & 0.01 & 0.00 & 0.02 & 0.01 & 0.00
\end{pmatrix} .$$

$$\widehat{P}_{12} = \begin{pmatrix}
0.00 & 0.95 & 0.01 & 0.00 & 0.01 & 0.01 & 0.00 & 0.00 & 0.01 & 0.00 \\
0.01 & 0.00 & 0.93 & 0.01 & 0.03 & 0.00 & 0.00 & 0.00 & 0.00 & 0.00 \\
0.00 & 0.00 & 0.00 & 0.94 & 0.00 & 0.00 & 0.00 & 0.03 & 0.00 & 0.01 \\
0.01 & 0.01 & 0.00 & 0.00 & 0.92 & 0.03 & 0.01 & 0.01 & 0.01 & 0.00 \\
0.01 & 0.00 & 0.03 & 0.01 & 0.00 & 0.92 & 0.01 & 0.00 & 0.02 & 0.00 \\
0.01 & 0.00 & 0.01 & 0.00 & 0.00 & 0.00 & 0.96 & 0.01 & 0.00 & 0.00 \\
0.01 & 0.01 & 0.00 & 0.00 & 0.00 & 0.00 & 0.00 & 0.93 & 0.00 & 0.03 \\
0.00 & 0.00 & 0.00 & 0.01 & 0.02 & 0.00 & 0.00 & 0.00 & 0.95 & 0.00 \\
0.00 & 0.00 & 0.00 & 0.03 & 0.01 & 0.00 & 0.01 & 0.00 & 0.00 & 0.94 \\
0.95 & 0.01 & 0.00 & 0.01 & 0.01 & 0.01 & 0.00 & 0.01 & 0.01 & 0.00
\end{pmatrix} .$$

6.6 Exercises

1. Consider the continuous multi-extremal optimization problem in Exercise 5.1. Add noise to the performance function S, and apply the CE method, for example as implemented in the program `cecont.m` of Appendix A.3. In particular, add $\mathsf{U}(-0.1, 0.1)$, $\mathsf{N}(0, 0.01)$ and $\mathsf{N}(0, 1)$ noise. Observe how $\widehat{\gamma}_t$, $\widehat{\mu}_t$ and $\widehat{\sigma}_t$ behave. Formulate an appropriate stopping criterion, for example based on $\widehat{\sigma}_t$.

2. Consider the Rosenbrock experiment of Table 5.1 and the corresponding Matlab code in Appendix A.5. Add $\mathsf{N}(0, 0.01)$ noise to the objective function `RosPen` by appending the line `S = S + 0.1*randn(N,1);` to the end of the code. Observe the convergence behavior of the CE algorithm for 10000 iterations, using the original parameters. Compare this with the behavior when $N = 200$, $N^{\text{elite}} = 5$ and $\beta = 3.5$.

3. Add random noise to the data from the URL

`http://www.iwr.uni-heidelberg.de/groups/comopt/software/TSPLIB95/atsp/`

and obtain a table similar to Table 6.15.

7

Applications of CE to COPs

In this chapter we show how CE can be employed to solve combinatorial optimization problems (COPs) from various fields. Our goal is to illustrate the flexibility and efficiency of the CE algorithm. We believe that the CE approach could open up a whole range of possible applications, for example in computational biology, graph theory, and scheduling. Section 7.1 deals with an application of the CE method concerning the alignment of DNA sequences; this section is based partly on [98]. In Sections 7.2–7.4, adapted from [113], the CE method is used to solve the single machine total weighted tardiness problem (SMTWTP), the single machine common due date problem (SMDDP), and the capacitated vehicle routing problem (CVRP). Finally, in Section 7.5 we consider various CE approaches for solving the maximum clique and related graph problems.

7.1 Sequence Alignment

Sequence alignment is a frequently encountered theme in computational biology. Many biologically important molecules are linear arrangements of subunits and can therefore be characterized as sequences. For example, a protein consists of amino acid residues linked by peptide bonds in a specific order known as its *primary structure*. A protein can alternatively be characterized as a sequence of larger subunits called *secondary structures*. Yet another characterization of a protein is its *tertiary structure*: the sequence of spatial positions and orientations taken by each of its amino acid residues. In order to study the structural, functional, and evolutionary relationships amongst biologically similar molecules, it is often useful to first align the corresponding sequences.

There are various forms of sequence alignment. Alignments can be made between sequences of the same type (for example, between the primary structures of proteins) or between sequences of different type (for example, alignment of a DNA sequence to a protein sequence, or of a protein to a three-dimensional structure). *Pairwise alignment* involves only two sequences, whereas *multiple sequence alignment* involves more than two sequences (although the term sometimes encompasses pairwise alignment also). *Global* alignment aligns whole sequences, whereas *local* alignment aligns only parts of sequences. In this section we consider only pairwise global alignments.

Algorithms for sequence alignment have been studied extensively. The inaugural paper on the subject is [122] and a useful reference is [74]. Other examples (see also [128]) include the dynamic programming approach initiated by [122], and the polyhedral approach initiated by [97], the hidden Markov model approach [104], and the Gibbs sampler approach [105]. The latter two involve randomized optimization algorithms.

Alignments

Consider two sequences of numbers $1, \ldots, n_1$ and $1, \ldots, n_2$. An *alignment* is an arrangement of the two sequences into two stacked rows, possibly including "spaces." Two opposite spaces not allowed. An example for $n_1 = 10$ and $n_2 = 6$ is given in Table 7.1.

Table 7.1. An example of an alignment.

{	1	2	-	3	4	5	6	7	8	9	10
	1	-	2	3	4	-	-	5	6	-	-

The significance of an alignment becomes clear when we assign a meaning to it. In particular, the two sequences of numbers could be associated with the positions of characters in a DNA or protein sequence. For example, the sequence $1, 2, \ldots, 10$ could be associated with the DNA string `AGTGCAGATA`, and the sequence $1, 2, \ldots, 6$ with another DNA string `ACTGGA`. Each alignment of the sequences $1, 2, \ldots, 10$ and $1, 2, \ldots, 6$ corresponds therefore with an "alignment" of the DNA strings. For example the alignment in Table 7.1 corresponds to the string alignment in Table 7.2.

Table 7.2. String alignment corresponding to the alignment in Table 7.1.

```
A G - T G C A G A T A
A - C T G - - G A - -
```

To be useful in applications, an alignment of two sequences should reflect in some way the commonalities of the sequences. Some alignments are therefore better than others. This concept is formalized using a *performance function* (often called a scoring function) to assign a value to an alignment. The most elementary way to do this is to use the *edit distance* between the aligned strings. In this case the performance of an alignment is derived from the aligned strings such that each character *mismatch* (for example A opposite to C) increases the performance by 1 and each *insertion* (space) also increases the performance by one. The objective is to find an alignment for which the edit distance is minimal. Table 7.3 gives two examples. Note that in this case the minimal edit distance is 5.

Table 7.3. Two possible alignments, with the corresponding edit distance.

```
A G T - G C A G A T A          A G - T G C A G A T A
A C T G - - G - A - -          A - C T G - - G A - -
   Performance: 8                 Performance: 6
```

Alignment Graph, Alignment Vector

A convenient way of looking at alignments is to associate with each alignment a unique *alignment path* through an *alignment graph*. An example of an alignment graph for the sequences 1,...,10 and 1,...,6 is given in Figure 7.1. The alignment path corresponding to Table 7.1 is also depicted. In particular, starting at $(0,0)$ the path jumps "to the right" whenever a space is inserted in the first sequence, and it jumps "downwards" whenever a space is inserted in the sequence; otherwise it jumps diagonally. Note that each path ends at (n_1, n_2).

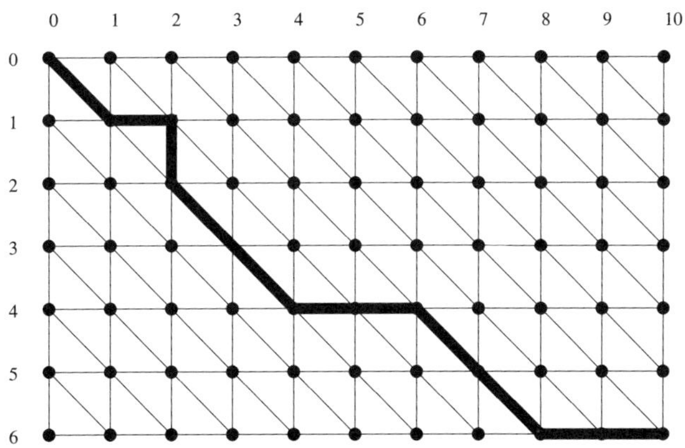

Fig. 7.1. Alignment graph. Directed edges from (i,j) to $(i+1,j)$, $(i+1,j+1)$, and $(i,j+1)$.

We can characterize the alignment path also by its *alignment vector* $\mathbf{x} = (x_1, \ldots, x_\tau)$, which is a vector of 0s, 1s, and 2s, where x_i denotes the "direction" the path takes at the i-th traversed node: $0 =$ horizontal, $1 =$ diagonal, and $2 =$ vertical. As an example, the alignment in Table 7.1 corresponds to the path in Figure 7.1 and the alignment vector of length $\tau = 11$ in Table 7.4.

Table 7.4. Each alignment vector corresponds to an alignment.

$\mathbf{x}$	1	0	2	1	1	0	0	1	1	0	0
	1	2	-	3	4	5	6	7	8	9	10
	1	-	2	3	4	-	-	5	6	-	-

Since there is a one-to-one correspondence between alignments and alignment paths it should not cause confusion if the same symbol $\mathbf{x}$ is used to represent both objects, and the same symbol $\mathcal{X}$ is used to represent the corresponding spaces. Moreover, any performance function for alignments may be regarded as a performance function for alignment paths. Now, let $S(\mathbf{x})$ denote the performance of the alignment corresponding to alignment vector $\mathbf{x}$. An optimal global sequence alignment is then an alignment $\mathbf{x}$ which solves

$$\min_{\mathbf{x} \in \mathcal{X}} S(\mathbf{x}) . \tag{7.1}$$

Remark 7.1. When using the edit distance, the performance depends in an "additive" way on the path. As a consequence, the optimal performance can efficiently be determined by *dynamic programming* (DP) [122, 161, 74, 84]. In practice, more complicated performance/scoring functions are used, but DP still works in many cases. When the performance function depends on the *whole* path, then the COP (7.1) is NP-complete. In particular, this holds for *structure* alignments, where the alignment corresponds to the spatial position of a protein residue (character). For such problems the CE method can provide a viable alternative. Moreover, deterministic algorithms only give one possible alignment and say nothing about the *distribution* of optimal alignments. The CE method can address this issue without much effort.

Trajectory Generation and Updating Formulas for CE

Using the edit distance, the optimal sequence alignment problem can be formulated in terms of a shortest path problem through the alignment graph; see Figure 7.1. The shortest path corresponds to the minimal edit distance. Thus, in the formulation of (7.1), we need to find an alignment path $\mathbf{x}$ through the alignment graph for which the alignment score $S(\mathbf{x})$ is minimal. We wish to solve (7.1) using the CE method. As usual, we need to specify:

1. How we generate the trajectories in $\mathcal{X}$ (that is, alignment vectors),
2. The probability updating formulas.

We generate our trajectories $\mathbf{x} \in \mathcal{X}$ by running a Markov chain $\mathbf{X} = (X_1, X_2, \ldots, X_\tau)$ on the alignment graph, starting at $(0,0)$ and ending at (n_1, n_2), with the following one-step transition probabilities in Table 7.5 (for all $0 \leqslant i \leqslant n_1 - 1$ and $0 \leqslant j \leqslant n_2 - 1$):

Table 7.5. The Transition Probabilities.

from	to	with prob.
(i,j)	$(i+1,j)$	$r(i,j)$
(i,j)	$(i,j+1)$	$d(i,j)$
(i,j)	$(i+1,j+1)$	$1 - r(i,j) - d(i,j)$

(Note that here r stands for *right* and d for *down*.) Moreover, for $j = 0, \ldots, n_2 - 1$ the transition probability from (n_1, j) to $(n_1, j+1)$ is 1. Similarly, for $i = 0, \ldots, n_1 - 1$ the transition probability from (i, n_2) to $(i+1, n_2)$ is 1. Finally, (n_1, n_2) is an absorbing state.

Parameter Update

Let us gather all parameters $\{r(i,j), d(i,j), 0 \leqslant i \leqslant n_1 - 1, 0 \leqslant j \leqslant n_2 - 1\}$ into a single parameter vector $\mathbf{v}$. For each such $\mathbf{v}$ we have

$$\begin{aligned} f(\mathbf{x}; \mathbf{v}) = \mathbb{P}_{\mathbf{v}}(\mathbf{X} = x) = \prod_{i=0}^{n_1-1} \prod_{j=0}^{n_2-1} \Big(& r(i,j)\, I_{\mathcal{X}_r(i,j)}(\mathbf{x}) \\ & + d(i,j)\, I_{\mathcal{X}_d(i,j)}(\mathbf{x}) \\ & + (1 - r(i,j) - d(i,j))\, I_{\mathcal{X}'(i,j)}(\mathbf{x}) \Big) . \end{aligned} \tag{7.2}$$

Here, $\mathcal{X}_r(i,j)$ is the set of all paths $\mathbf{x}$ making the transition from (i,j) to $(i+1,j)$; similarly, $\mathcal{X}_d(i,j)$ is the set of all paths $\mathbf{x}$ making the transition from (i,j) to $(i,j+1)$, and $\mathcal{X}(i,j) - \mathcal{X}_r(i,j) - \mathcal{X}_d(i,j)$ is abbreviated to $\mathcal{X}'(i,j)$. It follows that

$$\begin{aligned} \mathbb{E}_{\mathbf{u}} I_{\{S(\mathbf{X}) \leqslant \gamma\}} \ln f(\mathbf{X}; \mathbf{u}, \mathbf{v}) = \sum_{i=0}^{n_1-1} \sum_{j=0}^{n_2-1} \mathbb{E}_{\mathbf{u}} I_{\{S(\mathbf{X}) \leqslant \gamma\}} \Big(& \ln(r(i,j))\, I_{\mathcal{X}_r(i,j)}(\mathbf{X}) \\ + \ln(d(i,j))\, I_{\mathcal{X}_d(i,j)}(\mathbf{X}) + \ln(1 - r(i,j) - d(i,j))\, I_{\mathcal{X}'(i,j)}(\mathbf{X}) \Big) & . \end{aligned} \tag{7.3}$$

In view of the fundamental CE formula (4.7) we wish to maximize (7.3) with respect to $r(i,j)$ and $d(i,j)$ for all i and j, which amounts to differentiating the expression above with respect to $r(i,j)$ and $d(i,j)$ and equating it to zero. In the usual way this gives the deterministic updating formulas

$$r(i,j) = \frac{\mathbb{E}_{\mathbf{u}}\, I_{\{S(\mathbf{X}) \leqslant \gamma\}}\, I_{\{\mathbf{X} \in \mathcal{X}_r(i,j)\}}}{\mathbb{E}_{\mathbf{u}}\, I_{\{S(\mathbf{X}) \leqslant \gamma\}}\, I_{\{\mathbf{X} \in \mathcal{X}(i,j)\}}} \tag{7.4}$$

and

$$d(i,j) = \frac{\mathbb{E}_{\mathbf{u}}\, I_{\{S(\mathbf{X}) \leqslant \gamma\}}\, I_{\{\mathbf{X} \in \mathcal{X}_d(i,j)\}}}{\mathbb{E}_{\mathbf{u}}\, I_{\{S(\mathbf{X}) \leqslant \gamma\}}\, I_{\{\mathbf{X} \in \mathcal{X}(i,j)\}}} \, . \tag{7.5}$$

In a similar way we derive the formulas for the estimators $\widehat{r}_t(i,j)$ and $\widehat{d}_t(i,j)$ at stage t of Algorithm 4.2.1 as

$$\widehat{r}_t(i,j) = \frac{\sum_{k=1}^N I_{\{S(\mathbf{X}_k) \leqslant \widehat{\gamma}_t\}}\, I_{\{\mathbf{X}_k \in \mathcal{X}_r(i,j)\}}}{\sum_{k=1}^N I_{\{S(\mathbf{X}_k) \leqslant \widehat{\gamma}_t\}}\, I_{\{\mathbf{X}_k \in \mathcal{X}(i,j)\}}} \, , \tag{7.6}$$

and

$$\widehat{d}_t(i,j) = \frac{\sum_{k=1}^N I_{\{S(\mathbf{X}_k) \leqslant \widehat{\gamma}_t\}}\, I_{\{\mathbf{X}_k \in \mathcal{X}_d(i,j)\}}}{\sum_{k=1}^N I_{\{S(\mathbf{X}_k) \leqslant \widehat{\gamma}_t\}}\, I_{\{\mathbf{X}_k \in \mathcal{X}(i,j)\}}} \, , \tag{7.7}$$

where $\mathbf{X}_1, \ldots, \mathbf{X}_N$ are independent alignment paths generated according to the transition probabilities $\widehat{r}_{t-1}(i,j)$ and $\widehat{d}_{t-1}(i,j)$. These estimators have an easy interpretation. For example, to obtain $\widehat{r}_t(i,j)$ we count the number of paths (out of N) going from (i,j) to $(i,j+1)$ that have a performance less than or equal to $\widehat{\gamma}_t$, and divide this number by the total number of paths passing through (i,j) that have a performance less than or equal to $\widehat{\gamma}_t$. The estimator for $\widehat{d}_t(i,j)$ has a similar natural interpretation. The CE algorithm for sequence alignment follows directly from Algorithm 4.2.1 and is given below for convenience.

Algorithm 7.1.1 (Sequence Alignment CE Algorithm)

1. *Choose some initial vector of transition probabilities* $\widehat{\mathbf{v}}_0$, *for example with* $\widehat{r}_0(i,j) = \widehat{d}_0(i,j) = 1/3$. *Set* $t = 1$.
2. *Generate* N *paths* $\mathbf{X}_1, \ldots, \mathbf{X}_N$ *of the Markov process* $\mathbf{X}$, *using the transition probabilities specified in* $\widehat{\mathbf{v}}_{t-1}$ *and compute the sample* ϱ*-quantile of the performances.*
3. *Using the same sample* $\mathbf{X}_1, \ldots, \mathbf{X}_N$ *update the transition probabilities via* (7.6) *and* (7.7), *for each* (i,j).
4. *If for some* $t \geqslant d$, *say* $d = 5$, *stopping criterion* (4.10) *is met then stop; otherwise set* $t = t + 1$ *and reiterate from Step 2.*

Remark 7.2 (Modifications). Various modifications of Algorithm 7.1.1 are possible. In particular, we mention the modification in [98] where $\mathbf{X}$ is allowed to start at any position along the upper border $\{(0,j), j = 0, 1, \ldots, n_2\}$ or the left border $\{(i,0), i = 0, 1, \ldots, n_1\}$. The updating formulas for the starting probabilities are similar to the ones for the transition probabilities. The advantage of the latter approach is that a larger part of the alignment graph will be explored at an early stage in the procedure. As a consequence a considerable speed-up of the algorithm can be achieved.

Example 7.3. As an illustration, we provide the outcome of Algorithm 7.1.1 when applied to two protein sequences from Escherichia coli: Nitrogen Regulatory Protein P-II 1 (database: gi121386) and Nitrogen Regulatory Protein P-II 2 (database: gi1707971); see [34]. The two protein sequences are shown below:

```
MKKIDAIIKPFKLDDVREALAEVGITGMTVTEVKGFGRQKGHTELYRGAEYMVDFLPK
VKIERTAQTGKIGDGKIFVFDVARVIRIRTGEEDDAAI

MKLVTVIIKPFKLEDVREALSSIGIQGLTVTEVKGFGRQKGHAELYRGAEYSVNFLPK
VKIDVAIADDQLDEVIDIVSKAAYTGKIGDGKIFVAELQRVIRIRTGEADEAAL
```

The CE algorithm was implemented with a smoothing parameter $\alpha = 0.9$ and an adaptive updating/selection of the parameters such as in the FACE modification of Section 5.3, with $\varrho_0 = 0.01$ and $N_0 = 40000$. The algorithm found the following alignment:

```
MKKIDAIIKPFKLDDVREALAEVGITGMTVTEVKGFGRQKGHTELYRGAEYMVDFLPKVKI
MKLVTVIIKPFKLEDVREALSSIGIQGLTVTEVKGFGRQKGHAELYRGAEYSVNFLPKVKI

__________E___R__TAQ_TGKIGDGKIFVFDVARVIRIRTGEEDDAAI
DVAIADDQLDEVIDIVSKAAYTGKIGDGKIFVAELQRVIRIRTGEADEAAL
```

which gives an optimal performance of 39. Note that the edit distance does not impose gap penalties. A more sophisticated performance function would favor a single big gap to multiple small gaps. A typical evolution of the algorithm in given in Table 7.6. The CPU time was 94 seconds.

Table 7.6. Typical evolution of the sequence alignment CE algorithm.

t	$S_{t,(1)}$	$\widehat{\gamma}_t$
1	133	146
2	106	125
3	89	102
4	63	77
5	49	59
6	42	48
7	40	42
8	39	40
9	39	39
10	39	39
11	39	39

7.2 Single Machine Total Weighted Tardiness Problem

The single machine total weighted tardiness problem (SMTWTP) is a scheduling problem with the following description.

A set $V = \{1, \ldots, n\}$ of jobs with their corresponding processing times $\{L_1, \ldots, L_n\}$ are given. A due date d_j and a weight w_j (denoting the cost of one time unit of tardiness) are associated with each job j. The aim is to schedule the jobs without interruption on a single machine that can handle no more than one job at a time, so that the total cost is minimized.

Mathematically, the SMTWTP can be formulated as follows: Let $\mathcal{X}$ be the set of all the permutations of V. Each permutation $\mathbf{x} = (x_1, \ldots, x_n) \in \mathcal{X}$ defines the order in which the jobs are processed. Thus, for a 5-job problem $\mathbf{x} = (3, 1, 2, 5, 4)$ means that job 3 is processed first, then job 1, then 2, etc. Define f_k as a finishing date of a job which is assigned to the place k. This depends of course on the schedule $\mathbf{x}$. Specifically,

$$f_k = \sum_{l=1}^{k} L_{x_l} \; . \tag{7.8}$$

So the SMTWTP is the problem of finding the solution of the program

$$\min_{\mathbf{x} \in \mathcal{X}} S(\mathbf{x}) = \min_{\mathbf{x} \in \mathcal{X}} \sum_{k=1}^{n} w_{x_k} \max\{f_k - d_{x_k}, 0\} \; , \tag{7.9}$$

where f_k is defined by (7.8). Several techniques have been applied to the SMTWTP: branch and bound algorithms, simulated annealing, tabu search, genetic algorithms, and ant colony optimization; see for example [131, 3, 40, 27]. We apply here, of course, the CE algorithm.

We can generate the permutations $\mathbf{x}$ either according to Algorithm 4.7.1 or according to Algorithm 4.7.2 with the associated matrices $P = (p_{ij})$ and $P = (p_{(i,j)})$ (as in (4.39)), respectively. It turns out that with the second representation the method works better.

To illustrate the main steps of the CE algorithm for the SMTWTP consider the following simple example.

Example 7.4 (SMTWTP with 3 Jobs). Consider the single machine total weighted tardiness problem with 3 jobs and assume the following parameters are given:

Table 7.7. Example parameter for the SMTWTP.

j	L_j	d_j	w_j
1	1	1	1
2	1	2	1
3	1	2.5	1

Let P_t be the matrix corresponding to stage t of Algorithm 4.7.2. Thus,

$$P_t = \begin{pmatrix} p_{t,(1,1)} & p_{t,(1,2)} & p_{t,(1,3)} \\ p_{t,(2,1)} & p_{t,(2,2)} & p_{t,(2,3)} \\ p_{t,(3,1)} & p_{t,(3,2)} & p_{t,(3,3)} \end{pmatrix}, \tag{7.10}$$

where $p_{t,(i,j)}$ corresponds to the arrangement of the job i to the place j. Assume that P_0 is the matrix with equal probabilities 1/3. The six possible schedules (permutations) are: $(1,2,3)$; $(1,3,2)$; $(2,1,3)$; $(2,3,1)$; $(3,1,2)$; $(3,2,1)$. Because of the equal probabilities in P_0, each of these 6 solutions is generated initially with probability 1/6. It is easily verified that the corresponding objective values are

$$S((1,2,3)) = 0.5\ ;\ S((1,3,2)) = 1.0\ ;\ S((2,1,3)) = 1.5\ ;$$

$$S((2,3,1)) = 2.0\ ;\ S((3,1,2)) = 2.0\ ;\ S((3,2,1)) = 2.0\ .$$

Assume next that $\varrho = 1/3$. It follows that the $(1-\varrho)$-quantile of the performances γ_1 is equal to 1.0. Therefore,

$$I_{\{S((1,2,3))\leq\gamma_1\}} = 1\ ;\ I_{\{S((1,3,2))\leq\gamma_1\}} = 1\ ;\ I_{\{S((2,1,3))\leq\gamma_1\}} = 0\ ;$$

$$I_{\{S((2,3,1))\leq\gamma_1\}} = 0\ ;\ I_{\{S((3,1,2))\leq\gamma_1\}} = 0\ ;\ I_{\{S((3,2,1))\leq\gamma_1\}} = 0\ .$$

In the first iteration we have from the deterministic version of (4.41) (note that we use here $\mathbf{X}$ and S instead of $\mathbf{Y}$ and $\tilde{S}$, because there is no need to distinguish the node placement from the node transition algorithm here):

$$p_{1,(1,1)} = \frac{\mathbb{E}_{P_0} I_{\{S(\mathbf{X})\leq\gamma_1,\, X_1=1\}}}{\mathbb{E}_{P_0} I_{\{S(\mathbf{X})\leq\gamma_1\}}} = \frac{2/6}{2/6} = 1\ ,$$

and similarly $p_{1,(2,2)} = p_{1,(2,3)} = p_{1,(3,2)} = p_{1,(3,3)} = 1/2$, and for all other i, j we have $p_{1,(i,j)} = 0$, so that

$$P_1 = \begin{pmatrix} 1 & 0 & 0 \\ 0 & 1/2 & 1/2 \\ 0 & 1/2 & 1/2 \end{pmatrix}.$$

Using P_1 we can generate only two solutions:

$$(1,2,3) \quad \text{and} \quad (1,3,2)\ .$$

Obviously, each of the above solutions is generated with probability 1/2. Noting that $\varrho = 1/3$, it follows that $\gamma_2 = 0.5$. Therefore,

$$I_{\{S((1,2,3)\leq\gamma_2\}} = 1\ ;\ I_{\{S((1,3,2)\leq\gamma_2\}} = 0\ ,$$

and the probabilities for the second iteration are given by

$$P_2 = \begin{pmatrix} 1\,0\,0 \\ 0\,1\,0 \\ 0\,0\,1 \end{pmatrix} .$$

For example,

$$p_{2,(3,3)} = \frac{\mathbb{E}_{P_1} I_{\{S(\mathbf{X}) \leq \gamma_2,\, X_3=3\}}}{\mathbb{E}_{P_1} I_{\{S(\mathbf{X}) \leq \gamma_2\}}} = \frac{1/2}{1/2} = 1 .$$

Thus, the CE method indeed has converged to the optimal solution $(1, 2, 3)$.

Below we present numerical results with the CE Algorithm 4.2.1 for the SMTWTP for the set of benchmark problems taken from the site

`http://mscmga.ms.ic.ac.uk/jeb/orlib/wtinfo.html`

The set contains 5 problems with 40 jobs, 5 problems with 50 jobs, and 5 problems with 100 jobs. Table 7.8 presents the performance of the algorithm using the following parameters: $N = 10\,n^2$, $\varrho = 0.05$, $\alpha = 0.5$. The data were averaged over 10 independent replications. In Table 7.8 $\bar{T}$ denotes the average total number of iterations needed before stopping, $\bar{\gamma}_1$ and $\bar{\gamma}_T$ denote, respectively, the average of the initial and the final estimates of the optimal solution, γ^* denotes the best known solution, $\bar{\varepsilon}$ denotes the average relative experimental error, ε_* and ε^* denote the worse and the best relative experimental error among the 10 generated optimal solutions, and finally CPU denotes the average CPU time in seconds.

Table 7.8. The average performance of the Algorithm 4.2.1 for the SMTWTP based on 10 independent replications.

Probl.	n	$\bar{T}$	$\bar{\gamma}_1$	$\bar{\gamma}_T$	γ^*	$\bar{\varepsilon}$	ε_*	ε^*	CPU
1	40	21.4	3091.4	926.6	913.0	0.015	0.000	0.019	47
2	40	26.2	2255.0	1240.6	1225.0	0.013	0.001	0.022	56
3	40	18.4	2243.8	573.0	537.0	0.067	0.067	0.067	43
4	40	22.8	3269.8	2107.2	2094.0	0.006	0.006	0.009	49
5	40	11.0	2044.8	990.0	990.0	0.000	0.000	0.000	23
1	50	18.8	3774.4	2134.0	2134.0	0.000	0.000	0.000	88
2	50	18.0	3929.8	1998.0	1996.0	0.001	0.001	0.001	93
3	50	15.2	4631.2	2583.0	2583.0	0.000	0.000	0.000	71
4	50	14.0	5500.2	2691.0	2691.0	0.000	0.000	0.000	66
5	50	20.8	4700.2	1580.4	1518.0	0.041	0.023	0.057	97
1	100	34.4	25387.6	5988.0	5988.0	0.000	0.000	0.000	2111
2	100	32.0	24230.8	6304.2	6170.0	0.022	0.015	0.024	1947
3	100	33.4	19855.2	4311.0	4267.0	0.010	0.008	0.013	2011
4	100	53.6	19460.0	5042.8	5011.0	0.006	0.002	0.009	3320
5	100	44.4	21333.6	5283.0	5283.0	0.000	0.000	0.000	2743

Table 7.9 presents a typical evolution of the CE Algorithm for the SMTWTP with the indicator function for benchmark Problem 1 with 50 jobs. The parameters were as given above, in particular $N = 25000$. In the table t is the iteration number, $S_{t,(1)}$ the best solution obtained at the t-th iteration, $\widehat{\gamma}_t$ the worst solution of the elite samples, and $P_t^{mm} = \min_r \max_s \widehat{p}_{t,(r,s)}$.

Table 7.9. Typical evolution of the FACE algorithm for the SMTWTP.

t	$S_{t,(1)}$	$\widehat{\gamma}_t$	P_t^{mm}
1	3392	6850	0.030
2	2669	4482	0.035
3	2452	3224	0.080
4	2243	2621	0.121
5	2163	2392	0.148
6	2163	2340	0.159
7	2134	2335	0.201
8	2134	2247	0.306
9	2134	2184	0.286
10	2134	2134	0.342
11	2134	2134	0.428
12	2134	2134	0.512
13	2134	2134	0.631

7.3 Single Machine Common Due Date Problem

The description of the single machine common due date problem (SMCDDP) is similar to that of the SMTWTP in the previous section. Again, we are given a set $V = \{1, \dots, n\}$ of jobs and a set of corresponding processing times $\{L_1, \dots, L_n\}$. As before, the jobs have to be processed on one machine. However, now all the jobs have a *common* due date d. For each job j an earliness a_j and tardiness b_j penalty is given. These are incurred if the job is finished before or after the common due date d, respectively. The goal is to find a schedule for the n jobs that minimizes the sum of earliness and tardiness penalties.

As for the SMTWTP, $\mathcal{X}$ denotes the set of all the permutations of V; a permutation $\mathbf{x} = (x_1, \dots, x_n)$ corresponds to the order in which the jobs are processed. Let f_k as in (7.8) denote the finishing time of a job which is assigned to place k in the processing order. The program to be solved is:

$$\min_{\mathbf{x} \in \mathcal{X}} S(\mathbf{x}) = \min_{\mathbf{x} \in \mathcal{X}} \sum_{k=1}^{n} \left\{ a_{x_k} \max\{d - f_k, 0\} + b_{x_k} \max\{f_k - d, 0\} \right\} . \quad (7.11)$$

We can use the same permutation generation algorithms as in the TSP and SMTWTP. Numerical results for various benchmark problems from

`http://mscmga.ms.ic.ac.uk/jeb/orlib/schinfo.html`

are presented in [113].

7.4 Capacitated Vehicle Routing Problem

The capacitated vehicle routing problem (CVRP) presents the basic version of the more general vehicle routing problem (VRP), first proposed by Dantzig and Ramser [42]. The CVRP can be described as follows (see Figure 7.2 for an illustration): A set of n customers, labeled $\{1, 2, \ldots, n\}$, must be served from a single depot, labeled 0. The transit cost from i to j is given by c_{ij} for each $0 \leqslant i, j \leqslant n$. We assume that the cost structure is symmetric, that is, $c_{ij} = c_{ji}$. We also assume $c_{ii} = 0$. Each customer i has a demand d_i of goods and a vehicle of capacity D is available to deliver goods. Since the vehicle capacity is limited, the vehicle has to periodically return to the depot for reloading. It is forbidden to split customer delivery. The goal is to find the set of tours of minimal total cost, such that each tour begins and ends in the depot, each customer is served by exactly one tour and the total capacity of each tour is at most D. Various references on the (C)VRP may be found in [35] and [133].

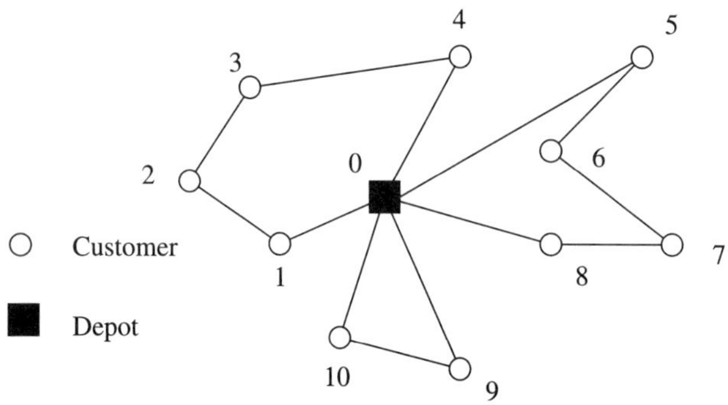

Fig. 7.2. An illustration of the CVRP.

To solve the CVRP using CE we use an algorithm similar to the TSP. In particular, we first assume that the graph is fully connected, assigning costs ∞ to "nonexisting" edges. We then generate tours $\mathbf{X}$ in exactly the same way as for the TSP, by representing them as random permutations $(X_0, \ldots, X_n)$ of $(0, 1, \ldots, n)$. To see how a tour/permutation $\mathbf{x} = (x_0, \ldots, x_n)$ and the d_i and D uniquely define a vehicle route, consider Figure 7.3. Here $\mathbf{x} = (0, 1, \ldots, n)$ corresponds to the tour $0 \to 1 \to \cdots \to n \to 0$. This tour only visits the depot

(0) at the beginning and at the end. However, when at a certain node k we have $d_1 + d_2 + \cdots + d_k \leqslant D$ and $d_1 + d_2 + \cdots + d_k + d_{k+1} > D$, the vehicle must return to the depot between nodes k and $k+1$. Suppose that this is the case for $k = 4$. For the second part of the vehicle route we can likewise determine exactly when the vehicle needs returning to the depot, for example after having visited node 8. We can see that for a given demand structure $\{d_i\}$ and capacity D the tour $\mathbf{x}$ in Figure 7.3 could correspond uniquely to the vehicle route given in Figure 7.2.

The only difference with the TSP is in calculating the performance of a tour $\mathbf{x}$, which is given by the total cost of the vehicle route determined by $\mathbf{x}$. For example the performance of $\mathbf{x} = (x_0, \ldots, x_n) = (0, 1, \ldots, n)$ in Figure 7.3 is given by

$$S(\mathbf{x}) = \left\{ \sum_{i=0}^{n-1} c_{x_i, x_{i+1}} + c_{x_n, x_0} \right\} + c_{40} + c_{05} - c_{45} + c_{80} + c_{09} - c_{89} \, ,$$

where the terms in brackets correspond to the usual TSP performance, as in (4.32).

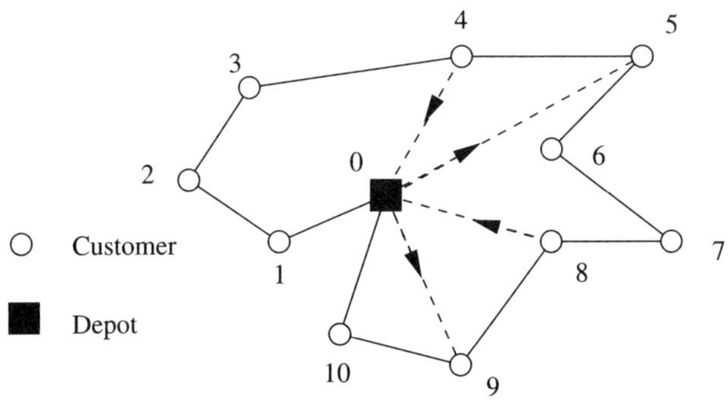

Fig. 7.3. Describing the routes via a tour/permutation.

Various benchmark problems for the (symmetric) capacitated vehicle routing problem can be found at the following URL:

`http://www.iwr.uni-heidelberg.de/groups/comopt/software/TSPLIB95`

Table 7.10 presents the performance of Algorithm 4.2.1 for the CVRP. The parameters are: $N = 10\,n^2$, $\varrho = 0.05$, $\alpha = 0.5$, and the data were averaged over 10 independent replications. In Table 7.10 we use the same notation as in Table 7.8.

Table 7.10. The average performance of the Algorithm 4.2.1 for the CVRP based on 10 independent replications.

name	n	$\bar{T}$	$\bar{\gamma}_1$	$\bar{\gamma}_T$	γ^*	$\bar{\varepsilon}$	ε_*	ε^*	CPU
`eil22`	22	39.0	611.5	377.4	375.0	0.006	0.001	0.012	24
`eil23`	23	30.5	1005.4	569.0	569.0	0.000	0.000	0.001	24
`eil30`	30	54.8	1095.0	515.4	-	-	-	-	127
`eil33`	33	78.0	1335.3	852.4	834.0	0.022	0.020	0.025	142
`eil51`	51	107.2	1352.4	530.0	-	-	-	-	980

Table 7.11 represents a typical evolution of the FACE Algorithm for the CVRP for benchmark problem `eil22`. The parameters are $N = 4840$, $\varrho = 0.05$, and $\alpha = 0.5$. The relative experimental error is $\varepsilon = 0.001$.

Table 7.11. Typical evolution of the FACE algorithm for the CVRP.

t	$S_{t,(1)}$	$\widehat{\gamma}_t$	P_t^{mm}
1	597.8	699.7	0.065
2	566.2	643.5	0.090
3	521.9	595.4	0.087
4	482.1	563.4	0.094
5	473.3	527.5	0.108
6	451.6	507.9	0.117
7	441.1	497.4	0.147
8	445.1	486.1	0.136
9	424.7	471.6	0.130
10	403.3	458.6	0.138
11	400.8	447.7	0.160
12	375.3	428.8	0.148
13	376.5	406.8	0.178
14	375.3	389.0	0.226
15	375.3	375.3	0.357
16	375.3	375.3	0.397
17	375.3	375.3	0.438
18	375.3	375.3	0.500

7.5 The Clique Problem

The maximum clique (MC) problem is a classical combinatorial optimization problem with important applications in different fields, such as cluster analysis, information retrieval, mobile networks, and computer vision. The MC problem is highly intractable and it presents one of the first problems that has been proven to be NP-complete.

Let $G = (V, E)$ be an arbitrary undirected graph, where $V = \{1, 2, \ldots, n\}$ is the vertex set of G, and $E \subseteq V \times V$ is the edge set of G. The total number of vertices (here n) is called the *order* of G; the total number of edges the *size* of G, [29]. The symmetric matrix $A_G = (a_{ij})$ where $a_{ij} = 1$ if $(i, j) \in E$ is an edge of G, and $a_{ij} = 0$ if $(i, j) \notin E$, is called the *adjacency matrix* of G. A graph $G = (V, E)$ is said to be *complete* if all its vertices are pairwise adjacent, that is, $\forall i, j \in V$ with $i \neq j$ we have $(i, j) \in E$. For any subset $\mathcal{C}$ of V the graph that contains all edges of G that join two vertices of $\mathcal{C}$ is called the *subgraph of G spanned by* $\mathcal{C}$; notation $G(\mathcal{C})$. A *clique* $\mathcal{C}$ is a subset of V such that $G(\mathcal{C})$ is complete. The *clique number* of G, denoted by $\omega(G)$, is the order (the number of vertices) of the maximum clique. The MC problem asks for cliques of maximum order $|\mathcal{C}|$:

$$\omega(G) = \max |\mathcal{C}| \ : \ \mathcal{C} \text{ is a clique in } G.$$

An example is given in Figure 7.4.

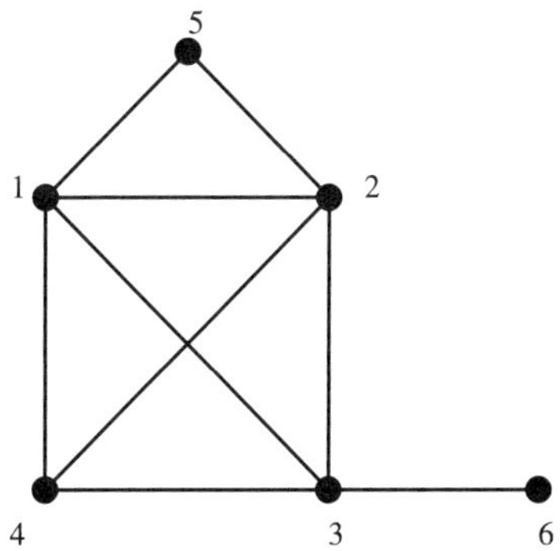

Fig. 7.4. A graph with maximum clique $\{1, 2, 3, 4\}$.

It is readily seen that the maximum clique is $\{1, 2, 3, 4\}$, which has clique value (cardinality, order) 4. Hence, the clique number of the graph is 4. Note that the corresponding (6×6) adjacency matrix $A = (a_{ij})$ is given by

$$A = \begin{pmatrix} 0\,1\,1\,1\,1\,0 \\ 1\,0\,1\,1\,1\,0 \\ 1\,1\,0\,1\,0\,1 \\ 1\,1\,1\,0\,0\,0 \\ 1\,1\,0\,0\,0\,0 \\ 0\,0\,1\,0\,0\,0 \end{pmatrix} . \tag{7.12}$$

We should distinguish a *maximum* clique from a *maximal* clique. A **maximal** clique is a clique that is not a proper subset of any other clique. A **maximum** clique is a maximal clique that has maximum order.

We shall represent a clique $\{x_1, \ldots, x_\tau\}$ via a "clique vector" $\mathbf{x} = (x_1, x_2, \ldots, x_\tau)$, where $x_i \in \{1, \ldots, n\}$, for $i = 1, \ldots, \tau \leqslant n$, and $x_1 \neq x_2 \neq \cdots \neq x_\tau$. Let $\mathcal{X}$ denote the space of all such vectors. Examples of clique vectors $\mathbf{x} \in \mathcal{X}$ for the graph in Figure 7.4 are $(1), (5), (1,2), (3,2,1)$, and $(3,2,4,1)$ — but not $(1,2,3,4,5), (2,6)$, and $(1,5,2,4)$. For each vector $\mathbf{x} = (x_1, \ldots, x_\tau)$ the corresponding clique value (cardinality of the clique) is given by

$$S((x_1, \ldots, x_\tau)) = \tau \ .$$

With this notation, the maximum clique problem is of the form (4.2).

Remark 7.5 (Related problems). The MC problem is related to various other problems in graph theory. In particular, it is closely associated with the maximum independent set and the minimum vertex cover problems.

An *independent set* of a graph $G = (V, E)$ is a subset of V whose elements are pairwise nonadjacent. The maximum independent set problem is to find an independent set in G of maximum order (cardinality). The order of this set is called the *stability number* of G, denoted by $\sigma(G)$. For the graph in Figure 7.4 we have $\sigma(G) = 3$, where the maximum independent set is $\{4, 5, 6\}$.

A *vertex cover* of a graph $G = (V, E)$ is a subset of V such that every edge $(i, j) \in E$ has at least one endpoint i or j in the subset. The minimum vertex cover problem is to find a vertex cover of minimum order. For example, in the graph in Figure 7.4 a possible minimum vertex cover is $\{5, 6\}$.

The relation with the maximum clique problem is the following: Let $\bar{G} = (V, \bar{E})$ be the *complement graph* of $G = (V, E)$, that is, the graph with edge set $\bar{E} = \{(i, j) \,:\, i, j \in V, i \neq j \text{ and } (i, j) \notin E\}$; see for example Figure 7.5. Then, $\mathcal{C}$ is a clique of G if and only if $\mathcal{C}$ is an independent set of $\bar{G}$, and if and only if $V \setminus \mathcal{C}$ is a vertex cover of $\bar{G}$. Hence $\omega(G) = \sigma(\bar{G})$. In the example $\{1, 2, 3, 4\}$ is a maximum independent set of $\bar{G}$, and $\{5, 6\}$ is a minimum vertex cover of $\bar{G}$. Any result obtained for one of the above problems has its equivalent forms for the other problems.

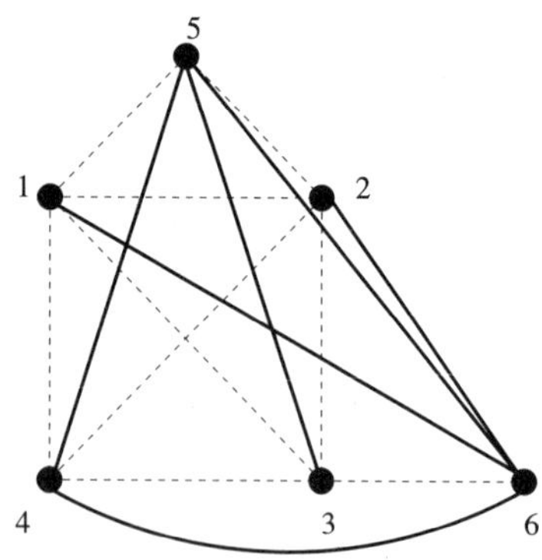

Fig. 7.5. The complement graph of the graph in Figure 7.4 (the dashed edges correspond to the edges in the original graph).

Trajectory Generation for Cliques

To explain the trajectory generation, consider the graph in Figure 7.4. It will be convenient to define an additional node, with label 0, to the graph, which is adjacent to all other nodes. By doing so, we obtain an expanded 7-node network instead of the original 6-node network. It is readily seen that in CE framework the MC problem is a SEN-type problem, and the trajectory generation is determined by probability matrix P. For example, for our 7-node network, we associate with A the following 7-state probability matrix

$$P = \begin{pmatrix} 0 & p_{01} & p_{02} & p_{03} & p_{04} & p_{05} & p_{06} \\ 0 & 0 & p_{12} & p_{13} & p_{14} & p_{15} & 0 \\ 0 & p_{21} & 0 & p_{23} & p_{24} & p_{25} & 0 \\ 0 & p_{31} & p_{32} & 0 & p_{34} & 0 & p_{36} \\ 0 & p_{41} & p_{42} & p_{43} & 0 & 0 & 0 \\ 0 & p_{51} & p_{52} & 0 & 0 & 0 & 0 \\ 0 & 0 & 0 & p_{63} & 0 & 0 & 0 \end{pmatrix} . \tag{7.13}$$

For a given $P = (p_{ij})$ we generate the clique vector $\mathbf{X} = (X_1, \ldots, X_\tau)$ in the following way:

Algorithm 7.5.1 (Naive Search)

1. *Generate X_1 according to the distribution formed by the 0-th row of P, that is, $(p_{01}, \ldots, p_{0n})$. Let $k = 1$.*
2. *Generate X_{k+1} according to the distribution formed by the X_1-st row of P.*
3. *Proceed with the trajectory generation through the nodes (states) $X_1, X_2, \ldots, X_\tau$ until some node, say node $X_{\tau+1}$ is reached, which does not belong to the clique.*
4. *Deliver the clique value $S(\mathbf{X}) = \tau$.*

Remark 7.6. In contrast to the TSP we do not require that all off-diagonal entries of P be nonzero. Indeed, we always assume that $p_{ij} = 0$ if $a_{ij} = 0$. The total number of nodes connected with a node k is called the *degree* of the node, written $\delta(k)$. Note that the maximum clique value associated with a node k is always less than or equal to its degree.

Remark 7.7. While proceeding from state X_k to state X_{k+1} in Step 3 of Algorithm 7.5.1 one can either set the probability p_{X_{k+1},X_k} to zero and then renormalize the remaining probabilities in the row X_{k+1} of P or reject the outcome X_k when generating from the X_{k+1}-st row of P. This avoids duplication, that is, the possibility that $X_{k+1} = X_k$.

Remark 7.8. If during the course of Algorithm 7.5.1 one obtains a clique value k, then one should set to zero all columns of P associated with nodes that have degree less than or equal to k, because these nodes can never be part of the current clique. If, for example, in our 6-node network $X_1 = 1$, then one can immediately eliminate node 6 (with $\delta(6) = 1$) from the generation algorithm by setting the 6th column of P to 0 and normalizing the rows. If subsequently $X_2 = 3$, we may repeat the same elimination procedure for node 5, because $\delta(5) = 2$.

Let $\widehat{P}_t = (\widehat{p}_{t,ij})$ denote the transition matrix at iteration t of the main CE Algorithm 4.2.1. As usual, we start from $p_{0,0i} = 1/n, \ i = 1, \ldots, n$, and we choose the nonzero elements in each of the other rows equal, that is, $1/\delta(k)$ for each row k. For the $\widehat{p}_{t,ij}$ we use the same updating rules (4.37) as for the TSP (with $\leqslant$ replaced with $\geqslant$). Note that even if there is a unique optimal clique, there will be various possible ODTMs. For example, two possible ODTM for the optimal clique $\{1, 2, 3, 4\}$ are

$$P^{*(1)} = \begin{pmatrix} 0 & 1 & 0 & 0 & 0 & 0 & 0 \\ 0 & 0 & p_{12} = 1 & 0 & 0 & 0 & 0 \\ 0 & 0 & 0 & p_{23} = 1 & 0 & 0 & 0 \\ 0 & 0 & 0 & 0 & p_{34} = 1 & 0 & 0 \\ 0 & p_{41} = 1 & 0 & 0 & 0 & 0 & 0 \\ 0 & 0 & 0 & 0 & 0 & 0 & 0 \\ 0 & 0 & 0 & 0 & 0 & 0 & 0 \end{pmatrix} \tag{7.14}$$

and

$$P^{*(2)} = \begin{pmatrix} 0 & 0 & 0 & 0 & 1 & 0 & 0 \\ 0 & 0 & 0 & 0 & p_{14} = 1 & 0 & 0 \\ 0 & p_{21} = 1 & 0 & 0 & 0 & 0 & 0 \\ 0 & 0 & p_{32} = 1 & 0 & 0 & 0 & 0 \\ 0 & 0 & 0 & p_{43} = 1 & 0 & 0 & 0 \\ 0 & 0 & 0 & 0 & 0 & 0 & 0 \\ 0 & 0 & 0 & 0 & 0 & 0 & 0 \end{pmatrix} . \tag{7.15}$$

Note that in (7.14) and (7.15) states 5 and 6 are redundant.

Next, we shall enhance Algorithm 7.5.1. This modification will lead to low bias clique generation and is called the *deep search* algorithm to distinguish it from Algorithm 7.5.1, which is called the *naive search* algorithm. In the deep search algorithm we consider the clique value $S_{kr} = r$ obtained by Algorithm 7.5.1 only as a possible (worst) alternative.

Algorithm 7.5.2 (Deep Search)

1-3. *As in Steps 1–3 of Algorithm 7.5.1.*

4*. *Return to the previous node X_τ and try from node X_τ again (randomly or deterministically) until either a larger clique is found or not. In the latter case stop and go to Step 5* below. In the former case continue with the trajectory $X_1, X_2, \ldots, X_\tau$ (generated at Step 3 of Algorithm 7.5.1) proceeding through the nodes $X_{\tau+1}, X_{\tau+2}, \ldots, X_{\tau_1+1}$, where X_{τ_1+1} corresponds to the node which does not belong to the clique passing through the* *nodes $X_1, \ldots, X_{\tau_1}$. Proceed with the loop based on Step 3 – Step 4*, that is, return to Step 3 of Algorithm 7.5.1 and proceed from node X_{τ_1} again until either a larger clique is found or not. In the latter case, stop and go to Step 5* below.*

5*. *Deliver the clique value τ_b where b corresponds to the total number of successful loops through Step 3 and Step 4* until stopping.*

Remark 7.9 (Even Deeper Search). Let $S_{t,(N-q+1)}, \ldots, S_{t,(N)}$ be the q largest (elite) clique values from the sample performances (clique values) $S_{t,1}, \ldots,$ $S_{t,N}$ obtained at the t-th iteration of Algorithm 7.5.2. One can try to enlarge the cliques from the set $\mathcal{E} = \{S_{t,(N-q+1)}, \ldots, S_{t,(N)}\}$ by arguing as follows: Choose any node contained in a particular clique with a clique value in $\mathcal{E}$, label that node as X_τ and proceed directly to Step 4* of Algorithm 7.5.2. If, for example, in our 6-node network the clique value 3 belongs to the elite set $\mathcal{E}$, and is associated, say, with the nodes 1, 2, 3, then we could select (randomly or deterministically) any node from the collection $\{1, 2, 3\}$ and proceed directly to Step 4* of Algorithm 7.5.2, hopefully obtaining a larger (the largest) clique value 4 associated with the nodes 1, 2, 3, 4. Again, such a policy of selecting a node (nodes) from the collection {1, 2, 3} can be applied to all three nodes 1, 2, 3 in turn.

Remark 7.10. Let τ be a clique associated with the starting node k and let $\delta_{(k1)}, \ldots, \delta_{(k\tau)}$ be the degrees of the nodes in the clique, arranged in increasing order. Clearly, if $\tau = \delta_{(k1)}$, then Algorithm 7.5.2 will be unable to enlarge the clique value τ starting at any node associated with τ. It also follows that if $\tau < \delta_{(k1)}$ then in order to attempt to enlarge τ it is more efficient (computationally) to proceed with Step 4* of Algorithm 7.5.2 from the node associated with the smallest row cardinality $\delta_{(k1)}$.

Remark 7.11 (Reducing Randomness). In order to avoid additional randomness and to speed up the algorithm, one can proceed directly to Step 2 of Algorithm 7.5.2 deterministically by ignoring Step 1. Indeed, for a given sample size N and given vector $(p_{t,0k},\ k = 1, \ldots, n)$, generate for each k a total of $N\, p_{t,0k}$ trajectories (in fact, the integer part $N\, p_{t,0k}$) starting at node k, $k = 1, \ldots, n$. In our 6-node example, let $N = 10$ and let $(p_{t,0k},\ k = 1, \ldots, 6) = (1/5, 1/5, 1/5, 1/5, 1/10, 1/10)$. It follows that in this case $(N\, p_{t,0k},\ k = 1, \ldots, 6) = (2,\ 2,\ 2,\ 2,\ 1,\ 1)$. This procedure is called *stratification.*

Remark 7.12 (Faster Trajectory Generation). Step 2 of the CE algorithm requires generating a random variable from a discrete distribution $\mathbf{p} = (p_1, \ldots, p_n)$, which can be quite time consuming. Similar to Algorithm 4.11.4 we can speed up the trajectory generation by dividing the p_i into two groups (in Algorithm 4.11.4 there are three): one group consisting of the c highest probabilities, with c typically between 5 or 10, and the other group consisting of the rest. We apply the inverse-transform method to draw from the first group, while the elements in the second group are drawn from a discrete uniform distribution.

We ran Algorithm 4.2.1 with $N = 3\,n^2$ for different case studies given in the URL

`ftp://dimacs.rutgers.edu/pub/challenge/graph/benchmarks/clique`

(needs to be accessed via anonymous ftp under Unix). Problems `brock200_1`, `brock200_2`, `brock200_3`, `brock200_4` and `brock400_4` were attempted and in all cases we obtained the relative experimental error $\varepsilon = 0$, that is Algorithm 4.2.1 performed perfectly. All runs were made on a 600MHz PIII computer with sample size $N = 3\,n^2$ and smoothing factor $\alpha = 0.7$. The algorithm stopped if the best sampled solution did not change for 5 consecutive iterations. The network sizes were up to $n = 400$ nodes and the maximum clique value was up to 33 (for `brock400_4`). A typical evolution (for `brock200_1`), using the naive search, is given in Table 7.12. The following optimal clique (with size 21) was found: $\{4, 26, 32, 41, 46, 48, 83, 100, 103, 104, 107, 120, 122, 132, 137, 138, 144, 175, 180, 191, 199\}$. A similar result for `brock400_4` is given in Table 7.13. In five repetitions the exact optimal clique $\{$7, 8, 17, 19, 112, 135, 147, 154, 157, 161, 186, 197, 202, 211, 241, 242, 245, 247, 266, 267, 270, 294, 324, 334, 340, 343, 353, 362, 380, 389, 393, 394, 396$\}$ was always found.

Table 7.12. Typical evolution of the clique CE algorithm.

t	1	2	3	4	5	6	7	8	9	10
$\widehat{\gamma}_t$	17	18	19	20	20	20	20	20	20	21

Table 7.13. Typical evolution of the CE algorithm for `brock400_4` .

t	1	2	3	4	5	6
$\widehat{\gamma}_t$	21	21	22	33	33	33

We also ran the improved trajectory generation algorithm of Remark 7.12 and compared it with the standard version. For the improved version we set the number of first class elements to $c = 5$. The improved version of the algorithm was as accurate as the naive version, while being approximately twice as fast. Table 7.14 shows the optimal solution (size of the maximum clique) and the execution time range for each instance. The execution time range was obtained by executing each algorithm on each instance for five times.

Table 7.14. Optimal solutions and typical execution times for the examined instances.

Instance	Maximum clique size	Execution time (seconds)			
		Naive		Improved	
		Min	Max	Min	Max
`brock200_1`	21	314	375	189	227
`brock200_2`	12	135	144	91	92
`brock200_3`	15	189	197	121	123
`brock200_4`	17	235	249	143	146
`brock400_4`	33	2874	3414	1857	2698

7.6 Exercises

1. **Clique problem**. Apply the CE Algorithm 4.2.1 to the clique problem of Section 7.5.
   a) First run the algorithm on a synthetic problem, using the naive clique generation Algorithm 7.5.1.
   b) When the synthetic problem is working, run the algorithm on a problem from the clique benchmark cases from the URL in Section 7.5, again using the naive clique generation.
   c) To speed up the trajectory generation apply the procedure outlined in Remark 7.12.
   d) Finally, run the FACE Algorithm 5.3.1.

2. **CE for Independent Set and Vertex Cover**. Follow the instructions for the clique problem, design your own trajectory generation algorithm for the independent set and vertex cover problems, and run Algorithm 4.2.1 first on a synthetic problem and then on a benchmark problem from the Web. Introduce noise to your problem by adding $\mathsf{U}(-a, a)$ noise to the objective function, run the noisy version of Algorithm 4.2.1 and present 2–3 tables (for different values a) similar to Table 6.14.

3. **CE for Vertex Coloring and Node Coloring**. Similar to the independent set and vertex cover problem, design your own trajectory generation algorithm for vertex coloring and node coloring problems and run Algorithm 4.2.1 first on a synthetic problem and then on a benchmark problem from the Web.

4. **CE for Permutation Flow Shop Problem (PFSP)**. In the permutation flow shop sequencing problem n jobs have to be processed (in the same order) on m machines. The objective is to find the permutation of jobs that will minimize the *makespan*, that is, the time at which the last job is completed on machine m. Let $t(i, j)$ be the processing time for job

i on machine j and let $\mathbf{x} = (x_1, x_2, ..., x_n)$ be a job permutation. Then the completion time $C(x_i, j)$ for job i on machine j can be calculated as follows:

$$\begin{aligned}
C(x_1, 1) &= t(x_1, 1) \\
C(x_i, 1) &= C(x_{i-1}, 1) + t(x_i, 1), \forall i = 2, ..., n \\
C(x_1, j) &= C(x_1, j-1) + t(x_1, j), \forall j = 2, ..., m \\
C(x_i, j) &= \max\{C(x_{i-1}, j), C(x_i, j-1)\} + t(x_i, j) \, , \\
&\qquad\qquad \text{for all } i = 2, ..., n; \quad j = 2, ..., m \, .
\end{aligned}$$

The objective is to minimize $S(\mathbf{x}) = C(x_n, m)$. The trajectory generation for the PFSP is exactly the same as in the TSP.

a) Follow the instructions at the beginning of Appendix A and run Algorithm 4.2.1 first for a synthetic problem and then for a benchmark problem from the Internet.
b) To speed up the trajectory generation apply the procedure given in the Appendix of Chapter 4. Add noise to your problem and present a table similar to Table 6.14.

5.* **CE Versus the EM Algorithm**. Suppose we have some statistical data $y_1, \ldots, y_n$ and we wish to fit the data to a mixture of two pdfs. Specifically, we assume that $y_1, \ldots, y_n$ are the outcomes of i.i.d. random variables $Y_1, \ldots, Y_n$ with density

$$f(y; a, \theta^0, \theta^1) = (1 - a) \, f_0(y; \theta^0) + a \, f_1(y; \theta^1) \, , \tag{7.16}$$

for some unknown a, θ^0 and θ^1. For example

$$f_j(y; \theta^j) = \frac{1}{\sqrt{2\pi\sigma_j^2}} \, \mathrm{e}^{-\frac{1}{2}\frac{(y-\mu_j)^2}{\sigma_j^2}} \, , \quad j = 0, 1 \, , \tag{7.17}$$

with $\theta^i = (\mu_i, \sigma_i^2)$, $i = 0, 1$. A straightforward way of estimating the parameters from the data $\mathbf{y} = (y_1, \ldots, y_n)$ is to choose the estimates such that the likelihood function

$$\mathcal{L}(a, \theta^0, \theta^1; \mathbf{y}) := \prod_{i=1}^{n} f(y_i; a, \theta^0, \theta^1) \tag{7.18}$$

is maximized. However, finding these *maximum likelihood* estimates is in general not easy for mixture models, since the likelihood function $\mathcal{L}(a, \theta^0, \theta^1)$ is typically multi-extremal.
In the well-known *EM method* [115], the likelihood is optimized via an iterative procedure. Specifically, consider the mixture model (7.16). We

may generate the random variables Y_i via a two-step procedure: first draw a random variable $X_i \sim \mathsf{Ber}(a)$ and then draw Y_i from f_{X_i}. Using this point of view, we can interpret the data $y_1, \ldots, y_n$ as only a part of the true data. The values of the 0-1 variables $x_1, \ldots, x_n$ — which indicate whether the y_i's were drawn from f_0 or f_1 — are *hidden*. Now, if $\mathbf{x} = (x_1, \ldots, x_n)$ were known, then the MLEs of the parameters could be easily estimated, namely as

$$\widehat{a} = n^{-1} \sum_{i=1}^{n} x_i \,, \tag{7.19}$$

$$\widehat{\mu}_j = n^{-1} \sum_{i:x_i=j} y_i, \quad j = 0, 1 \,, \tag{7.20}$$

$$\widehat{\sigma_j^2} = n^{-1} \sum_{i:x_i=j} (y_i - \widehat{\mu}_j)^2, \quad j = 0, 1 \,. \tag{7.21}$$

On the other hand, if the parameters a, θ^0 and θ^1 were known, then estimating the distribution of X_i from the data would be easy, using Bayes's formula:

$$p_i = \mathbb{E}(X_i \,|\, Y_i = y_i) = \frac{f_1(y_i; \theta^1)\, a}{f(y_i; a, \theta^0, \theta^1)}, \quad i = 1, \ldots, n \,. \tag{7.22}$$

Since we do not know the x_i we cannot use formulas (7.19)–(7.21). However, instead of maximizing the logarithm of the likelihood function in (7.18), we can maximize the *expected* log-likelihood function

$$\mathbb{E} \ln \mathcal{L}(a, \theta^0, \theta^1; \mathbf{X}, \mathbf{y}) = \mathbb{E} \ln \left[\prod_{i:X_i=0} (1-a)\, f_0(y_i; \theta^0) \times \prod_{i:X_i=1} a\, f_1(y_i; \theta^1) \right], \tag{7.23}$$

where $\mathbf{X} \sim \mathsf{Ber}(\mathbf{p})$, with $\mathbf{p} = (p_1, \ldots, p_n)$ as defined above. This leads to the following algorithm:

Algorithm 7.6.1 (EM algorithm)

1. *Choose initial estimates (guesses)* $\widehat{\theta}^i_0$, $i = 0, 1$. *Set* $t = 1$.
2. *(E-step) For each* i *define*

$$\widehat{p}_{t,i} = \frac{f_1(y_i; \widehat{\theta}^1_{t-1})\, \widehat{a}_{t-1}}{f(y_i; \widehat{a}_{t-1}, \widehat{\theta}^0_{t-1}, \widehat{\theta}^1_{t-1})} \,,$$

   and determine the expected log-likelihood function in (7.23), *with* $\mathbf{X} \sim \mathsf{Ber}(\widehat{\mathbf{p}}_t)$.
3. *(M-step) Maximize, with respect to* a, θ^0 *and* θ^1 *the expected log-likelihood function obtained in the E-step. Call the maximizing parameters* $\widehat{a}, \widehat{\theta}^0_t$ *and* $\widehat{\theta}^1_t$. *Specifically, for the case* (7.17) *it can be shown that*

$$\widehat{a}_t = n^{-1} \sum_{i=1}^{n} \widehat{p}_{t,i} \tag{7.24}$$

$$\widehat{\mu}_{t,0} = \frac{\sum_{i=1}^{n} (1 - \widehat{p}_{t,i})\, y_i}{\sum_{i=1}^{n} (1 - \widehat{p}_{t,i})} \tag{7.25}$$

$$\widehat{\mu}_{t,1} = \frac{\sum_{i=1}^{n} \widehat{p}_{t,i}\, y_i}{\sum_{i=1}^{n} \widehat{p}_{t,i}} \tag{7.26}$$

$$\widehat{\sigma^2}_{t,0} = \frac{\sum_{i=1}^{n} (1 - \widehat{p}_{t,i})(y_i - \widehat{\mu}_{t,0})^2}{\sum_{i=1}^{n} (1 - \widehat{p}_{t,i})} \tag{7.27}$$

$$\widehat{\sigma^2}_{t,1} = \frac{\sum_{i=1}^{n} \widehat{p}_{t,i}\, (y_i - \widehat{\mu}_{t,1})^2}{\sum_{i=1}^{n} \widehat{p}_{t,i}} . \tag{7.28}$$

4. *If some stopping criterion is met, then stop; otherwise set* $t := t + 1$ *and reiterate from Step 2.*

Note that EM is a local search procedure and therefore there is no guarantee that it converges to the global maximum. As an alternative to EM consider the CE Algorithm. There are two different approaches. First, we can view the maximization of (7.18) as a continuous multi-extremal optimization with respect to a and the θ^i; see Section 5.1. Alternatively, we can seek to maximize the function

$$S(\mathbf{x}) = \prod_{i:x_i=0} (1 - a)\, f_0(y_i; \theta^0) \times \prod_{i:x_i=1} a\, f_1(y_i; \theta^1) \tag{7.29}$$

over all $\mathbf{x}$. Note that in this case, we have to estimate S as we go along, since we do not know the true parameters. We can do this by using estimates for the parameters a and θ^j, $j = 0, 1$.

a) Present an explicit CE algorithm for the maximization of (7.18) as a function of a and the θ^i.
b) Present an explicit CE algorithm for the maximization of (7.29) as a function of $\mathbf{x}$.
c) Run the EM algorithm and both versions of the above CE algorithms, and compare their efficiencies on a mixture of 10 Normal pdfs with $\mu_j = j$ and $\sigma^2 = j,\ j = 1, 2, \ldots, 10$.

8

Applications of CE to Machine Learning

In this chapter we apply the CE method to several problems arising in machine learning, specifically with respect to optimization. In Section 8.1, adapted from [50], we apply CE to the well-known mastermind game. Section 8.2, based partly on [112], describes the application of the CE method to Markov decision processes. Finally, in Section 8.3 the CE method is applied to clustering problems. In addition to its simplicity, the advantage of using the CE method for machine learning is that it does not require direct estimation of the gradients, as many other algorithms do (for example, the stochastic approximation, steepest ascent, or conjugate gradient method). Moreover, as a global optimization procedure the CE method is quite robust with respect to starting conditions and sampling errors, in contrast to some other heuristics, such as simulated annealing or guided local search.

8.1 Mastermind Game

In the well-known *mastermind* game the objective is to decipher a hidden "code" of colored pegs through a series of guesses. After each guess new information on the true code is provided in the form of black and white pegs. A black peg is earned for each colored peg that is in exactly the right place, and a white peg for each peg that is in the solution, but in the wrong position; see Figure 8.1. To get a feel for the game, one can visit

`http://www.javaonthebrain.com/java/mastermind`

and play its java implementations.

Consider a mastermind game with m colors, numbered $\{1, \ldots, m\}$ and n pegs (positions). The hidden solution and a guess can be represented by a row vector of length n with numbers in $\{1, \ldots, m\}$. For example, for $n = 5$ and $m = 7$ the solution could be $\mathbf{y} = (4, 2, 4, 7, 3)$, and a possible guess $\mathbf{x} = (4, 3, 4, 2, 5)$. Let $\mathcal{X}$ be the space of all possible guesses. Note that the

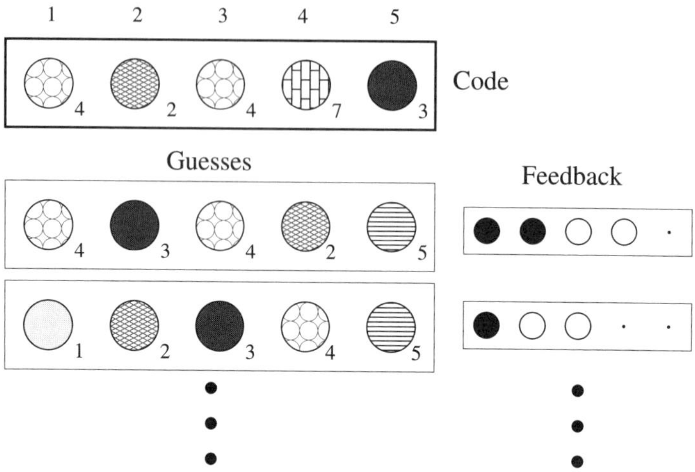

Fig. 8.1. The mastermind game.

total number of possibilities is m^n. On $\mathcal{X}$ we define a performance function S which returns for each guess $\mathbf{x}$ the "pegscore"

$$S(\mathbf{x}) = 2 \times N_{\text{BlackPegs}} + N_{\text{WhitePegs}} \;, \tag{8.1}$$

where $N_{\text{BlackPegs}}$ and $N_{\text{WhitePegs}}$ are the number of black and white pegs returned after the guess. We have assumed, somewhat arbitrarily, that a black peg is worth twice as much as a white peg. As an example, for the solution $\mathbf{y} = (4, 2, 4, 7, 3)$ and guess $\mathbf{x} = (4, 3, 4, 2, 5)$ above, one gets 2 black pegs (for the first and third pegs in the guess) and 2 white pegs (for the second and fourth pegs in the guess, which match the fifth and second pegs of the solution but are in the wrong place). Hence the score for this guess is 6.

There are some specially designed algorithms that efficiently solve the problem for different numbers of colors and pegs. For examples see

`http://www.mathworks.com/contest/mastermind.cgi/home.html`

With the above performance function, the problem can be formulated as an optimization problem of the form (4.2), and hence we could apply the CE method to solve it. In order to apply the CE method we first need to generate random guesses $\mathbf{X} \in \mathcal{X}$. We can do this via an $n \times m$ matrix P. Each element p_{ij} of P describes the probability that we choose the j-th color for the i-th peg (location). Since only one color may be assigned to one peg we have that $\sum_{j=1}^{m} p_{ij} = 1$ for all i. At each iteration t of the CE algorithm we independently sample for each row (that is, each peg) a color using probability matrix $P = \widehat{P}_t$ and calculate the score S according to (8.1). The updating of the elements $\widehat{p}_{t,ij}$ of the probability matrix $\widehat{P}_t$ is performed exactly as in (4.14). Note that in this case the numerator in (4.14) simply counts the number of times the color index j has been assigned to peg i for the elite samples.

8.1.1 Numerical Results

Table 8.1 represents a typical evolution of the CE algorithm for the mastermind test problem denoted on the site

`http://www.mathworks.com/contest/mastermind.cgi/home.html`

as "problem (5.2.3)," with the matrix P of size $n \times m = 36 \times 33$. We used $N = 5\,m\,n = 5940$, $\varrho = 0.01$ and $\alpha = 0.7$. It took 34 seconds of CPU time to find the true solution. The notation in the table is the same as in Section 4.9, with $S_{t,(N_t)}$ abbreviated to S_t^*. Table 8.2 represents data similar to Table 8.1 for the FACE Algorithm 5.3.1. The initial sample size is $N^{\min} = 594$ and the elite sample size is 297. The true solution was found in 18 seconds.

Table 8.1. Evolution of Algorithm 4.2.1 for the mastermind problem with $m = 33$ colors and $n = 36$ pegs.

t	S_t^*	$\widehat{\gamma}_t$	$p\uparrow_{\max}$	$p\downarrow_{\min}$
1	31	26	0.1138	0.0330
2	34	28	0.2382	0.0308
3	38	32	0.3574	0.0304
4	42	35	0.5348	0.0304
5	48	40	0.6457	0.0304
6	55	45	0.7756	0.0306
7	60	51	0.8312	0.0304
8	64	57	0.9016	0.0309
9	70	63	0.9426	0.0316
10	72	68	0.9804	0.8247
11	72	72	0.9961	0.9649

Table 8.2. Evolution of FACE Algorithm 5.3.1 for the mastermind problem with $m = 33$ colors and $n = 36$ pegs.

t	S_t^*	$\widehat{\gamma}_t$	N_t	$p\uparrow_{\max}$	$p\downarrow_{\min}$
1	30	21	594	0.05	0.03
2	32	24	594	0.10	0.03
3	33	28	5212	0.17	0.03
4	35	29	594	0.23	0.03
5	36	29	594	0.28	0.03
6	39	31	594	0.30	0.03
7	42	32	594	0.37	0.03
8	43	34	594	0.45	0.03
9	48	36	594	0.52	0.03
10	49	40	1600	0.62	0.03

t	S_t^*	$\widehat{\gamma}_t$	N_t	$p\uparrow_{\max}$	$p\downarrow_{\min}$
11	51	42	594	0.71	0.03
12	54	44	594	0.76	0.03
13	55	46	606	0.79	0.03
14	59	49	594	0.84	0.03
15	62	52	594	0.87	0.03
16	65	54	594	0.91	0.03
17	66	57	594	0.93	0.03
18	67	60	730	0.96	0.03
19	72	63	594	0.98	0.03
20	72	70	11880	0.99	0.82

Table 8.3 compares the efficiencies (the relative experimental errors and the CPU times) of CE Algorithm 4.2.1 and FACE Algorithm 5.3.1 for different mastermind games. The results are averaged over 10 different runs. We see that the algorithms have similar accuracies, and that the FACE algorithm is typically two times faster than its CE algorithm counterpart.

Table 8.3. The efficiency of CE Algorithm 4.2.1 versus the FACE algorithm for different mastermind games.

m	n	CE time (sec.)	$\varepsilon(CE)$	FACE time (sec.)	$\varepsilon(FACE)$
18	36	13.667	0.0000	7.909	0.0221
23	32	15.497	0.0000	8.720	0.0000
30	41	41.006	0.0000	20.392	0.0000
42	41	66.695	0.0000	35.525	0.0000
43	45	90.788	0.0000	42.469	0.0000

8.2 The Markov Decision Process and Reinforcement Learning

The *Markov decision process* (MDP) model is standard in machine learning, operations research, and related fields. We review briefly some of the basic definitions and concepts in MDP. For details see for example [25, 132, 174]. This section is based partly on [112]. For another example of the power of the CE method in the MDP context see [117].

An MDP is defined by a tuple $(\mathcal{Z}, \mathcal{A}, \mathcal{P}, r)$, where

1. $\mathcal{Z} = \{1, \ldots, n\}$ is a finite set of states.
2. $\mathcal{A} = \{1, \ldots, m\}$ is the set of possible actions by the decision maker. We assume it is the same for every state — to ease notations.
3. $\mathcal{P}$ is the transition probability matrix with elements $\mathcal{P}(z'|z, a)$ presenting the transition probability from state z to state z', when action a is chosen.
4. $r(z, a)$ is the reward for performing action a in state z (r may be random).

At each time instance k the decision maker observes the current state z_k, and determines the action to be taken, say a_k. As a result, a reward $r(z_k, a_k)$, or shortly r_k, is received and a new state z' is chosen according to $\mathcal{P}(z'|z_k, a_k)$. A *policy* or *strategy* π is a rule that determines, for each history $H_k = z_1, a_1, \ldots, z_{k-1}, a_{k-1}, z_k$ of states and actions, the probability distribution of the decision maker's actions at time k. A policy is called *Markov* if each action is deterministic and depends only on the current state z_k. A Markov policy is called *stationary* if it does not depend on the time k. The goal of the decision maker is to maximize a certain reward function.

The following are standard reward criteria:

1. *Finite horizon reward.*
   This applies when there exists a finite time τ (random or deterministic) at which the process terminates. The objective is to maximize the total reward
   $$S(\pi) = \mathbb{E}_\pi \sum_{k=0}^{\tau-1} r_k \,. \tag{8.2}$$
   Here $\mathbb{E}_\pi$ denotes the expectation with respect to some probability measure induced by the strategy π.
2. *Infinite horizon discounted reward.*
   The objective is to find a strategy π that maximizes
   $$S(\pi) = \mathbb{E}_\pi \sum_{k=0}^{\infty} \beta^k \, r_k \,, \tag{8.3}$$
   where $0 < \beta < 1$ is the discount factor.
3. *Average reward.*
   The objective is to maximize
   $$S(\pi) = \liminf_{\tau \to \infty} \frac{1}{\tau} \, \mathbb{E}_\pi \sum_{k=0}^{\tau-1} r_k \,.$$

In the next section we will restrict our attention to the *stochastic shortest path* MDP, where it is assumed that the process starts from a specific initial state $z_0 = z^{\text{start}}$, and terminates in an absorbing state z^{fin} with zero reward. For the finite horizon reward criterion, τ is the time at which z^{fin} is reached (which we will assume will happen eventually). It is well known [132] that for the shortest path MDP there exists a stationary Markov policy which maximizes $S(\pi)$, for each of the reward functions above.

If both r and $\mathcal{P}$ are known, then several efficient methods, such as *value iteration* and *policy iteration*, can be used to find the optimal policy [132, 173, 174]. However, if the transition probability or the reward function are *unknown*, the problem is much more difficult, and is referred to as a *learning* problem. A well-known framework for learning algorithms is *reinforcement learning* (RL), where an agent learns the behavior of the system through trial and error in an unknown dynamic environment; see [92].

There are several approaches to RL, which can be roughly divided into the following three classes: model-based, model-free, and policy search. In the model-based approach, first a model of the environment is constructed. The estimated MDP is then solved using standard tools of dynamic programming [96]. In the model-free approach one learns a utility function, instead of learning the model. The optimal policy is to choose at each state an action that maximizes the expected utility. The popular Q-learning algorithm [171] is an example of this approach. In the policy search approach a subspace of the policy space is searched, and the performance of policies is evaluated based on their empirical performance [19, 163]. An example of a gradient-based policy search method is the REINFORCE algorithm [175]. A detailed account of policy gradient methods can be found in [20]. For an approach which uses a direct search in policy space see [139]. The CE algorithm in the next section can be viewed as a policy search approach.

Remark 8.1 (Stochastic Approximation). Many RL algorithms are based on the classic *stochastic approximation* (SA) algorithm. To explain SA, assume that we need to find the unique solution x^* of some nonlinear equation $S(x) = 0$, where instead of $S(x)$ only an estimate $\widehat{S}(x)$ is available, with $\mathbb{E}\widehat{S}(x) = S(x)$. The SA algorithm for estimating x^* involves the iteration

$$x_{t+1} = x_t + \beta_t \, \widehat{S}(x_t) \, ,$$

where $\{\beta_t, t = 1, 2, \ldots\}$ is a positive sequence satisfying

$$\sum_{t=1}^{\infty} \beta_t = \infty, \ \sum_{t=1}^{\infty} \beta_t^2 < \infty \, . \tag{8.4}$$

The connection between SA and Q-learning is given in [165]. This work has made an important impact on the entire field of RL. Unfortunately, SA is known as a very slow converging procedure, because of (8.4). Even if β_k remains bounded away from 0, (and thus convergence is not guaranteed) it is still required that β_k is small in order to ensure convergence to a reasonable neighboring solution [30]. We shall employ the CE method instead of SA and shall demonstrate its high efficiency.

8.2.1 Policy Learning via the CE Method

We consider a CE learning algorithm for the *shortest path* MDP, where it is assumed that the process starts from a specific initial state z_0, and that there is an absorbing state z^{fin} with zero reward. The objective is given in (8.2), with τ being the stopping time at which z^{fin} is reached, which we will assume will always happen.

To put this problem in the CE framework, consider the maximization problem (8.2). Recall that for the shortest path MDP an optimal stationary strategy exists. We can represent each stationary strategy as a vector $\mathbf{x} = (x_1, \ldots, x_n)$ with $x_i \in \{1, \ldots, m\}$ being the action taken when visiting state i. Writing the expectation in (8.2) as

$$S(\mathbf{x}) = \mathbb{E}_{\mathbf{x}} \sum_{k=0}^{\tau-1} r(Z_k, A_k) \, , \tag{8.5}$$

where $Z_0, Z_1, \ldots$ are the states visited, and $A_0, A_1, \ldots$ the actions taken, we see that the optimization problem (8.2) is of the form (4.2). We shall also consider the case where $S(\mathbf{x})$ is measured (observed) with some noise, in which case we have a *noisy* optimization problem. The idea now is to combine the random policy generation and the random trajectory generation in the following way: At each stage of the CE algorithm we generate random policies and random trajectories using an auxiliary $n \times m$ matrix $P = (p_{za})$, such that for each state z we choose action a with probability p_{za}. Once this "policy matrix" P is defined, each iteration of the CE algorithm comprises the following two standard phases:

1. Generation of N random trajectories $(Z_0, A_0, Z_1, A_1, \ldots, Z_\tau, A_\tau)$ using the auxiliary policy matrix P. The cost of each trajectory is computed via
$$\widehat{S}(\mathbf{X}) = \sum_{k=0}^{\tau-1} r(Z_k, A_k) . \tag{8.6}$$
2. Updating of the parameters of the policy matrix (p_{za}) on the basis of the data collected in the first phase.

The matrix P is typically initialized to a uniform matrix $(p_{ij} = 1/m)$. We describe both the trajectory generation and updating procedure in more detail below. We shall show that in calculating the associated sample performance, one can take into account the Markovian nature of the problem to speed up the Monte Carlo process.

Generating Random Trajectories

Generation of random trajectories for MDP is straightforward and is given for convenience only. All one has to do is to start the trajectory from the initial state $z_0 = z^{\text{start}}$ and follow the trajectory by generating each new state according to the probability distribution of P, until the absorbing state z^{fin} is reached at time τ, say.

Algorithm 8.2.1 (Trajectory Generation for MDP)
Input: P auxiliary policy matrix.

1. *Start from the given initial state $Z_0 = z^{\text{start}}$, set $k = 0$.*
2. *Generate an action A_k according to the Z_k-th row of P, calculate the reward $r_k = r(Z_k, A_k)$ and generate a new state Z_{k+1} according to $\mathcal{P}(\cdot \mid Z_k, A_k)$. Set $k = k + 1$. Repeat until $Z_k = z^{\text{fin}}$.*
3. *Output the total reward of the trajectory $(Z_0, A_0, Z_1, A_1, \ldots, Z_\tau)$, given by (8.6).*

Updating rules

Given the N strategies $\mathbf{X}_1, \ldots, \mathbf{X}_N$ and their scores, $\widehat{S}(\mathbf{X}_1), \ldots, \widehat{S}(\mathbf{X}_N)$, one can update the policy matrix (p_{za}) using the CE method, namely as per

$$\widehat{p}_{t,za} = \frac{\sum_{k=1}^{N} I_{\{\widehat{S}(\mathbf{X}_k) \geqslant \widehat{\gamma}_t\}} I_{\{\mathbf{X}_k \in \mathcal{X}_{za}\}}}{\sum_{k=1}^{N} I_{\{\widehat{S}(\mathbf{X}_k) \geqslant \widehat{\gamma}_t\}} I_{\{\mathbf{X}_k \in \mathcal{X}_z\}}}, \tag{8.7}$$

where $\{\mathbf{X}_k \in \mathcal{X}_z\}$ is the event that the trajectory generated by strategy $\mathbf{X}_k$ contains a visit to state z, and $\{\mathbf{X}_k \in \mathcal{X}_{za}\}$ is the event that this trajectory contains a visit to state z in which action a was taken.

We now explain how to take advantage of the Markovian nature of the problem. Let us think of a maze where a certain trajectory starts badly, that is, the path is not efficient in the beginning, but after some time it starts moving quickly towards the goal. According to (8.7), all the updates are performed in a similar manner in every state in the trajectory. However, the actions taken in the states that were sampled near the target were successful, so one would like to "encourage" these actions. Using the Markov property one can substantially improve the above algorithm by considering for each state the part of the reward from the visit to that state onwards. We therefore use the same trajectory and simultaneously calculate the performance for every state in the trajectory separately. The idea is that each choice of action in a given state affects the reward from that point on, disregarding the past.

The sampling Algorithm 8.2.1 does not change in Steps 1 and 2. The difference is in Step 3. Given a policy $\mathbf{X}$ and trajectory $(Z_0, A_0, Z_1, A_1, \ldots, Z_\tau)$, we calculate the performance from every state until termination. For every state $z = Z_j$ in the trajectory the (estimated) performance is $\widehat{S}_z(\mathbf{X}) = \sum_{k=j}^{\tau-1} r_k$. The updating formula for $\widehat{p}_{za}$ is similar to (8.7), however *each state z is updated separately* according to the (estimated) performance $\widehat{S}_z(\mathbf{X})$ obtained from state z onwards.

$$\widehat{p}_{t,za} = \frac{\sum_{k=1}^{N} I_{\{\widehat{S}_z(\mathbf{X}_k) \geqslant \widehat{\gamma}_{t,z}\}} I_{\{\mathbf{X}_k \in \mathcal{X}_{za}\}}}{\sum_{k=1}^{N} I_{\{\widehat{S}_z(\mathbf{X}_k) \geqslant \widehat{\gamma}_{t,z}\}} I_{\{\mathbf{X}_k \in \mathcal{X}_z\}}}. \tag{8.8}$$

A crucial point here is to understand that in contrast to (8.7) the CE optimization is carried for every state separately and a different threshold parameter $\widehat{\gamma}_{t,z}$ is used for every state z, at iteration t. This facilitates faster convergence for "easy" states where the optimal strategy is easy to find. The trajectory sampling method above can be viewed as a variance reduction method. Numerical results indicate that the CE algorithm with updating (8.8) is much faster than with updating (8.7).

Remark 8.2 (2-dependence). It is possible to further improve the efficiency of the algorithm by remembering the last two steps of the path, rather than the last step. Mathematically, this amounts to replacing the Markov property of the trajectories with a similar "2-dependence" property.

Remark 8.3 (Different reward criteria). The sampling and updating procedures for the discounted and the average reward criteria are somewhat more involved, but not fundamentally different from those discussed above. For a more detailed discussion see [112].

8.2.2 Numerical Results

The CE with the updating rule (8.8) and trajectory generation according to Algorithm 8.2.1 was implemented for a maze problem, which presents a two-dimensional grid. We assume the following:

1. The moves in the grid are allowed in four possible directions with the goal to move from the upper-left corner to the lower-right corner.
2. The maze contains obstacles ("walls") into which movement is not allowed.
3. The reward for every allowed movement until reaching the goal is -1.

In addition we introduce:

- A small (failure) probability not to succeed moving in an allowed direction.
- A small probability of succeeding moving in the forbidden direction ("moving through the wall").
- A high cost for the moves in a forbidden direction.

In Figure 8.2 we present the results for a 20×20 maze. We set the following parameters: $N = 1000$, $\varrho = 0.03$, $\alpha = 0.7$. The initial policy was uniformly random. The cost of the moves were assumed to be random variables uniformly distributed between 0.5 and 1.5 and uniformly distributed between 25 and 75 for the allowed and forbidden moves, respectively. Note that the expected cost for the allowed and forbidden moves are equal to 1 and 50, respectively. The success probabilities in the allowed and forbidden states were taken to be 0.95 and 0.05, respectively. The arrows $z \to z'$ in Figure 8.2 indicate that at the current iteration the probability of going from z to z' is at least 0.01. In other words, if a corresponds to the action that will lead to state z' from z, then we will plot an arrow from z to z' — provided $p_{za} > 0.01$.

In all our experiments CE found the target exactly, within 5–10 iterations. The CPU time was less than one minute (on a 500MHz Pentium processor). Note that the successive iterations of the policy in Figure 8.2 quickly converge to the optimal policy. We have also run the algorithm for several other mazes and the optimal policy was always found.

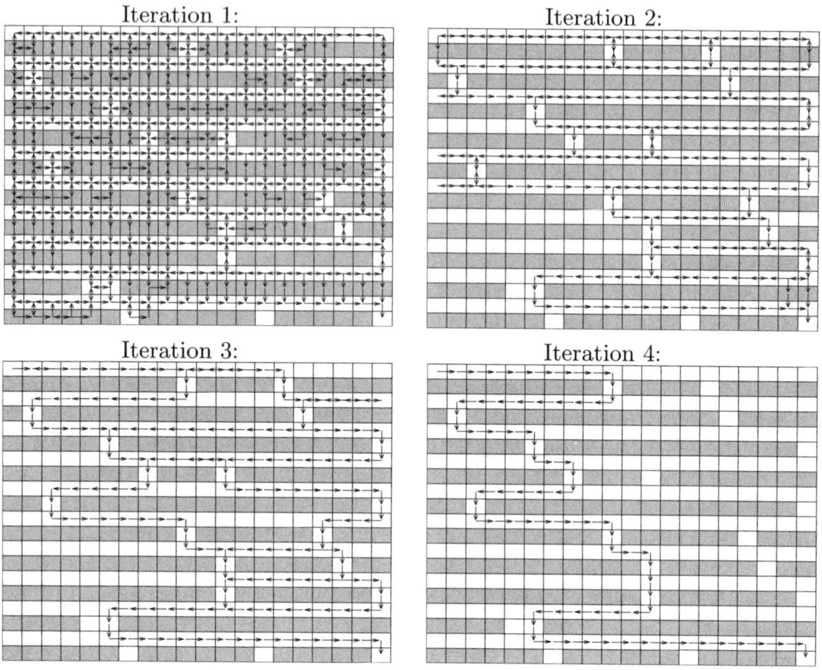

Fig. 8.2. Performance of the CE Algorithm for the 20 × 20 maze. Each arrow indicates a probability > 0.01 of going in that direction.

8.3 Clustering and Vector Quantization

The clustering problem reads as follows: given a dataset $\mathscr{Z} = \{\mathbf{z}_1, \ldots, \mathbf{z}_n\}$ of points in some d-dimensional Euclidean space, partition the data into K "clusters" $R_1, \ldots, R_K$ (with $R_i \cap R_j = \emptyset$, for $i \neq j$, and $\cup_j R_j = \mathscr{Z}$), such that some empirical *loss function* (performance measure) is minimized. A typical loss function (see, for example, [172]) is:

$$\sum_{j=1}^{K} \sum_{\mathbf{z} \in R_j} ||\mathbf{z} - \mathbf{c}_j||^2 , \tag{8.9}$$

where

$$\mathbf{c}_j = \frac{1}{|R_j|} \sum_{\mathbf{z} \in R_j} \mathbf{z} \tag{8.10}$$

presents the cluster center or *centroid* of cluster R_j. Denoting by $\mathbf{x} = (x_1, \ldots, x_n)$ the vector with $x_i = j$ when $\mathbf{z}_i \in R_j$, and letting $\mathbf{z}_{ij} = I_{\{x_i = j\}}\, \mathbf{z}_i$, we can write (8.9) as

$$\sum_{j=1}^{K}\sum_{i=1}^{n} I_{\{x_i=j\}} \, ||\mathbf{z}_{ij} - \mathbf{c}_j||^2 \, , \tag{8.11}$$

where the centroids can be written as

$$\mathbf{c}_j = \frac{1}{n_j}\sum_{i=1}^{n} \mathbf{z}_{ij} \, ,$$

with $n_j = \sum_{i=1}^{n} I_{\{x_i=j\}}$ the number of points in the j-th cluster.

As we mentioned, our goal is to partition the set of points $\mathscr{Z}$ into K not necessarily equal sized clusters R_j, such that (8.9) is minimized. In other words, we want to find a vector of centroids $(\mathbf{c}_1, \ldots, \mathbf{c}_K)$ and the corresponding partition $\{R_j\}$ that minimize (8.9). This definition also combines both the encoding and decoding steps in *vector quantization* [172]. Namely, we wish to "quantize" or "encode" the vectors in $\mathscr{Z}$ in such a way that each vector is represented by one of K *source vectors* $\mathbf{c}_1, \ldots, \mathbf{c}_K$, such that the loss (8.9) of this representation is minimized. Most well-known clustering and vector quantization methods update the vector of centroids, starting from some initial choice $(\mathbf{c}_{0,1}, \ldots, \mathbf{c}_{0,K})$ and using iterative (typically gradient-based) procedures. It is important to realize that in that case (8.9) is seen as a function of the centroids, where each point $\mathbf{z}$ is assigned to the nearest centroid, thus determining the clusters. It is well known that these type of problems — optimization with respect to the centroids — are multi-extremal and, depending on the initial value of the clusters, the gradient-based procedures converge to a *local minimum* rather than global minimum. A standard heuristic to minimize (8.10) is the *K-means algorithm* [172] which consists of the following steps:

1. Initialize by assigning (randomly or deterministically) to each point in $\mathscr{Z}$ a cluster number in $\{1, \ldots, K\}$.
2. Calculate the centroids $\mathbf{c}_1, \ldots, \mathbf{c}_K$ of the clusters.
3. (Re)assign each point to the nearest centroid.
4. Repeat Steps 2 and 3 until convergence is reached, for example if the clusters no longer change.

A useful modification is the *fuzzy K-means* algorithm [28]. A detailed description of various types of clustering methods may be found in [52] and the accompanying [162].

Some useful URLs for clustering analysis:

```
http://www.pitt.edu/~csna/

http://www.astro.psu.edu/statcodes/sc_multclass.html

http://www.ph.tn.tudelft.nl/prtools/
```

We next present a CE approach to solve the clustering problem by viewing it as a continuous multi-extremal optimization problem where, in analogy to the K-means method, the centroids $\mathbf{c}_1, \ldots, \mathbf{c}_K$ are the decision variables. In short, we consider the program

$$\min_{\mathbf{c}_1,\ldots,\mathbf{c}_K} S(\mathbf{c}_1, \ldots, \mathbf{c}_K) = \min_{\mathbf{c}_1,\ldots,\mathbf{c}_K} \sum_{j=1}^{K} \sum_{\mathbf{z}\in R_j} ||\mathbf{z} - \mathbf{c}_j||^2 , \tag{8.12}$$

where $R_j = \{\mathbf{z} : ||\mathbf{z} - \mathbf{c}_j|| < ||\mathbf{z} - \mathbf{c}_k||,\ k \neq j\}$. That is, R_j is the set of data points that are closer to $\mathbf{c}_j$ than to any other centroid.

For better insight and easy reference we consider the program (8.12) with $K = 2$ clusters and present the main steps of the CE method while using normal pdfs for updating the centroids $\mathbf{c}_j,\ j = 1, 2$ and assuming that each $\mathbf{z}_i \in \mathbb{R}^2$. We associate with the program (8.12) two 2-dimensional normal distributions $\mathsf{N}(\boldsymbol{\mu}_1, \Sigma_1)$ and $\mathsf{N}(\boldsymbol{\mu}_2, \Sigma_2)$, where Σ_1, Σ_2 are the corresponding covariance matrices. As in a typical CE application for a continuous multi-extremal optimization we set the initial matrices Σ_1, Σ_2 to be diagonal (with quite large variances at the diagonals, say 100) and then we proceed as follows:

1. Choose deterministically or randomly the initial vectors $\boldsymbol{\mu}_1$ and $\boldsymbol{\mu}_2$.
2. Generate $K = 2$ sequences of centroids (for cluster 1 and 2, respectively)

$$\mathbf{Y}_{11}, \ldots, \mathbf{Y}_{1N} \quad \text{and} \quad \mathbf{Y}_{21}, \ldots, \mathbf{Y}_{2N} ,$$

   with $\mathbf{Y}_{jk} \sim \mathsf{N}(\boldsymbol{\mu}_j, \Sigma_j),\ j = 1, 2$, independently. For each $k = 1, \ldots, N$ calculate the objective function as in (8.9), with $\mathbf{c}_j$ replaced by $\mathbf{Y}_{jk},\ j = 1, 2$.
3. Apply the CE Algorithm 4.2.1, (say with $\varrho = 0.01$ and $\alpha = 0.7$) and update the parameters $(\boldsymbol{\mu}_1, \boldsymbol{\mu}_2)$ and (Σ_1, Σ_2), accordingly.
4. Stop according to stopping rule (4.10) and accept the resulting parameter vector $(\boldsymbol{\mu}_{1T}, \boldsymbol{\mu}_{2T})$ (at the final T-th iteration) as the estimate of the true optimal solution $(\mathbf{c}_1^*, \mathbf{c}_2^*)$ of the program (8.12).

Using Remark 2.8 and (3.66), (3.67) we see that the means and variances for each centroid are updated simply as the corresponding sample mean and sample variance of the $N^{\text{elite}} = \lceil \varrho N \rceil$ elite samples. Specifically, if $\mathbf{X}_1, \ldots, \mathbf{X}_{N^{\text{elite}}}$ are the elite samples corresponding to a specific centroid (for cluster 1 or 2), then the related $\boldsymbol{\mu}$ and Σ are updated as

$$\widehat{\boldsymbol{\mu}} = \frac{1}{N^{\text{elite}}} \sum_{i=1}^{N^{\text{elite}}} \mathbf{X}_i$$

and

$$\widehat{\Sigma} = \frac{1}{N^{\text{elite}}} \sum_{i=1}^{N^{\text{elite}}} (\mathbf{X}_i - \widehat{\boldsymbol{\mu}})(\mathbf{X}_i - \widehat{\boldsymbol{\mu}})^T .$$

Note that we have not assumed independent components for each centroid distribution $\mathsf{N}(\boldsymbol{\mu}, \Sigma)$. In the 2-dimensional case we need to update therefore 5 parameters for each centroid. For a d-dimensional normal distribution the number of parameters is $d+(d+1)d/2$. However, if we use *independent* components for each $\mathsf{N}(\boldsymbol{\mu}, \Sigma)$ centroid distribution, then the number of distributional parameters is $2\,d$, because only the means and variances need to be updated; the off-diagonal elements of Σ are equal 0. It follows that for K clusters the total number of decision variables is $2\,d\,K$, when using independent components. Henceforth we will only consider the case with independent components.

Remark 8.4 (Starting Positions). One advantage of the CE method is that, as a global optimization method, it is very robust with respect to the initial positions of the centroids. Provided that standard deviation σ is chosen large enough, the initial mean μ has little or no effect on the accuracy and convergence speed of the algorithm. In general we choose the μ and σ such that the initial sampling distribution is fairly "uniform" over the smallest rectangle that contains the data points. Practically, this means that the initial σs should be not too small, say equal to the width or height of this "bounding box."

For the K-means method, however, a correct choice of starting positions is essential. A well-known data-dependent initialization method is to generate the starting positions independently, drawing each centroid from a d-dimensional Gaussian distribution $\mathsf{N}(\boldsymbol{\mu}, \Sigma)$, where $\boldsymbol{\mu}$ is the sample mean of the data and Σ the sample covariance matrix of the data.

8.3.1 Numerical Results

In this section we present numerical experiments using the CE Algorithm 4.2.1 as well as three well-known clustering heuristics, namely K-means, fuzzy K-means (FKM), and linear vector quantization (LVQ). The Matlab code for these last three algorithms was taken from the Matlab *classification toolbox* [162]; see also [52].

Two well-known types of 2-dimensional data sets were used from [162]:

(a) Banana data: Points are scattered around a segment of a circle.
(b) 3-Gaussian mixture data: Points are generated from a mixture of three 2-dimensional Gaussian distributions.

Generation of these data sets is straightforward. For convenience a banana data generation algorithm is included in Appendix A.7.

Tables 8.4–8.6 present a comparative study of CE Algorithm 4.2.1 and the traditional clustering ones for $n = 200$ and various cluster sizes K, on the models (a) and (b). In all experiments $\alpha = 0.7$ and $\varrho = 0.025$. In all cases the sample size $N = 800$ is taken, so that the number of elite samples is $N^{\text{elite}} = \varrho N = 20$. All initial standard deviations are 14 for the banana data and 6 for the 3-Gaussian data, corresponding to the width/height of the bounding box for the data. The initial means are chosen uniformly over this bounding box. The starting positions for the other algorithms are chosen according to the standard initialization procedure for the K-means algorithm discussed in Remark 8.4. We stop the CE algorithm when the performance no longer changes in two decimal places.

Each method was repeated 10 times for both data sets (a) and (b). We use the following notations in the tables: T denotes the average total number of iterations; $\bar{\gamma}_T$ denotes the averaged solution over 10 runs; $\gamma^\dagger$ is the best known solution; $\bar{\varepsilon}$ denotes the average relative experimental error (based on 10 runs) with respect to the best known solution $\gamma^\dagger$. That is,

$$\bar{\varepsilon} = \frac{\bar{\gamma}_T - \gamma^\dagger}{\gamma^\dagger} . \tag{8.13}$$

Similarly, ε_* and ε^* denote the largest and smallest relative experimental errors. Finally, CPU denotes the average CPU time in seconds on a 1.6GHz PC.

In order to identify accurately the global minimum, we repeated our experiments many times, using different ϱ, N, and smoothing parameter α (see Remark 5.2). The smallest value found from these experiments is given by $\gamma^\dagger$ for each case. It was found that the smallest CE performance of the 10 runs gives a reliable estimate for the true global minimum.

Table 8.4. Performance of the four different methods for the data sets (a) and (b), with $n = 200$, $K = 5$, $N = 800$, $N^{\text{elite}} = 20$, $\alpha = 0.7$.

Approach	T	$\bar{\gamma}_T$	$\gamma^\dagger$	$\bar{\varepsilon}$	ε_*	ε^*	CPU
(a) - Banana data set							
CE	49.6	288.49	288.11	0.00	0.00	0.01	26.67
K-Means	9.3	294.31	288.11	0.02	0.01	0.04	0.09
FKM	80.6	290.19	288.11	0.01	0.01	0.01	0.14
LVQ	17.7	302.81	288.11	0.05	0.01	0.19	0.07
(b) - 3 Gaussian mixture							
CE	44.2	69.15	69.11	0.00	0.00	0.00	28.53
K-Means	7.8	81.68	69.11	0.18	0.02	0.96	0.11
FKM	43.9	69.92	69.11	0.01	0.01	0.01	0.09
LVQ	6.4	83.75	69.11	0.21	0.06	0.96	0.03

Table 8.5. Performance of the four different methods for the data sets (a) and (b), with $n = 200$, $K = 10$, $N = 800$, $N^{\text{elite}} = 20$, $\alpha = 0.7$.

Approach	T	$\bar{\gamma}_T$	$\gamma^{\dagger}$	$\bar{\varepsilon}$	ε_*	ε^*	CPU
(a) - Banana data set							
CE	75.8	197.22	195.87	0.01	0.00	0.02	64.25
K-Means	9	221.49	195.87	0.13	0.03	0.18	0.20
FKM	86.1	199.36	195.87	0.02	0.01	0.03	0.20
LVQ	14	210.85	195.87	0.08	0.01	0.20	0.08
(b) - 3 Gaussian mixture							
CE	82.4	49.15	48.16	0.02	0.00	0.04	94.93
K-Means	10.9	58.38	48.16	0.21	0.07	0.44	0.42
FKM	63.5	49.04	48.16	0.02	0.00	0.05	0.15
LVQ	8.4	54.54	48.16	0.13	0.05	0.24	0.05

Table 8.6. Performance of the four different methods for the data sets (a) and (b), with $n = 200$, $K = 20$, $N = 800$, $N^{\text{elite}} = 20$, $\alpha = 0.7$.

Approach	T	$\bar{\gamma}_T$	$\gamma^{\dagger}$	$\bar{\varepsilon}$	ε_*	ε^*	CPU
(a) - Banana data set							
CE	142.1	138.06	135.80	0.02	0.00	0.03	261.92
K-Means	10.1	169.03	135.80	0.24	0.12	0.31	1.20
FKM	385.2	141.26	135.80	0.04	0.02	0.07	1.32
LVQ	13.1	160.84	135.80	0.18	0.12	0.32	0.13
(b) - 3 Gaussian mixture							
CE	159.8	31.88	31.29	0.02	0.01	0.03	284.98
K-Means	10.9	45.32	31.29	0.45	0.26	0.66	2.15
FKM	108.8	32.94	31.29	0.05	0.02	0.08	0.38
LVQ	8.3	42.73	31.29	0.37	0.26	0.58	0.07

We see that the CE algorithm, although significantly slower, is more accurate and consistent than the other algorithms. Among the fast algorithms the FKM is by far the best. Observe also from Tables 8.4–8.6 that as K increases, the efficiency (in terms of $\bar{\varepsilon}$, ε_*, ε^*) of CE increases relative to their counterparts K-means, FKM and LVQ. We found this in general to be the case. This can explained by arguing as follows:

1. The number of minima of the objective function in (8.12) increases with K.
2. The CE method, which presents a global optimization method typically avoids the local minima and as a result settles down in the global one.
3. The alternatives, K-means, FKM, and LVQ, which present local optimization methods are typically trapped in local minima.

It is clear that the "classical" K-means method with an average relative experimental error of 10–100% compares poorly to the CE method, which is slower but yields a vastly superior relative error of less than 1%. To formally compare the CE approach with some other optimization method one could use criterion (4.1).

Figures 8.3 and 8.4 illustrate for the banana and 3-Gaussian data, respectively, the difference in the placement of the centroids for CE (circles) and KM (crosses). Note that for the 3-Gaussian data the K-means algorithm has (wrongly) placed *two* centroids in the lower left-hand cluster.

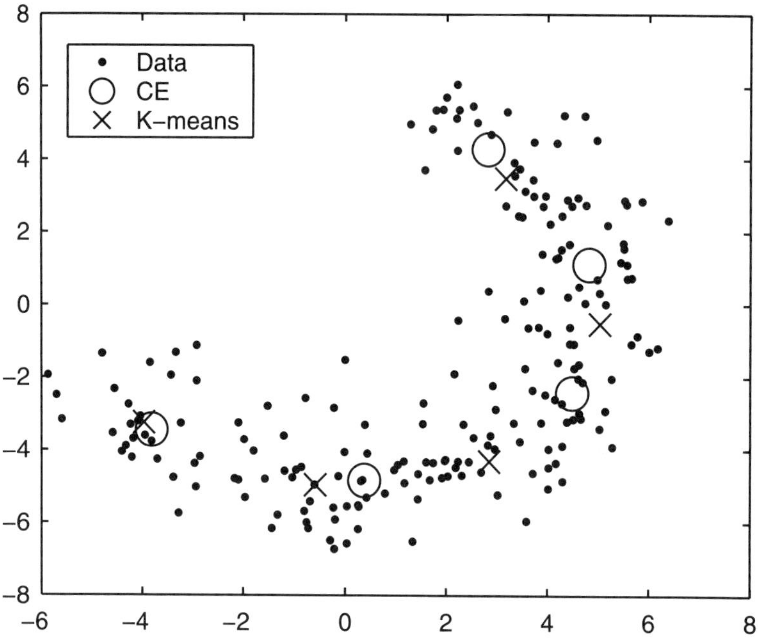

Fig. 8.3. The CE results for vector quantization of the 2-D *banana* data set. Circles designate the final cluster centers (centroids) produced by CE. Crosses are cluster centers of the K-means algorithm.

Finally, Table 8.7 represents the evolution of Algorithm 4.2.1 for the banana problem with $n = 200$, and $K = 5$. Here σ_t^* denotes the largest standard deviation at iteration t. We found that the CE method for clustering works well even if the data set is noisy. This is in contrast to most clustering methods that often require stochastic approximation procedures, which are very slow.

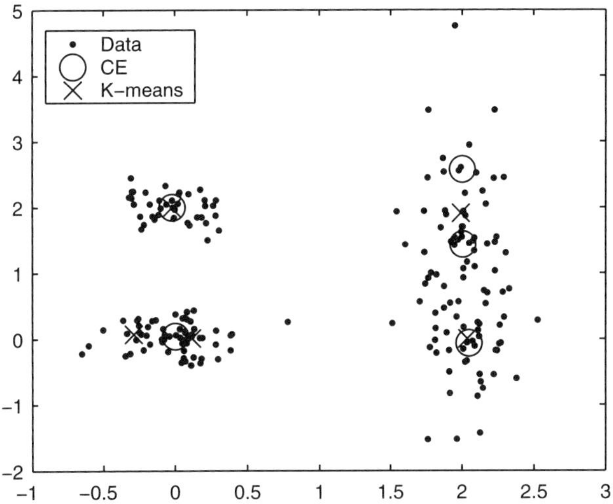

Fig. 8.4. The CE results for vector quantization of the 2-D 3-Gaussian data set. Circles designate the final cluster centers (centroids) produced by CE. Crosses are cluster centers of the K-means algorithm.

t	$\widehat{\gamma}_t$	$S_{t,(1)}$	σ_t^*
1	452.18	377.37	3.00000
2	434.09	395.01	3.20577
3	420.91	366.87	3.38746
4	403.82	356.78	3.16696
5	374.37	336.55	3.30599
6	364.62	333.48	2.94838
7	344.82	325.18	2.53459
8	333.42	313.58	2.22936
9	317.22	302.15	1.58735
10	305.09	295.90	1.15077
11	296.10	292.34	0.65408
12	291.75	290.31	0.45115
13	289.66	288.55	0.26106
14	288.80	288.50	0.18901
15	288.46	288.27	0.11668
16	288.26	288.18	0.07314
17	288.18	288.14	0.05173
18	288.14	288.13	0.03390
19	288.12	288.12	0.02360
20	288.11	288.11	0.01730
21	288.11	288.11	0.01013
22	288.11	288.11	0.00705

Table 8.7. Evolution of Algorithm 4.2.1 for the banana problem with $n = 200$, $d = 2$, $K = 5, N = 800, N^{\text{elite}} = 20$ and $\alpha = 0.7$.

8.4 Exercises

Markov Decision Processes

1. Consider the MDP problem of Section 8.2. Run Algorithm 4.2.1 for the 20×20 maze problem and obtain a figure similar to that in Figure 8.2.

Clustering

2. Generate banana data using the Matlab code in Appendix A.7. Repeat the experiments for the banana data in Tables 8.4 - 8.6 for various choices for n and K, for example $n = 2000$ and $K = 5$. As a rule of thumb take $N = 20\,K$, $\varrho = 0.025$ and $\alpha = 0.7$. Note that N is 5 times the number of parameters that has to be estimated. Does modified smoothing (Remark 5.2) help to bring the solution closer to the global minimum?

3. An alternative CE approach to clustering is to view the problem as a combinatorial minimal cut (min-cut) problem with n nodes and K partitions. In particular, each partition $R_1, \ldots, R_K$ is represented via a partition vector $\mathbf{x} \in \mathscr{X} = \{1, \ldots, n\}^K$ as in (8.9). The "trajectory generation" of the CE Algorithm 4.2.1 consists of drawing random $\mathbf{X} \in \mathscr{X}$ according to an n-dimensional discrete distribution with independent marginals, such that $\mathbb{P}(X_i = j) = p_{ij}$, $i = 1, \ldots, n$, $j = 1, \ldots, K$. For $K = 2$, we may, alternatively, use $\mathbf{X} \sim \mathsf{Ber}(\mathbf{p})$, as in the ordinary $(K = 2)$ max-cut problem. With the performance $S(\mathbf{x})$ given in (8.11), the updating rules for the p_{ij} are of the form (4.14), noting that we have here a minimization problem. This approach may be useful when the performance is a complicated function of the data. For example, the data could represent a collection of proteins, each of which has a list of characteristics. In this case the performance function is not merely the sum of the Euclidean distances but some complicated function of these characteristics.
   Compare the min-cut approach with the continuous multi-extremal approach for the *banana* data set and obtain figures similar to Figure 8.3. Run also the FACE Algorithm 5.3.1. Compare the results of the FACE Algorithm 5.3.1 with those obtained via the main CE Algorithm 4.2.1.

Image Analysis

4. A graph-theoretic approach to *image analysis* is to view a digital image as a set of nodes $1, \ldots, n$, say, representing the pixels. With each node i is associated a node "feature" $\mathbf{y}_i$, indicating for example the gray level or the RGB components of the pixel. The spatial position of node i in the image is denoted by $\mathbf{z}_i$. The "similarity" between two nodes i and j is specified via an edge (i, j), with weight or cost c_{ij}. For example, the cost

$$c_{ij} = \begin{cases} e^{-\frac{\|\mathbf{y}_i - \mathbf{y}_j\|^2}{\sigma_y^2}} \times e^{-\frac{\|\mathbf{z}_i - \mathbf{z}_j\|^2}{\sigma_z^2}} & \text{if } \|\mathbf{z}_i - \mathbf{z}_j\| < r, \\ 0 & \text{otherwise,} \end{cases} \tag{8.14}$$

where σ_y^2, σ_z^2 and r are constants, captures both the feature similarity and the spatial proximity of nodes i and j.

In *image segmentation* the objective is to partition the nodes into two or more "similar" groups. The goodness of the image partition can be measured via many different performance functions [51]. For the bipartition case Shi and Malik [157] suggested the following *normalized cut* performance function:

$$S(\mathbf{x}) = \frac{\sum_{i \in V_1(\mathbf{x}), j \in V_2(\mathbf{x})} c_{ij}}{\sum_{i \in V_1(\mathbf{x})} d_i} + \frac{\sum_{i \in V_1(\mathbf{x}), j \in V_2(\mathbf{x})} c_{ji}}{\sum_{j \in V_2(\mathbf{x})} d_j}, \tag{8.15}$$

where $d_i = \sum_j c_{ij}$ denotes the total cost from i to all other nodes, and $\{V_1(\mathbf{x}),\ V_2(\mathbf{x})\}$ is the partition corresponding to the binary cut vector $\mathbf{x} = (x_1, \ldots, x_n)$, as in the max-cut problem.

a) Segment a sample image into two segments (black and white), by maximizing (8.15) via the CE method.
b) Compare the outcomes for a variety of CE parameters and model parameters in (8.14).

5.* Bayesian inference is a rich source of complicated optimization problems, which may be tackled via the CE method. Consider the following example in image analysis: An observed digital image consists of n "gray levels" $\mathbf{y} = (y_1, \ldots, y_n) \in \mathbb{R}^n$. Assume that the original image consists of only two gray levels, μ_0 and μ_1. However, the observed image $\mathbf{y}$ is contaminated by random and independent Gaussian noise with zero mean and variance σ^2 (see Figure 8.5). In order to recover the original image we must infer from $\mathbf{y}$ the original gray levels (μ_0 or μ_1) for each pixel i.
Solving this problem is equivalent to two-class labeling of the observed image $\mathbf{y}$. For each pixel i we assign a label $x_i \in \{0, 1\}$ that designates one of two gray levels μ_k, $k = 0, 1$ of the original uncorrupted image.
A Bayesian approach to this problem proceeds in three steps. First, the *a priori* information about the unknown $\mathbf{x}$ is summarized by some density $f(\mathbf{x})$. Second, given an original image $\mathbf{x}$, the *likelihood* of obtaining data $\mathbf{y}$ is described by the conditional density $f(\mathbf{y} \mid \mathbf{x})$. Third, by Bayes's formula, the *posterior* information about $\mathbf{x}$ given the data $\mathbf{y}$ is given by

$$f(\mathbf{x} \mid \mathbf{y}) = c\, f(\mathbf{y} \mid \mathbf{x})\, f(\mathbf{x}),$$

where c is a normalization constant. The objective is to maximize this posterior probability. Defining the *energies* $U(\mathbf{x}) = -\ln f(\mathbf{x})$, $U(\mathbf{x} \mid \mathbf{y}) = -\ln f(\mathbf{x} \mid \mathbf{y})$ and $U(\mathbf{y} \mid \mathbf{x}) = -\ln f(\mathbf{y} \mid \mathbf{x})$, we see that maximizing the posterior probability with respect to $\mathbf{x}$ is equivalent to minimizing the posterior energy

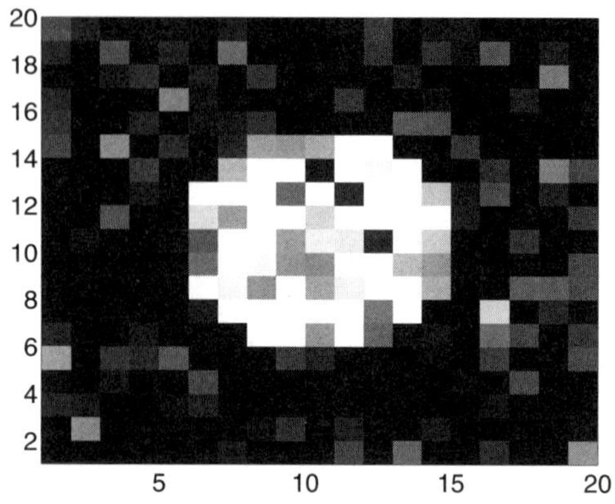

Fig. 8.5. A gray scale image corrupted by noise.

$$S(\mathbf{x}) = U(\mathbf{x} \,|\, \mathbf{y}) = U(\mathbf{y} \,|\, \mathbf{x}) + U(\mathbf{x}) \, . \tag{8.16}$$

It thus remains to specify the likelihood $f(\mathbf{y} \,|\, \mathbf{x})$ and the a-priori density $f(\mathbf{x})$, or equivalently the energies $U(\mathbf{y} \,|\, \mathbf{x})$ and $U(\mathbf{x})$. An often used model for the likelihood of the observations is to assume that given $x_i = k$ the corrupted pixel value Y_i has a Gaussian distribution with mean μ_k and variance σ^2. Thus,

$$f_i(y_i \,|\, x_i = k) = \frac{1}{\sqrt{2\pi\sigma^2}} \exp\left(-\frac{1}{2\sigma^2}(y_i - \mu_k)^2\right) . \tag{8.17}$$

Assuming independent noise, the joint pdf for the entire image $\mathbf{Y}$ is given by $f(\mathbf{y} \,|\, \mathbf{x}) = \prod_i f_i(y_i \,|\, x_i)$, which implies

$$U(\mathbf{y} \,|\, \mathbf{x}) = -\ln f(\mathbf{y} \,|\, \mathbf{x}) = \sum_{k=0}^{1} \sum_{i=1}^{n} \left(\frac{1}{2\,\sigma^2}(y_i - \mu_k)^2\right) \delta_{ik} + \text{const}, \tag{8.18}$$

where δ_{ik} equals 1 when the pixel i is assigned to class k, and 0 otherwise. A well-known choice for the a-priori energy is

$$U(\mathbf{x}) = \sum_{i=1}^{n} \sum_{j \in \mathcal{N}_i} \beta_j (x_i - x_j)^2 \, , \tag{8.19}$$

where $\mathcal{N}_i$ is the set of neighborhood pixels of i, and the β_j are constants, independent of i.

a) Select or construct a black and white test image.
b) Corrupt the image with independent Gaussian noise.
c) Reconstruct the image by solving (8.16) using the CE method. Consider two cases: (a) μ_0 and μ_1 are known in advance, and (b) μ_0 and μ_1 are unknown and have to be estimated.

A

Example Programs

In this appendix we give various Matlab implementations of CE algorithms. Devising and implementing CE algorithms for COPs is sometimes more of an art than a science. We would like to suggest a few strategies and "rules of practice" when making CE programs:

1. Clearly specify the trajectory generation and updating rules for each problem. If several alternative trajectory generation algorithms are found it is suggested to describe at least two of them. Present also the pseudocodes and the programs (in Matlab or C).
2. Repeat all runs at least 5–10 times (starting each run from a different stream of random numbers) and present summary data similar to the tables in this book.
3. Run, in addition, simulated annealing or/and genetic algorithms and compare the results with CE ones.
4. Run the CE algorithm first on a simple artificial (synthetic) problem, where the solution is known in advance, and only then run more complicated cases, such as case studies on the Web.
5. For each problem, associate a relevant artificial problem of size n =4–5 and using Algorithm 4.2.2 — the deterministic version of Algorithm 4.2.1 — calculate $\{(\gamma_t, \mathbf{v}_t)\}$ *analytically* step by step, as in Example 4.7.
6. To assess whether more complicated reward functions (see Section 5.2) are preferred, run CE Algorithm 4.2.1 using the standard indicator $I_{\{S(\mathbf{X}) \geqslant \gamma_t\}}$ and with the alternative reward $I_{\{S(\mathbf{X}) \geqslant \gamma_t\}} S(\mathbf{X})$. Compare the results.
7. Discuss briefly your results and make conclusions.

Parameter setting for COPs

For the main CE Algorithm 4.2.1 we suggest selecting

$$\varrho = \begin{cases} \dfrac{\ln n}{n} \text{ if } n < 100 \,, \\ \\ 0.01 \text{ if } n \geq 100 \,, \end{cases}$$

and taking $\alpha \in (0.3,\ 0.8)$. Depending on whether the problem is of SNN- or SEN-type, take the sample size $N = C\,n$ and $C\,n^2$, $(5 \leq C \leq 10)$ respectively.

To define C and α more accurately we suggest (for quite large networks) running an associated one of smaller size. For example, in a TSP problem with $n = 200$ cities, one could

1. associate with the original TSP problem an auxiliary one with, say, 20 out of 200 randomly chosen cities, keeping the distance matrix between these 20 cities the same as in the original problem;
2. run the auxiliary problem for several combinations of C and α;
3. adopt the best C and α (corresponding to the most accurate version) to the original problem.

For the noisy version of Algorithm 4.2.1 (see Chapter 6) the sample size needs to be increased, e.g., from $N = 5\,n$ and $N = 5\,n^2$, for SNN and SEN to $N = 25\,n$ and $25\,n^2$, respectively. We keep again $\varrho = 0.01$ and $\alpha \in (0.3, 0.8)$. Finally, for the FACE modification of CE, Chapter 5, follow Algorithm 5.3.1.

A.1 Rare Event Simulation

The Matlab function below can be used to reproduce the results of the first toy example of the tutorial; see Section 2.2.1.

```
function toy1(u)
% this function takes a vector of means, u, and outputs
% an estimate of the probability that the min path length
% exceeds gamma, and gives a table showing the parameters
% at each step


tic
v=u; % initialize v
gamma=2;
N = 10^3; % number of samples for a normal step
N1= 10^5; % number of samples for the final step
n=length(u);
rho=0.1; % required later on


g = 0;
```

```
while (g < gamma0)
  y=-log(rand(N,n));
  for k=1:N,
y(k,:)=v.*y(k,:);
  end
  % calculate S
  S = S_len(y);

  [SS,SSidx]=sort(S);
  eidx=round((1-rho)*N);
  g=SS(eidx); % the (1-rho) quantile of S
  if g>=gamma,
g=gamma;
while SS(eidx)>=g, % succesively decrease eidx until we go below gamma
  eidx=eidx-1;
end
eidx=eidx+1; % add one, and we are at the lowest eidx
  end

  W=ones(N,1);
  for j=1:n,
  W=W.*exp(-y(:,j)*(1/u(j) - 1/v(j)))*(v(j)/u(j));
  end

  for j=1:n, % update v
v(j)=sum(W(SSidx(eidx:N)).*(y(SSidx(eidx:N),j)))/sum(W(SSidx(eidx:N)));
  end
  disp([g,v])

end

y=-log(rand(N1,n));
for k=1:N1,
  y(k,:)=v.*y(k,:);
end
S = S_len(y); %min(S1,S2);

W=ones(N1,1);
for j=1:n,
  W=W.*exp(-y(:,j)*(1/u(j) - 1/v(j)))*(v(j)/u(j));
end

l=sum((S>=g).*W)/N1;

std = sqrt((sum((S>=g).*W(:).^2)/N1 - l*l)/N1);
RE = std/l;

disp(['RE: ',num2str(RE)])
```

```
disp(['l:  ',num2str(l)])
fprintf('95 perc. CI : (%g,%g)\n', l - l*RE*1.96, l+l*RE*1.96)

toc
return;


function s=S_len(X)
% Current S(X)
s1 = min(X(:,1) + X(:,4),X(:,1)+X(:,3)+X(:,5));
s2 = min(X(:,2)+X(:,5),X(:,2)+X(:,3)+X(:,4));
s = min(s1,s2);
```

A.2 The Max-Cut Problem

The following Matlab program has been used to produce the results of the synthetic max-cut problem in Table 2.4.1 and Figures 2.3 and 2.4.

```
% A Matlab demonstration of the CE method for
% solving the synthetic MAXCUT problem

clear
% Setting parameters
m = 200; % (2m nodes total)
N = 1000; % sample size
rho = 0.1;
maxIters=23; % Number of iterations

% set random seed
rand('state',1234);
% first construct a random cost matrix
Z11 = rand(m);
for i = 1:m,
  for j =i:m,
  Z11(j,i) = Z11(i,j);
  end;
  Z11(i,i) = 0;
end

Z22 = rand(m);
for i = 1:m,
  for j =i:m,
  Z22(j,i) = Z22(i,j);
  end;
  Z22(i,i) = 0;
end

B12  = ones(m);
```

```
B21 = ones(m);

C = [Z11 B12; B21 Z22]; % cost matrix


% Calculating optimal score

optp = [ones(1,m),zeros(1,m)]
y = rand(1,2*m);
x = (y <optp);  % generate cut vector, according to p
s = scut(C,x);
fprintf('Best score %f\n',s)

%Initializing
p = 1/2 * ones (1,2*m);
p(1) = 1;
gammas = zeros(1,maxIters);
curbests = zeros(1,maxIters);
ps = zeros(maxIters,2*m);
pdist = zeros(1,maxIters);

tic
curbest=0.0;
for j=(1:maxIters) % main CE loop
j % output interation number
% generate matrix X of cutvectors
     Y = rand(N,2*m);
     X = (Y < ones(N,1)*p);
     g = zeros(N,1);
     for i=1:N
           g(i,:) = scut(C,X(i,:));
     end
     [sortcut,sortidx] = sort(g);
                         % sortidx(1) contains index of the smallest
                         % cutvector
     % update gamma
     eidx = ceil((1-rho)*N);  % smallest index of best
     gamma = sortcut(eidx);
     curbest = max(curbest,sortcut(end)) %keep track of best result
     gammas(j)=gamma;
 curbests(j) = curbest;
     pdist(j) = min(norm(p -optp),norm(p - ~optp));

     % update p on basis of elite vectors, that is, the vectors
     % X(sortidx(eidx),:) to X(sortidx(N),:)
     for i=1:2*m
       p(i)= sum(X(sortidx(eidx:N),i))/(N-eidx+1);
     end
    ps(j,:) = p;
```

```
end
toc  %end timing

% gammas is the calculated gamma
% pdist is the distance from the optimal p
finalp = p

fprintf('Calculated best score %f\n',curbest);
figure(1);
plot(1:maxIters,gammas);
xlabel('t');ylabel('\gamma_t');
figure(4);
plot(1:maxIters,pdist);
xlabel('Number of iteration');ylabel('||p-p_o_p_t ||_2');
figure(2);
subplot(10,1,1); bar(1/2*ones(1,2*m));
for j=1:9
   subplot(10,1,j+1); bar(ps(j,:));
end
figure(3);
for j=1:10
   subplot(10,1,j); bar(ps(j+9,:));
end
for i = 1:maxIters
  fprintf('%d   %f  %f  %f\n',i,gammas(i),curbests(i), pdist(i));
end


function [perf] = scut(C,x)

V1 = find(x); % {V1,V2} is the partition
V2 = find(~x);
perf = sum(sum(C(V1,V2))); % the size of the cut
```

A.3 Continuous Optimization via the Normal Distribution

The program below illustrates how to optimize a general continuous multi-extremal function via the CE method, using "normal" updating rules. Typical use: `cecont(-6,10,0.1,0.7,100,0.05,'S')`, with `S.m` a certain function, for example as below; see also Example 5.1.

```
% 1. mu    - initial mean
% 2. sigma - initial standard deviation
% 3. rho   - rarity parameter
% 4. alpha - smoothing parameter
% 5. N     - sample size
% 6. S     - performance function
```

```
% 7. eps   - tolerance
%
function cecont(mu,sigma,rho,alpha,N,eps,FUN)
tic
t=0; % iteration counter
while sigma > eps
  t=t+1;
  x= mu + sigma*randn(N,1);
  SX=feval(FUN,x); % Compute the performance.
  sortSX=sortrows([x SX],2);
  mu1=mean(sortSX((1-rho)*N:N,1));
  mu=alpha.*mu1+(1-alpha).*mu; % smoothed updating of mu
  sigma1 = std(sortSX((1-rho)*N:N,1));
  sigma=alpha*sigma1+(1-alpha)*sigma; % smoothed updating of sigma
  fprintf('%g  %3.4f  %3.4f  %3.4f \n', t, feval(FUN,mu),mu, sigma)
end
toc

function res=S(x);
  res =exp(-(x-2).^2) + 0.8*exp(-(x+2).^2);
```

A.4 FACE

The Matlab program `q.m` below is an example of a FACE implementation for the n-queen problem of Exercise 6 of Chapter 2. To run, simply type `q` in the Matlab shell, or call the program with your own selection of parameters.

```
function B=q(Ne,Nmin,Nmax,alpha,d,c,n)
% A program to solve the Queen Problem
% via the Fully Adaptive Cross-Entropy method
%
% Syntax: B=q(Ne,Nmin,Nmax,alpha,d,c,n)
%
% Inputs -
%   Ne:    Number of elite samples to use
%   Nmin:  Minimum number of samples to use (must be >= Ne)
%   Nmax:  Maximum number of samples to use (must be >= Ne)
%   alpha: Smoothing Parameter
%
% Extra Inputs -
%   d:  number of consec. S_t^*'s  with no gamma_hat_t improvement
%   c:  number of Nt's=Nmax in a row
%   n:  n queens on an nxn board
%
% Output -
%   B:     An nxn matrix with queen's denoted as 1's, and blank
%          squares denoted as 0's
```

```
% some default parameters if they are not supplied
if nargin<7,n=8;end
if nargin<6,c=4;end
if nargin<5,d=6;end
if nargin <4,alpha=0.7;end
if nargin ==0,Ne = 50;end
if nargin <2,Nmin = 10*Ne;Nmax = 50*Ne;end
if (Nmin<Ne)||(Nmax<Ne)||(Nmax<Nmin),
    fprintf('Please give a valid (Ne,Nmin,Nmax) tuple.\n')
    B=0;return;
end
N=Nmin; % start sampling with the minimum allowed
st=rand(d,1)+n*n;
nt=rand(c,1);
gt=100;
v=(1/n)*ones(n); % initial probabilities
iii=1;
ind=1;
count=1;
ccc=0;
while iii~=0,
  if (N+ccc)>=Nmax,
    if (sum(nt>=Nmax)==c), % we have reached the limit here
      fprintf('unreliable solution!\n')
      break;
    end
  end
  s=[]; % set up score vector
  for kk=1:N, % calculate and score N_t queen layouts
    B=genq(v,n);
    S=scoreq(B,n);
    s=[s;S];
    V(:,:,kk)=B;
  end
  [s,I]=sort(s); % Sort the scores
  gtold=gt;
  gt=s(Ne);
  V=V(:,:,I(1:Ne)); % the elite samples
  if s(1)==0, % if we have a valid solution
    fprintf('t = %d , N_t = %d , g_hat_t = %g , S_star_t = %d\n',
            count,N,s(Ne),s(1))
    fprintf('valid solution!\n')
    break;
  end
  if s(Ne)<gtold,
    st=rand(d,1)+n*n; % reset counting for s_star_t
  end
  if (gt>=gtold)&&(s(1)>=st(d)),
    if (sum(st==s(1))==d),
```

```
    fprintf('reliable solution.\n')
    break;
  else
    ccc=0;
    s=s(1:Ne);
    fprintf('N_t: %d ',Nmin)
    while ind~=0,
      if Nmin==Nmax,break;end
      B=genq(v,n);
      S=scoreq(B,n);
      ccc=ccc+1;
      if S<s(Ne),
        s=[s;S];
        V(:,:,Ne+1)=B;
        [s,I]=sort(s);
        s=s(1:Ne);
        V=V(:,:,I(1:Ne));
        if (s(1)==0),
          fprintf('\nt = %d , N_t = %d , g_hat_t = %g ,
                   S_star_t = %d\n', count,N+ccc,s(Ne),s(1))
          fprintf('valid solution!\n')
          B=V(:,:,1);
          return;
        end
      end
      if mod(ccc,5*Nmin)==0,
        fprintf('\nN_t:')
      end
      if mod(ccc,Nmin)==0,
        fprintf(' %d ',N+ccc)
      elseif mod(ccc,round(Nmin/5))==0,
        fprintf('.')
      end
      if s(Ne)<gtold,
        st=rand(d,1)+n*n; % reset counting for s_star_t
        break;
      end
      if ((N+ccc)>=Nmax)||(s(1)<st(d)),
        break;
      end
    end
    gt=s(Ne);
    w=(1/Ne)*sum(V,3);
    v=alpha*w+(1-alpha)*v;
    fprintf('\nt = %d , N_t = %d , g_hat_t = %g , S_star_t = %d\n',
             count,N+ccc,s(Ne),s(1))
    N=Nmin;
    count = count+1;
    st(1:(d-1))=st(2:d);
```

```
      st(d)=s(1);
      nt(1:(c-1))=nt(2:c);
      nt(c)=N+ccc;
    end
  else
    if gt<gtold,
      nt=rand(c,1);
    end
    w=(1/Ne)*sum(V,3);
    v=alpha*w+(1-alpha)*v;
    fprintf('t = %d , N_t = %d , g_hat_t = %g , S_star_t = %d\n',
             count,N,s(Ne),s(1))
    N=Nmin;
    count = count+1;
    st(1:(d-1))=st(2:d);
    st(d)=s(1);
    nt(1:(c-1))=nt(2:c);
    nt(c)=N;
  end
end
B=V(:,:,1);

%%%%%%%%%%%%%%%%%%%%%%%%%%%%%%%%%%%%%%%%
function B=genq(v,n)
% generate an outcome based on probabilities v
% where v is an nxn matrix of probs.
% note: we require one queen per row
% each row of v MUST sum to > 0
Z=zeros(n);B=Z;
r=rand(n,1);
if mod(n,2)==1, %odd
    h=(n-1)/2;
else
    h=n/2;
end
for i=1:n, % see where to add the next queen
    for j=1:n,
        Z(i,j+1)=Z(i,j)+v(i,j);
        if ((Z(i,j)<=r(i)) && (r(i)<Z(i,j+1))),
            if((i==1)&&(j>(ceil(n/2)))),
                B(i,j-h)=1;
            else
                B(i,j)=1;
            end
        end
    end
end

%%%%%%%%%%%%%%%%%%%%%%%%%%%%%%%%%%%%%
```

```
function s=scoreq(B,n)
% Compute the score for an nxn chessboard,
% ie: Score is the number of "hits" between queens
s=0; % initial score
% score cols and not rows, because of the way we gen outcomes
a=sum(B,1);
s=s+sum((a(a>1)-1));
% score diagonal hits
a=sum(spdiags(B),1);
s=s+sum((a(a>1)-1));
a=sum(spdiags(B(n:-1:1,:)),1);
s=s+sum((a(a>1)-1));
```

A.5 Rosenbrock

The following Matlab script was used to find the global optimum of the Rosenbrock function in Section 5.4. The program uses a fixed number of (elite) samples to update the parameters of the normal distributions. It also implements the modified smoothing scheme in (5.1). The algorithm simply stops after a total of `t_max` iterations. To apply the program to the trigonometric function simply uncomment the relevant line below.

```
clear all
format short e
% t : iteration number
% t_max : total number of iterations
% N : sample size
% Gamma: vector of gammas
% Alf :  smoothing parameter;
% N_El : number of elite samples
% n : dimension
% Betta : parameter used in modified smoothing
% q : parameter used in modified smoothing
% D : vector of variances
% M : vector of means
% X : N x n matrix, containing a sample of N n-dimensional vectors
% SA : Performance of X
% S_Sort : vector of sorted performances
% S_Best : best performance
% V_Sort : vector of indices of sorted performances
% X_Gamma : performace of worst of elite samples
% X_Best : performance of the best of the elite samples
N = 1000,  n = 10, N_El = 20, Alf = 0.8 , Betta = 0.9, q = 6
t0 = 1, t_max = 2000;
M = -2 + 4*rand(1,n); % random initial Mean
M_Last = M;
D = 10000.0*ones(1,n);
D_Last = D;
```

```
X_Best_Last = zeros(1,n);
X_Gamma_Last = zeros(1,n);
tic
for t = t0:t_max,
  M = Alf*M + (1-Alf)*M_Last;
  B_Mod = Betta - Betta*(1-1/t)^q;
  D = B_Mod*D + (1-B_Mod)*D_Last ;
  R = randn(N,n);
  Std = D.^0.5  ;
  for v = 1:N,
    X(v,:) = M + Std.*R(v,:);
  end;
  SA = RosPen(N,n,X); % call the Rosenbrock function (with penalty)
 %SA = trigo(N,n,X); % call the trigonometic function
  [S_Sort,V_Sort] = sort(SA);

   X_Gamma = X(V_Sort(N_El),:);
  X_Best = X(V_Sort(1),:);
  M_Current(t,:) = M;
  D_Current(t,:) = D;
  Gamma(t) = S_Sort(N_El);
  S_Best(t) = S_Sort(1);
  S_Min(t) = min(S_Best);
  Glob_Min = S_Min(t);
  t_Last = t;
  M_Last = M;
  D_Last = D;

  X_Best_Last = X_Best;
  X_Gamma_Last = X_Gamma;
  X_El  =  X(V_Sort(1:N_El),:);
  M = mean(X_El);
  D = (std(X_El).^2);
  if mod(t,100)==0
    fprintf('%d   %5.3f  %5.3f' ,t,Gamma(t), S_Best(t));
    fprintf('  %6.2f',M)
    fprintf('\n')
  end;
end
toc
M
S_Best(t)

%%%%%%%%%%%%%%%%%%%%%%%%%%%%%%%%%%%%%%%%%%%%%%%%%%%%%%%%%%%%%

function S = RosPen(N,n,X) %Rosenbrock function with penalty
X_High = 2*ones(N,n); X_Low = -2*ones(N,n);
X_Pen = max(X,X_High) - X_High + X_Low - min(X_Low,X);
S = 1000000000.0*X_Pen(:,1);
```

```
for i=1:(n-1) ,
   S_i = 100*(X(:,(i+1))-X(:,i).^2).^2 + (X(:,i)-1).^2;
   S_Pen = 1000000000.0*X_Pen(:,(i+1));
   S = S + S_i + S_Pen;
end;

%%%%%%%%%%%%%%%%%%%%%%%%%%%%%%%%%%%%%%%%%%%%%%%%%%%%%%%%%%%%%%%%%

function S = trigo(N,n,X) %trigonometric function
S=ones(N,1);
for i=1:n ,
  S_i = 8*sin(7*(X(:,i)- 0.9).^2).^2 + ...
      + 6*sin(14*(X(:,i)- 0.9).^2).^2 + (X(:,i) - 0.9).^2;
  S=S + S_i;
end;
```

A.6 Beta Updating

The following program computes the value and derivative of the logarithm of the gamma function at x. Its use, in conjunction with the second program in this section, is illustrated in Exercise 5.4.

```
function [f,df]=slava(x);
if (x <= 0)
    f = inf;
    df = [];
end;
dx = 1.e-10;
f = gammaln(x);
if (x<1)
    y = x+1;
    df = (gammaln(y+dx)-gammaln(y))/dx-1/x;
elseif (x<2)
    df = (gammaln(x+dx)-gammaln(x))/dx;
elseif (x<200)
    df = 0;
    y = x;
    while(1)
        y = y-1;
        df = df+1/y;
        if (y<2)
            df = df+(gammaln(y+dx)-gammaln(y))/dx;
            break;
        end;
    end;
else
    df = log(x-1)+0.5/(x-1);
end;
```

The function below calculates the solution to (5.11) and (5.12), by implementing the *ellipsoid method*. The input parameters pp and qq are equal to minus the quotient of the sums in (5.11) and (5.12), respectively.

```
function [alpha,beta] = SlavaEll(pp,qq);
n = 2;
e = zeros(n,1);
p = zeros(n,1);
q = zeros(n,1);
blx = [1.e-6;1.e-6];
bux = [1.e12;1.e12];
x = 0.5*(blx+bux);
rad = 0.5*norm(blx-bux);
B = rad*eye(n);
rep.cc = -1;
frec = inf;
fmod = -inf;
df = zeros(2,1);
contr.print = 1;
contr.eps = 1.e-12;
for itr = 1:1000,
    flag_c = 1;
    aa = sum((blx>x).*(blx-x));
    if aa,
        e = -(x<blx);
        flag_c = 0;
    end;
    if flag_c,
        aa = sum((x>bux).*(x-bux));
        if aa,
            e = (x>bux);
            flag_c = 0;
        end;
    end;
    if flag_c,
        al = x(1);
        bt = x(2);
        [fsum,dfsum] = slava(al+bt);
        [fal,dfal] = slava(al);
        [fbt,dfbt] = slava(bt);
        df(1) = -dfsum + dfal + pp;
        df(2) = -dfsum + dfbt + qq;
        f = -fsum + fal + fbt + pp*x(1) + qq*x(2);
        if f<frec,
            alpha = al;
            beta = bt;
            frec = f;
            if contr.print,
               % disp(sprintf('itr# %3d rec = %.6e',itr,frec));
```

```
            end;
        end;
        aa = f-frec;
        e = df;
    end;
    p = B'*e;
    pn = norm(p);
    if pn == 0,
        break;
    end;
    if flag_c,
        fmod = max(fmod,f-pn);
    end;
    if frec-fmod<contr.eps*max(1,abs(frec)),
        break;
    end;

    p = p/pn;
    aln = aa/pn;
    if (aln>1)&(frec==inf),
        break;
    end;
    if aln> = 1,
        aln = 0.5;
    end;
    cfo = sqrt((1-aln^2)*n^2/(n^2-1));
    cfi = cfo*(1-sqrt((1-aln)*(n-1)/((1+aln)*(n+1))));
    shift = (1+n*aln)/(1+n);
    q = B*p;
    x = x - shift*q;
    B = cfo*B - cfi*q*p';
end;
[gsum,dgsum] = slava(alpha+beta);
[gal,dgal] = slava(alpha);
[gbt,dgbt] = slava(beta);
res1 = (dgsum-dgal-pp)/max(1,abs(pp));
res2 = (dgsum-dgbt-qq)/max(1,abs(qq));
%disp(sprintf('***Residuals: %.3e %.3e',res1,res2));
```

A.7 Banana Data

Banana shaped data sets occur frequently in clustering and classification test problems. Below is a simple Matlab function that generates such data.

```
function b = banana(n,sigma,radius,theta1,theta2)
% Generate a Banana-shaped data set
% n      - number of data points (default: 200)
% sigma  - move points according to a normal dist. with this
```

```
%           standard deviation in both x and y directions
%           (default: 1)
% radius - the radius of the circle (of which the "banana" is
%           an arc (default: 5)
% theta1 - starting angle of the arc (default: 9*pi/8)
% theta2 - ending angle of the arc (default: 19*pi/8)
if nargin<4,theta1=pi + pi/8;end
if nargin<5,theta2=(5*pi/2 - pi/8) - theta1;end
if nargin<3,radius=5;end
if nargin<2,sigma=1;end
if nargin<1,n=200;end

randn('seed', 123456789); %optional
rand('seed', 987654321);  %optional
angles=theta1 + rand(n,1)*theta2; % angles between pi/8 and 11*pi/8
b=radius.*[cos(angles),sin(angles)]; % transform these to random points
                                      % on an arc of a circle
b=b+sigma*randn(n,2); % shift these points off the arc ind.
                      % normally in x and y directions
```

References

1. E. H. L. Aarts and J. H. M. Korst. *Simulated Annealing and Boltzmann Machines.* John Wiley & Sons, Chichester, 1989.
2. E. H. L. Aarts and J. K. Lenstra. *Local Search in Combinatorial Optimization.* John Wiley & Sons, Chichester, 1997.
3. T. S. Abdul-Razaq, C. N. Potts, and L. N. Van Wassenhove. A survey of algorithms for the single machine total weighted tardiness scheduling problem. *Discrete Applied Mathematics*, 26:235–253, 1990.
4. W. A. T. W. Abdullah. Seeking global minima. *Journal of Computational Physics*, 110:320, 1994.
5. I. Adan and J. van der Wal. Monotonicity of the throughput in single server production and assembly networks with respect to the buffer sizes. In H. G. Perros and T. Altiok, editors, *Queueing Networks with Blocking*, pages 345–356. Elsevier Science, 1989.
6. R. K. Ahuja, T. L. Magnanti, and J. B. Orlin. *Network Flows Theory, Algorithms and Applications.* Prentice Hall, Englewood Cliffs, 1993.
7. V. M Aleksandrov, V. I. Sysoyev, and V. V. Shemeneva. Stochastic optimization. *Engineering Cybernetics*, 5:11–16, 1968.
8. S. M. Ali and S. D. Silvey. A general class of coefficients of divergence of one distribution from another. *Journal of the Royal Statistical Society, Series B*, 28:131–142, 1966.
9. G. Alon, D. P. Kroese, T. Raviv, and R. Y. Rubinstein. Application of the cross-entropy method to the buffer allocation problem in a simulation-based environment. *Annals of Operations Research*, Kluwer Academic, 2004.
10. S. Andradóttir. A method for discrete stochastic optimization. *Management Science*, 41:1946–1961, 1995.
11. S. Andradóttir. A global search method for discrete stochastic optimization. *SIAM J. Optimization*, 6:513–530, 1996.
12. S. Asmussen. *Applied Probability and Queues.* John Wiley & Sons, 1987.
13. S. Asmussen and K. Binswanger. Simulation of ruin probabilities for subexponential claims. *ASTIN Bulletin*, 27(2):297–318, 1997.
14. S. Asmussen, K. Binswanger, and B. Højgaard. Rare events simulations for heavy-tailed distributions. *Bernoulli*, 6:303–322, 2000.
15. S. Asmussen, D. P. Kroese, and R. Y. Rubinstein. Heavy tails, importance sampling and cross-entropy. Submitted, 2003.

16. S. Asmussen and R. Y. Rubinstein. The efficiency and heavy traffic properties of the score function method in sensitivity analysis of queueing models. *Advances of Applied Probability*, 24(1):172–201, 1992.
17. S. Asmussen and R. Y. Rubinstein. Response surface estimation and sensitivity analysis via the efficient change of measure. *Stochastic Models*, 9:313–339, 1993.
18. S. Asmussen and R. Y. Rubinstein. Complexity properties of steady-state rare-events simulation in queueing models. In *Advances in Queueing: Theory, Methods and Open Problems*, pages 429–462. CRC Press, 1995.
19. A. Barto, R. Sutton, and C. Anderson. Neuron-like adaptive elements that can solve difficult learning control problems. *IEEE Transactions on Systems, Man, and Cybernetics*, 13:834–846, 1983.
20. J. Baxter, P. L. Bartlett, and L. Weaver. Experiments with infinite-horizon, policy-gradient estimation. *Journal of Artificial Intelligence Research*, 15:351–381, 2001.
21. R. E. Bellman. On a routing problem. *Quarterly of Applied Mathematics*, 16(87–90), 1958.
22. A. Ben-Tal and A. Nemirovski. *Lectures on Modern Convex Optimizations.* SIAM, Philadelphia, 2001, pp. 342–352.
23. J. M. Bernardo and A. F. M. Smith. *Bayesian Theory.* John Wiley & Sons, New York, 1994.
24. D. P. Bertsekas and R. Gallager. *Data Networks.* Prentice Hall, Englewood Cliffs, 1992.
25. D.P. Bertsekas. *Dynamic Programming and Optimal Control*, volume I. Athena Scientific, 1995.
26. D. J. Bertsimas. A vehicle routing problem with stochastic demand. *Operations Research*, 40:574–585, 1992.
27. M. Den Besten, T. Stützle, and M. Dorigo. Ant colony optimization for the total weighted tardiness problem. *Parallel Problem Solving from Nature*, pages 611–620, 2000.
28. J. Bezdek. *Pattern Recognition with Fuzzy Objective Function Algorithms.* Plenum Press, New York, 1981.
29. B. Bollobás. *Graph Theory.* Springer-Verlag, Berlin, 1979.
30. V. S. Borkar and S. P. Meyn. The O.D.E. method for convergence of stochastic approximation and reinforcement learning. *SIAM J. Control Optim.*, 38(2):447–469, 2000.
31. P. Boyle, O. Broadie, and P. Glasserman. Simulation methods for security pricing. *J. Economic Dynamics and Control*, 21:1267–1321, 1997.
32. J. A. Bucklew. *Large Deviation Techniques in Decision, Simulation and Estimation.* John Wiley & Sons, New York, 1990.
33. J. A. Buzacott and J. G. Shanthikumar. *Stochastic Models of Manufacturing Systems.* Prentice Hall, Englewood Cliffs, 1993.
34. P. D. Carr, E. Cheah, P. M. Suffolk, S. G. Vasudevan, N. E. Dixon, and D. L. Ollis. X-ray structure of the signal transduction protein from Escherichia Coli at 1.9 A. *Acta Crystallogr. D*, 52:93–104, 1996.
35. M. Charikar, S. Huller, and B. Raghavachari. Algorithms for capacitated vehicle routing. *SIAM Journal on Computing*, 31:665–682, 2001.
36. K. Chepuri and T. Homem de Mello. Solving the vehicle routing problem with stochastic demands using the cross entropy method. *Annals of Operations Research*, Kluwer Academic, 2004.

37. I. Cohen, B. Golany, and A. Shtub. Managing stochastic finite capacity multi-project systems through the cross-entropy method. *Annals of Operations Research*, Kluwer Academic, 2004.
38. H. Cohn and M. Fielding. Simulated annealing searching for an optimal temperature schedule. *SIAM Journal of Optimization*, 9(3):779–802, 1999.
39. T. M. Cover and J. A. Thomas. *Elements of Information Theory.* John Wiley & Sons, New York, 1991.
40. H. A. J. Crauvels, C. N. Potts, and L. N. Van Wassenhove. Local search heuristics for the single machine total weighted tardiness scheduling problem. *INFORMS Journal of Computing*, 10(3):341–350, 1998.
41. Y. Dallery, Z. Liu, and D. Towsley. Equivalence, reversibility, symmetry and concavity properties in fork/join queueing networks with blocking. *Journal of the ACM*, 41(5):903–942, 1994.
42. G. B. Dantzig and R. H. Ramser. The truck dispatching problem. *Management Science*, 6:80–91, 1959.
43. P. T. de Boer. *Analysis and efficient simulation of queueing models of telecommunication systems.* PhD thesis, University of Twente, 2000.
44. P. T. de Boer, D. P. Kroese, S. Mannor, and R. Y. Rubinstein. A tutorial on the cross-entropy method. *Annals of Operations Research*, Kluwer Academic, 2004.
45. P. T. de Boer, D. P. Kroese, and R. Y. Rubinstein. Estimating buffer overflows in three stages using cross-entropy. In *Proceedings of the 2002 Winter Simulation Conference, San Diego*, pages 301–309, 2002.
46. P. T. de Boer, D. P. Kroese, and R. Y. Rubinstein. A fast cross-entropy method for estimating buffer overflows in queueing networks. *Management Science*, 2004.
47. L. Devroye. *Non-Uniform Random Variate Generation.* Springer-Verlag, New York, 1986.
48. E. W. Dijkstra. A note on two problems in connection with graphs. *Numerische Math.*, 1:269–271, 1959.
49. M. Dorigo, V. Maniezzo, and A. Colorni. The ant system: optimization by a colony of cooperating agents. *IEEE Transactions on Systems, Man, and Cybernetics—Part B*, 26(1):29–41, 1996.
50. U. Dubin. The cross-entropy method for combinatorial optimization with applications. Master's thesis, The Technion, Israel Institute of Technology, Haifa, June 2002.
51. U. Dubin. Application of the cross-entropy method for image segmentation. *Annals of Operations Research*, 2004. Submitted.
52. R. O. Duda, P. E. Hart, and D. G. Stork. *Pattern Classification.* John Wiley & Sons, New York, 2001.
53. T. Elperin, I. B. Gertsbakh, and M. Lomonosov. Estimation of network reliability using graph evolution models. *IEEE Transactions on Reliability*, 40(5):572–581, Dec 1991.
54. P. Embrechts and N. Veraverbeke. Estimates for the probability of ruin with special emphasis on the possibility of large claims. *Insurance Mathematics and Economics*, 1:55–72, 1982.
55. W. Feller. *An Introduction to Probability Theory and Its Applications*, volume I. John Wiley & Sons, 2nd edition, 1970.
56. L. R. Ford. Network flow theory. Technical Report P-923, RAND Corporation, Santa Monica, California, 1956.

57. M. R. Garey and D. S. Johnson. *Computers and Intractability: A guide to the theory of NP-completeness.* W. H. Freeman and Company, San Francisco, 1979.
58. M. J. J. Garvels and D. P. Kroese. A comparison of RESTART implementations. In *Proceedings of the 1998 Winter Simulation Conference*, pages 601–609, Washington, DC, 1998.
59. M. J. J. Garvels, D. P. Kroese, and J. C. W. van Ommeren. On the importance function in RESTART simulation. *European Transactions on Telecommunications*, 13(4), 2002.
60. S. B. Gelfand and S. K. Mitter. Simulated annealing with noisy or imprecise energy measurements. *JOTA*, 62:49–62, 1989.
61. S. B. Gershwin and J. E. Schor. Efficient algorithms for buffer space allocation. *Annals of Operations Research*, 93:117–144, 2000.
62. I. B. Gertsbakh. *Statistical Reliability Theory.* Marcel Dekker, Inc., New York, 1989.
63. P. Glasserman, P. Heidelberger, P. Shahabuddin, and T. Zajic. A look at multilevel splitting. In H. Niederreiter, editor, *Monte Carlo and Quasi Monte Carlo Methods 1996, Lecture Notes in Statistics*, volume 127, pages 99–108. Springer-Verlag, New York, 1996.
64. P. Glasserman, P. Heidelberger, P. Shahabuddin, and T. Zajic. A large deviations perspective on the efficiency of multilevel splitting. *IEEE Transactions on Automatic Control*, 43(12):1666–1679, 1998.
65. P. Glasserman, P. Heidelberger, P. Shahabuddin, and T. Zajic. Multilevel splitting for estimating rare event probabilities. *Operations Research*, 47(4):585–600, 1999.
66. F. Glover and M. Laguna. *Modern Heuristic Techniques for Combinatorial Optimization*, chapter Chapter 3: Tabu search. Blackwell Scientific Publications, Oxford, 1993.
67. P. W. Glynn. Likelihood ratio gradient estimation for stochastic systems. *Communications of the ACM*, 33(10):75–84, 1990.
68. P. W. Glynn and D. L. Iglehart. Importance sampling for stochastic simulations. *Management Science*, 35:1367–1392, 1989.
69. D. Goldberg. *Genetic Algorithms in Search, Optimization and Machine Learning.* Addison Wesley, 1989.
70. W. B. Gong, Y. C. Ho, and W. Zhai. Stochastic comparison algorithm for discrete optimization with estimation. In *Proceedings of the 31st IEEE Conference on Decision and Control*, pages 795–800, 1992.
71. C. Görg. Simulating rare event details of ATM delay time distributions with RESTART/LRE. In *Proceedings of the RESIM Workshop.* University of Twente, The Netherlands, March 1999.
72. C. Görg and O. Fuß. Comparison and optimization of RESTART run time strategies. *AEÜ*, 52(3):197–204, 1998.
73. G. R. Grimmett and D. R. Stirzaker. *Probability and Random Processes.* Oxford University Press, Oxford, 3rd edition, 2001.
74. D. Gusfield. *Algorithms on Strings, Trees and Sequences.* Cambridge University Press, Cambridge, 1997.
75. W. J. Gutjahr. A generalized convergence result for the graph-based ant system meta-heuristic. Technical Report 91-016, Dept. of Statistics and Decision Support Systems, University of Vienna, Austria, 2000.

76. W. J. Gutjahr. A graph-based ant system and its convergence. *Future Generations Computing*, 16:873–888, 2000.
77. W. J. Gutjahr. Ant algorithms with the guaranteed convergence to the optimal solution. Technical report, Dept. of Statistics and Decision Support Systems, University of Vienna, Austria, 2001.
78. W. J. Gutjahr and G. C. Pflug. Simulated annealing for noisy cost functions. *Journal of Global Optimisation*, 8:1–13, 1996.
79. Z. Haraszti and J. Townsend. Rare event simulation of delay in packet switching networks using DPR-based splitting. In *Proceedings of the RESIM Workshop, 11-12 March 1999*, pages 185–190. University of Twente, the Netherlands, 1999.
80. W. K. Hastings. Monte Carlo sampling methods using Markov chains and their applications. *Biometrika*, 57:92–109, 1970.
81. C. Heavey, H. Y. Papadopoulos, and J. Browne. The throughput rate of multistation unreliable production lines. *European Journal of Operation Research*, 68:69–89, 1993.
82. P. Heidelberger. Fast simulation of rare events in queueing and reliability models. *ACM Transactions on Modeling and Computer Simulation*, 5:43–85, 1995.
83. B. E. Helvik and O. Wittner. Using the cross-entropy method to guide/govern mobile agent's path finding in networks. In *3rd International Workshop on Mobile Agents for Telecommunication Applications - MATA'01*, 2001.
84. D. S. Hirschberg. Algorithms for the longest common subsequence problem. *J. ACM*, 24:664–675, 1977.
85. T. Homem-de-Mello and R. Y. Rubinstein. Rare event probability estimation for static models via cross-entropy and importance sampling. Submitted.
86. R. Horst, P. M. Pardalos, and N. V. Thoai. *Introduction to Global Optimization.* Kluwer Academic, 1996.
87. K-P. Hui, N. Bean, M. Kraetzl, and D. P. Kroese. The cross-entropy method for network reliability estimation. *Annals of Operations Research*, Kluwer Academic, 2004.
88. M. R. Jerrum and A. J. Sinclair. Approximating the permanent. *SIAM Journal on Computing*, 18:1149–1178, 1989.
89. M. R. Jerrum and A. J. Sinclair. Polynomial-time approximation algorithms for the Ising model. *SIAM Journal on Computing*, 22:1087–1116, 1993.
90. N. L. Johnson and S. Kotz. *Urn Models and Their Applications.* John Wiley & Sons, New York, 1977.
91. S. Juneja and P. Shahabuddin. Simulating heavy tailed processes using delayed hazard rate twisting. *ACM Transactions on Modeling and Computer Simulation*, 12:94–118, 2002.
92. L. P. Kaelbling, M. Littman, and A. W. Moore. Reinforcement learning - a survey. *Journal of Artificial Intelligence Research*, 4:237–285, May 1996.
93. H. Kahn and T. E. Harris. *Estimation of Particle Transmission by Random Sampling.* National Bureau of Standards Applied Mathematics Series, 1951.
94. Y. M. Kaniovski, A. J. King, and R. J.-B. Wets. Probabilistic bounds (via large deviations) for the solutions of stochastic programming problems. *Annals of Operations Research*, 56:189–208, 1995.
95. J. N. Kapur and H. K. Kesavan. *Entropy Optimization Principles with Applications.* Academic Press, New York, 1992.

96. M. Kearns and S. Singh. Near-optimal reinforcement learning in polynomial time. In *Proc. of the 15th Int. Conf. on Machine Learning*, pages 260–268, San Francisco, 1998. Morgan Kaufmann.
97. J. D. Kececioglu, H.-P. Lenhof, K. Mehlhorn, P. Mutzel, K. Reinert, and M. Vingron. A polyhedral approach to sequence alignment problems. *Disc. Appl. Math.*, 104:143–186, 2000.
98. J. Keith and D. P. Kroese. Sequence alignment by rare event simulation. In *Proceedings of the 2002 Winter Simulation Conference*, pages 320–327, San Diego, 2002.
99. A. I. Khinchin. *Information Theory.* Dover Publications, Inc., New York, 1957.
100. S. Kirkpatrick, C. D. Gelatt, and M. P. Vecchi. Optimization by simulated annealing. *Science*, 220:671–680, 1983.
101. I. Kovalenko. Approximations of queues via small parameter method. In J. H. Dshalalow, editor, *Advances in Queueing: Theory, Methods and Open Problems*, pages 481–509. CRC Press, New York, 1995.
102. V. Kriman and R. Y. Rubinstein. Polynomial time algorithms for estimation of rare events in queueing models. In J. Dshalalow, editor, *Frontiers in Queueing: Models and Applications in Science and Engineering*, pages 421–448, New York, 1995. CRC Press.
103. D. P. Kroese and R. Y. Rubinstein. The transform likelihood ratio method for rare event simulation with heavy tails. *Queueing Systems*, 2004.
104. A. Krogh, M. Brown, I. S. Mian, K. Sjolander, and D. Haussler. Hidden Markov models in computational biology: applications to protein modeling. *J. Mol. Biol.*, 235:1501–1531, 1994.
105. C. E. Lawrence, S. F. Altschul, M. S. Boguski, J. S. Liu, A. F. Neuwald, and J. C. Wootton. Detecting subtle sequence signals: A Gibbs sampling strategy for multiple alignment. *Science*, 262:208–214, 1993.
106. P. L'Ecuyer. A unified view of the IPA, SF, and LR gradient estimation techniques. *Management Science*, 36:1364–1384, 1990.
107. E. L. Lehmann. *Testing Statistical Hypotheses.* Springer-Verlag, New York, 1997.
108. D. Lieber. *Rare-events estimation via cross-entropy and importance sampling.* PhD thesis, William Davidson Faculty of Industrial Engineering and Management, Technion, Haifa, Israel, 1998.
109. D. Lieber, R. Y. Rubinstein, and D. Elmakis. Quick estimation of rare events in stochastic networks. *IEEE Transaction on Reliability*, 46:254–265, 1997.
110. Z. Liu, A. Doucet, and S. S. Singh. The cross-entropy method for blind multiuser detection. In *IEEE International Symposium on Information Theory*, Chicago, 2004.
111. L. Lovasz. Randomized algorithms in combinatorial optimization. *DIMACS Series in Discrete Mathematics and Theoretical Computer Science*, 25:153–179, 1995.
112. S. Mannor, R. Y. Rubinstein, and Y. Gat. The cross-entropy method for fast policy search. In *The 20th International Conference on Machine Learning (ICML-2003)*, Washington, DC, August 2003.
113. L. Margolin. Cross-entropy method for combinatorial optimization. Master's thesis, The Technion, Israel Institute of Technology, Haifa, July 2002.
114. L. Margolin. On the convergence of the cross-entropy method. *Annals of Operations Research*, Kluwer Academic, 2004.

115. G. McLachlan and T. Krishnan. *The EM Algorithm and Extensions.* John Wiley & Sons, New York, 1997.
116. L. E. Meester and J. G. Shanthikumar. Concavity of the throughput of tandem queueing systems with finite buffer storage space. *Advances in Applied Probability*, 22:764–767, 1990.
117. I. Menache, S. Mannor, and N. Shimkin. Basis Function Adaption in Temporal Difference Reinforcement Learning. *Annals of Operations Research*, Kluwer Academic, 2004.
118. M. Metropolis, A. W. Rosenbluth, M. N. Rosenbluth, A. H. Teller, and E. Teller. Equations of state calculations by fast computing machines. *J. of Chemical Physics*, 21:1087–1092, 1953.
119. G. A. Mikhailov. Calculation of system parameter derivatives of functionals of the solutions to the transport equations. *Journal of Computational Mathematics and Mathematical Physics*, 1976.
120. L. B. Miller. Monte carlo analysis of reactivity coefficients in fast reactors: general theory and applications. Technical report, Argonne National Laboratory, IL, 1967.
121. C. N. Morris. Natural exponential families with quadratic variance functions. *The Annals of Statistics*, 10:65–80, 1982.
122. S. B. Needleman and C. D. Wunsch. A general method applicable to the search for similarities in the amino-acid sequence of two proteins. *J. Mol. Biol.*, 48:443–453, 1970.
123. W. I. Norkin, G. C. Pflug, and A. Ruszczyński. A branch-and-bound method for stochastic global optimization. Working paper, International Institute for Applied System Analysis, Laxenburg, Austria, 1996.
124. I. H. Osman and G. Laporte. Metaheuristics: a bibliography. *Annals of Operations Research*, 63:513–523, 1996.
125. C. H. Papadimitriou and M. Yannakakis. Optimization, approximation, and complexity classes. *J. Comput. System Sci.*, 43:425–440, 1991.
126. H. T. Papadopoulos and G. A. Vouros. A model management system (MMS) for the design and operation of production lines. *International Journal of production Research*, 35(8):2213–2236, 1997.
127. R. G. Parker and R. L. Rardin. *Discrete Optimization.* Academic Press, San Diego, 1996.
128. P. A. Pevzner. Multiple alignment, communication cost and graph matching. *SIAM J. Appl. Math.*, 52:1763–1779, 1992.
129. J. D. Pinter. *Global Optimization in Action.* Kluwer Academic, Dordrecht, 1996.
130. G. Potamianos and J. Goutsias. Stochastic approximation algorithms for partition function estimation of Gibbs random fields. *IEEE Transactions on Information Theory*, 43(6):1948–1965, 1997.
131. C. N. Potts and L. N. van Wassenhove. Singe machine tardiness sequencing heuristics. *IEEE Transactions*, 23:346–354, 1991.
132. M. Puterman. *Markov Decision Processes.* Wiley-Interscience, 1994.
133. T. K. Ralphs, L. Kopman, W. R. Pulleyblank, and L. E. Trotter. On the capacitated vehicle routing problem. *Mathematical Programming*, (B) 94, 2003.
134. V. J. Rayward-Smith, I. H. Osman, C. R. Reeves, and G. D. Smith. *Modern Heuristic Search Methods.* John Wiley & Sons, Chichester, 1996.

135. C. R. Reeves. Modern heuristic techniques. In V. J. Rayward-Smith, I. H. Osman, C. R. Reeves, and G. D. Smith, editors, *Modern Heuristic Search Methods*, Chichester, 1996. John Wiley & Sons.
136. M. I. Reiman and A. Weiss. Sensitivity analysis for simulations via likelihood ratios. *Operations Research*, 37(5):830–844, 1989.
137. A. Ridder. Importance sampling simulations of Markovian reliability systems using cross-entropy. *Annals of Operations Research*, Kluwer Academic, 2004.
138. H. E. Romeijn and R. L. Smith. Simulated annealing for constrained global optimization. *Journal of Global Optimization*, 5:101–126, 1994.
139. M. T. Rosenstein and A. G. Barto. Robot weightlifting by direct policy search. In Bernhard Nebel, editor, *Proceedings of the Seventeenth International Joint Conference on Artificial Intelligence*, pages 839–846, Seattle, 2001. Morgan Kaufmann.
140. R. Y. Rubinstein. *Some problems in Monte Carlo Optimization.* PhD thesis, University of Riga, Latvia, 1969. In Russian.
141. R. Y. Rubinstein. A Monte Carlo method for estimating the gradient in a stochastic network. Manuscript, Technion, Israel, 1976.
142. R. Y. Rubinstein. *Simulation and the Monte Carlo Method.* John Wiley & Sons, New York, 1981.
143. R. Y. Rubinstein. *Monte Carlo Optimization Simulation and Sensitivity of Queueing Network.* John Wiley & Sons, New York, 1986.
144. R. Y. Rubinstein. Optimization of computer simulation models with rare events. *European Journal of Operational Research*, 99:89–112, 1997.
145. R. Y. Rubinstein. The cross-entropy method for combinatorial and continuous optimization. *Methodology and Computing in Applied Probability*, 2:127–190, 1999.
146. R. Y. Rubinstein. Combinatorial optimization, cross-entropy, ants and rare events. In S. Uryasev and P. M. Pardalos, editors, *Stochastic Optimization: Algorithms and Applications*, pages 304–358, Dordrecht, 2001. Kluwer.
147. R. Y. Rubinstein. The cross-entropy method and rare-events for maximal cut and bipartition problems. *ACM Transactions on Modelling and Computer Simulation*, 12(1):27–53, 2002.
148. R. Y. Rubinstein and B. Melamed. *Modern Simulation and Modeling.* Wiley Series in Probability and Statistics, New York, 1998.
149. R. Y. Rubinstein and A. Shapiro. *Discrete Event Systems: Sensitivity Analysis and Stochastic Optimization via the Score Function Method.* John Wiley & Sons, New York, 1993.
150. J. S. Sadowsky. On the optimality and stability of exponential twisting in Monte Carlo simulation. *IEEE Trans. Info. Theory*, IT–39:119–128, 1993.
151. F. Schreiber and C. Görg. A modified RESTART method using the LRE-algorithm. In North Holland, editor, *Proceedings of the 14th International Teletraffic Congress*, pages 787–796, 1994.
152. F. Schreiber and C. Görg. The RESTART/LRE method for rare event simulation. In *Proceedings of the 1996 Winter Simulation Conference*, pages 390–397, 1996.
153. P. K. Sen and J. M. Singer. *Large Sample Methods in Statistics.* Chapman & Hall/CRC, New York, 1993.
154. P. Shahabuddin. Rare event simulation of stochastic systems. In *Proceedings of the 1995 Winter Simulation Conference*, pages 178–185, Washington, D.C., 1995. IEEE Press.

155. J. G. Shanthikumar and D. D. Yao. Monotonicity and concavity properties in cyclic queueing networks with finite buffers. In H.G. Perros and T. Altiok, editors, *Queueing Networks with Blocking*, pages 325–344. Elsevier Science, 1989.
156. A. Shapiro. Simulation based optimization- convergence analysis and statistical inference. *Stochastic Models*, 12:425–454, 1996.
157. J. Shi and J. Malik. Normalized cuts and image segmentation. *IEEE Transactions on Pattern Analysis and Machine Intelligence*, 22(8):888–905, 2000.
158. L. Shi and S. Olafsson. Nested partitioning method for global optimization. *Operations Research*, 48(3):390–407, 2000.
159. L. Shi, S. Olafsson, and N. Sun. New parallel randomized algorithm for traveling salesman problem. *Computers and Operations Research*, 26:371–394, 1999.
160. D. Siegmund. Importance sampling in the Monte Carlo study of sequential tests. *Annals of Statistics*, 4:673–684, 1976.
161. T. F. Smith and M. S. Waterman. Identification of common molecular subsequences. *J. Mol. Biol.*, 147:195–197, 1981.
162. D. G. Stork and E. Yom-Tov. *Computer Manual to Accompany Pattern Classification*. John Wiley & Sons, 2004.
163. R. Sutton and A. Barto. Reinforcement learning: An introduction. Technical report, 1998.
164. F. Szidarovszky and S. Yakowitz. *Principles and Procedures of Numerical Analysis*. Plenum Press, New York, 1978.
165. J. N. Tsitsiklis. Asynchronous stochastic approximation and Q-learning. *Machine Learning*, 16:185–202, 1994.
166. M. Villén-Altamirano and J. Villén-Altamirano. RESTART: A method for accelerating rare event simulations. In J. W. Cohen and C. D. Pack, editors, *Proceedings of the 13th International Teletraffic Congress, Queueing, Performance and Control in ATM*, pages 71–76, 1991.
167. M. Villén-Altamirano and J. Villén-Altamirano. About the efficiency of RESTART. In *Proceedings of the RESIM'99 Workshop*, pages 99–128. University of Twente, the Netherlands, 1999.
168. G. A. Vouros and H. T. Papadopoulos. Buffer allocation in unreliable production lines using a knowledge based system. *Computer & Operations Research*, 25(12):1055–1067, 1998.
169. I. A. Wagner, M. Lindenbaum, and A. M. Bruckstein. ANTS: Agents on Networks, Trees and Subgraphs. *Future Generation Computer Systems*, 16(8):915–926, 2000.
170. A. J. Walker. An efficient method for generating discrete random variables with general distributions. *Assoc. Comput. Mach. Trans. Math. Software*, 3:253–256, 1977.
171. C. Watkins. *Learning from Delayed Rewards*. PhD thesis, Cambridge University, 1989.
172. A. Webb. *Statistical Pattern Recognition*. Arnold, London, 1999.
173. R. Wheeler and K. Narendra. Decentralized learning in finite Markov chains. *IEEE Trans. on Automatic Control*, 31:519–526, 1996.
174. D. White. *Markov Decision Process*. John Wiley & Sons, 1992.
175. R. J. Williams. Simple statistical gradient-following algorithms for connectionist reinforcement learning. *Machine Learning*, 8:229–256, 1992.

Index